Advances in Natural Medicines, Nutraceuticals and Neurocognition

Advances in Natural Medicines, Nutraceuticals and Neurocognition

Edited by
Con Stough and Andrew Scholey

CRC Press
Taylor & Francis Group
Boca Raton London New York

CRC Press is an imprint of the
Taylor & Francis Group, an **informa** business

CRC Press
Taylor & Francis Group
6000 Broken Sound Parkway NW, Suite 300
Boca Raton, FL 33487-2742

First issued in paperback 2019

© 2013 by Taylor & Francis Group, LLC
CRC Press is an imprint of Taylor & Francis Group, an Informa business

No claim to original U.S. Government works

ISBN-13: 978-1-4398-9360-9 (hbk)
ISBN-13: 978-0-367-38044-1 (pbk)

Visit the Taylor & Francis Web site at
http://www.taylorandfrancis.com

and the CRC Press Web site at
http://www.crcpress.com

Contents

PART III   Essential Fatty Acids and Neurocognition

PART IV   Herbal Medicines, Nutraceuticals and Neurocognition

Preface

The field of natural medicines and neurocognition has made huge advances in the last few years, both methodologically and theoretically. It is a testament to such developments that many papers in the area are now published in mainstream psychopharmacology, neuroscience, nutrition, and medical journals rather than in more niche publications. We hope that this book reflects the breadth and depth of advances in this field.

Part I discusses approaches to best measure cognitive processes directly relevant for herbal pharmacological studies. The first two chapters have been written by authors with a wealth of experience in developing and applying neurocognitive tests in the context of cognitive aging and psychopharmacology. Chapter 1 by Wesnes draws on decades of experience in psychopharmacological research and in applying scientifically valid cognitive batteries to capture small changes in cognition due to herbal and supplement administration. In Chapter 2, Pipingas and Camfield consider more recent methodological developments related to cognitive aging, an issue progressively important in Western societies.

Part II considers the effects of vitamins and nutrients on neurocognitive change. It includes chapters that examine the neurocognitive effects of isolated compounds, including N-acetylcysteine (Chapter 3 by Camfield) and lipoic acid (Chapter 5 by Maczurek, Ooi, Patel, and Münch). Chapters by Macpherson (Chapter 4) and by Haskell and Kennedy (Chapter 7) examine recent developments, suggesting that supplementation with multivitamins may benefit cognitive health. Chapter 6 by Owen looks at the impact of several agents that improve metabolic activity in the context of improving neurocognitive function.

Part III turns to a topic that has been very much on the public agenda over the past decade, namely, the extent to which essential fatty acids, and in particular omega-3s, can improve cognitive function. It is interesting to note how successful the marketing of omega-3 essential oils for cognitive change has been. Chapters 8 and 9 by Jackson and Bradbury, respectively, provide an excellent scientific context and analysis of the state of the empirical science in this area.

The book concludes with Part IV, which deals with the effects of a range of herbal medicines on cognition. Chapter 10 by Chang, Colagiuri, and Luo provides a fascinating insight into the application of Chinese medicine in the context of dementia. While this chapter focuses on herbal extracts, it also examines acupuncture and other traditional Chinese medicine (TCM) approaches to improving behavioral decline. Chapter 11 by Finn, Scholey, Pipingas, and Stough evaluates the effects of one component of green tea, epigallocatechin gallate (EGCG), on cognitive function. Green tea has a long history within TCM, and there is emerging data suggesting that neurocognitive health may be added to the list of benefits that may be attributed to green tea and its constituents. The next two chapters explore both mechanistic and efficacy studies associated with chronic administration of *Bacopa monnieri* (BM), an Indian herb that has been subjected to considerable scientific scrutiny in India and progressively in other countries. Chapter 12 by Stough, Cropley, Pase, Scholey and Kean examines some potential overlapping mechanisms between cognitive aging and the effect of BM (particularly the extract CDR-I08) on the human brain. The chapter also evaluates the body of research indicating a cognitive enhancing effect due to BM. Chapter 13 by Singh explores the beneficial behavioral effects of BM. *Bacopa* has been systematically evaluated in vivo, in preclinical trials, and more recently in a series of clinical trials. The extract provides an excellent case study of translation of usage in traditional medicine to high-level empirical study through to neurocognitive health applications. The main theme of this volume is to study nutraceutical effects in the context of cognitive enhancement while preventing cognitive decline. There is also a growing body of research that deals with the effects of such agents in psychiatric disorder. Chapter 14 by Sarris reviews evidence for the efficacy of herbal extracts in the treatment of anxiety disorders, depression, and insomnia. Chapter 15 by Zangara again reflects a potentially important drug discovery path focusing on the Chinese club moss alkaloid Huperzine A, its mechanisms of action, and its potential in the treatment of Alzheimer's disease and related conditions.

This part concludes with a chapter by Stockley who provides an in-depth exploration of wine and its components in the context of effects on mechanisms of neural function and decline.

We have assembled some of the best researchers in the world in the field of nutraceuticals and neurocognition in order to provide the reader with an insight into the current state of research in this field from a number of perspectives.

Con Stough
Andrew Scholey

Editors

Professor Con Stough is codirector of the Centre for Human Psychopharmacology (formerly the Brain Sciences Institute) at Swinburne University and professor of cognitive neuroscience. He has an extensive track record in research and consulting and has published more than 150 peer-reviewed international expert papers in the areas of cognition, psychopharmacology, and biological psychology. He has also attracted more than $15 million in research grants over the last six years from national and international government and industry sources. He previously was on the advisory panel for the International Society for Intelligence Research and is currently on the editorial board for the journal *Intelligence* and other journals. He is currently an invited panel member for the World Economic Forum (Neuroscience and Behaviour). Many of Professor Stough's publications concern psychopharmacology, particularly in the areas of illicit drugs and the neurocognitive effects of natural medicines. He has published some of the first human clinical trials on cognition for many natural medicines and drugs and with Professor Scholey coordinates the largest group of researchers internationally in the neurocognitive effects of natural supplements and herbal medicines. Examples of current studies involve a range of vitamins, herbal medicines, and illicit and smart drugs. He is currently the lead investigator in a large Australian Government–funded trial examining the effects of natural medicines on cognitive ageing. He has a particular interest in the Indian herb bacopa and is coordinating a large group of postdocs and PhD students studying the neurocognitive effects of this interesting herb.

Professor Andrew Scholey, BSc (Hons) PhD CPsychol AFBPsS, is codirector (with Con Stough) of the Centre for Human Psychopharmacology at Swinburne University, Melbourne, Victoria, Australia. He is a leading international researcher into the neurocognitive effects of natural products, supplements, and food components, having published over 100 peer-reviewed journal articles and numerous book chapters. In 1998, he established the Human Cognitive Neuroscience Unit at Northumbria University, United Kingdom, and was the Unit's director until joining the then Brain Sciences Institute at Swinburne University in 2007. Prof. Scholey has been lead investigator in a series of studies into the human biobehavioral effects of natural products and their neurocognition-enhancing and anti-stress/anxiolytic properties. He has attracted millions of dollars in research funding, including as chief investigator on national competitive grants from the United Kingdom, Europe, and Australia, as well as from many industry bodies in Europe, North America, Asia, and Australia. He has acted as advisor to numerous industry bodies, ILSI Europe and ILSI South East Asia, and has reported to the UK Parliamentary Forum on Diet and Health.

Contributors

Joanne Bradbury
School of Health and Human
    Sciences
Southern Cross University
Lismore, New South Wales, Australia

David Camfield
Centre for Human
    Psychopharmacology
Swinburne University of Technology
Melbourne, Victoria, Australia

Dennis Chang
Center for Complementary Medicine
    Research
School of Science and Health
University of Western Sydney
Penrith, New South Wales, Australia

Ben Colagiuri
School of Psychology
The University of Sydney
Sydney, New South Wales, Australia

Vanessa Cropley
Melbourne Neuropsychiatry Centre
The University of Melbourne
Melbourne, Victoria, Australia

Melissa Finn
Centre for Human Psychopharmacology
Swinburne University of Technology
Melbourne, Victoria, Australia

Crystal F. Haskell
Department of Psychology
Brain, Performance, and Nutrition
    Research Centre
Northumbria University
Newcastle upon Tyne, United Kingdom

Philippa Jackson
Faculty of Health and Life Sciences
Brain, Performance, and Nutrition
    Research Centre
Northumbria University
Newcastle upon Tyne, United Kingdom

James Kean
Centre for Human Psychopharmacology
Swinburne University of Technology
Melbourne, Victoria, Australia

David O. Kennedy
Department of Psychology
Brain, Performance, and Nutrition
    Research Centre
Northumbria University
Newcastle upon Tyne, United Kingdom

Rong Luo
Chengdu University of Traditional
    Chinese Medicine
Chengdu, Sichuan, People's Republic of
    China

Helen Macpherson
Centre for Human
    Psychopharmacology
Swinburne University of Technology
Melbourne, Victoria, Australia

Annette E. Maczurek
Department of Pharmacology
School of Medicine
University of Western Sydney
Penrith, New South Wales, Australia

Gerald Münch
Department of Pharmacology
and
Molecular Medicine Research Group
School of Medicine
University of Western Sydney
Penrith, New South Wales, Australia

Lezanne Ooi
Department of Pharmacology
School of Medicine
University of Western Sydney
Penrith, New South Wales, Australia

Lauren Owen
Centre for Human Psychopharmacology
Swinburne University of Technology
Melbourne, Victoria, Australia

Matthew Pase
Centre for Human
    Psychopharmacology
Swinburne University of Technology
Melbourne, Victoria, Australia

Mili Patel
Department of Pharmacology
School of Medicine
University of Western Sydney
Penrith, New South Wales, Australia

Andrew Pipingas
Centre for Human
    Psychopharmacology
Swinburne University of Technology
Melbourne, Victoria, Australia

Jerome Sarris
Faculty of Medicine
Department of Psychiatry
The Melbourne Clinic
The University of Melbourne
Melbourne, Victoria, Australia

Andrew Scholey
Centre for Human Psychopharmacology
Swinburne University of Technology
Melbourne, Victoria, Australia

Hemant K. Singh
Central Drug Research Institute
Lucknow, India

and

Lumen Research Foundation
Chennai, India

Creina S. Stockley
The Australian Wine Research Institute
Urrbrae, South Australia, Australia

Con Stough
Centre for Human
    Psychopharmacology
Swinburne University of Technology
Melbourne, Victoria, Australia

Keith A. Wesnes
Bracket Global
Goring-on-Thames, United Kingdom

and

Division of Psychology
Northumbria University
Newcastle, United Kingdom

and

Centre for Human
    Psychopharmacology
Swinburne University of Technology
Melbourne, Victoria, Australia

Andrea Zangara
Soho-Flordis International
Sydney, New South Wales, Australia

and

Centre for Human
    Psychopharmacology
Swinburne University of Technology
Melbourne, Victoria, Australia

Part I

Methodologies to Measure Cognition in Natural Medicine Trials

1 Natural Substances as Treatments for Age-Related Cognitive Declines

Keith A. Wesnes

CONTENTS

Cognitive function has long been known to decline with normal ageing, and recent findings indicate that this decline starts in early adulthood. While these declines are recognised, there is currently no regulatory acceptance to encourage the pharmaceutical industry to develop medicines to treat these normal age-related deteriorations; and the industry is therefore currently focused on Alzheimer's and other dementias, as well as prodromes for Alzheimer's disease including Mild Cognitive Impairment. Recent surveys have shown that students, various professional groups, and the military are using 'smart drugs' like modafinil off-label to promote cognitive function, and such use is producing much controversy, due in part to the possible safety risks associated with such use. However, a growing body of data is accumulating showing that naturally occurring substances can enhance cognitive function, even in young volunteers. This provides an alternative strategy for individuals who wish to optimise their mental performance and to attempt to correct age-related declines, i.e. by consuming naturally occurring substances which are more widely available. Accepting that naturally occurring substances can have the same range of health risks as prescription medicines, this paper considers research findings that could provide a rationale for self-medication of cognitive function with natural substances.

COGNITION ENHANCEMENT

Bostrom and Sandberg (2009) define cognition enhancement as "the amplification or extension of core capacities of the mind through improvement or augmentation of internal or external information processing systems." Aspects of cognitive function that are targets for enhancement include attention, vigilance, information processing, memory, planning, reasoning, decision making, and motor control. Bostrom and Sandberg argue that an intervention aimed at correcting a specific pathology or defect of a cognitive subsystem may be characterized as therapeutic, while enhancement is an intervention that improves a subsystem in some way other than repairing something that is broken or remedying a specific dysfunction. This distinction is interesting, and accurately characterizes the various compounds which are being developed and studied in this rapidly growing field.

Substances to enhance cognitive function are currently receiving a large amount of public interest and ethical debate (Cakic, 2009), due to the recognition of their widespread use by students, the military, and many professional groups (Sahakian and Morein-Zamir, 2007). In 2008, the journal *Nature* reported the results of an online poll, in which 20% of the 1400 respondents admitted that they had used "neuro-enhancers" to stimulate their focus, concentration, or memory (Maher, 2008). Although 96% of respondents felt that individuals with neuropsychiatric disorders who have severe memory and concentration problems should receive such substances, 80% of respondents felt that anyone who wanted such substances should be allowed access, and 69% said they would take one provided the side-effects were low. The high level of interest can be illustrated by articles in *The Times Online* (Bannerman, 2010) entitled "Bring smart drugs out of the closet, experts urge Government," and in *Time Magazine* entitled "Popping Smart Pills: The Case for Cognitive Enhancement" (Szalavitz, 2009).

MEASUREMENT OF HUMAN COGNITIVE FUNCTION

Cognitive function concerns mental abilities which enable us to conduct the activities of daily living. Some aspects of cognitive function are relatively stable and unaffected by, for example, aging, fatigue, drugs, or trauma; while other aspects such as attention and memory are variable by nature and highly susceptible to change. Tests of cognitive function assess how well various cognitive skills are operating in an individual at any particular time. Such evaluations require individuals to perform tasks which involve one or more cognitive domains. Thus if a researcher wished to assess memory, the test would involve the memorization of information and the outcome measure would reflect how well such information could be retrieved. Equally, to assess the ability to sustain attention, the test could involve monitoring a source of information in order to detect predefined target stimuli over a period of time, and the outcome measures would reflect the speed and accuracy of the detections. It is important to note that the only way to measure cognitive function directly is by assessing the quality of performance on cognitive tests or behavioral tasks. It is of interest to assess how the individual feels about his or her levels of cognitive function, but this is simply supportive evidence for the objective assessment of task performance. Similarly, various measures of brain activity (for example, electroencephalography and fMRI scanning) do not measure the quality of cognitive function directly, but rather provide us with independent but nonetheless hugely valuable information about the activation of certain brain areas as well as the interconnecting pathways between various areas which are crucial for successful completion of various cognitive operations.

It is important that the researcher in this field identifies the appropriate domain of cognitive function to investigate. While "cognition enhancement" is an acceptable generic term, as is "health promoting," both science and regulators require more specific targets, which respect the independence of different domains when considering specific claims. For example, why in medicine would a drug which helped pulmonary function be expected to help the liver? This illustrates the limitation of global scores of cognition for nutritional claims, and should guide researchers to seek assessments of specific target domains of function. There are a number of core cognitive domains which can be evaluated, including attention, information processing, reasoning, memory, motor control, problem solving, and executive function. Taking memory as an example, there are four major types: episodic or declarative memory, working memory, semantic memory, and procedural memory (see Budson and Price, 2005). As Budson and Price illustrate, relatively few conditions are associated with impairments to semantic memory and procedural memory, while working and episodic memory are impaired in a wide variety of neurological, psychiatric, surgical, and medical conditions. This creates a rationale for directing testing toward working and episodic memory as a more fruitful potential area to evaluate in novel conditions, and most test systems recognize this approach. Further, tests specific to particular domains are, when available, ideal, as this helps to facilitate the substantiation of any claims made on the basis of the research findings. The most specific tests are attentional tests, as well-designed tests of attention do not require aspects of memory or reasoning for task performance, and thus changes in performance can be relatively clearly attributable to effects on attentional processes. As attention is important for the performance of any task, when seeking to evaluate other domains, it is useful to also assess attention additionally in order that the relative contribution to any effects of changes to attention can be established. Most well-established test batteries include assessments of attention, working and episodic memory, motor control, and aspects of executive function.

AUTOMATION OF COGNITIVE TESTS

The automation of cognitive tests brings numerous advantages (e.g., Wesnes et al., 1999); the most relevant to the area of cognition enhancement is improving the signal-to-noise ratio. Noise, i.e., unwanted variability, is decreased by the standardization such testing can bring to test administration and the reduction of errors in scoring. However, the signal can also be increased due to the extra precision in assessment which millisecond resolution of response times can bring. Furthermore, aspects of cognitive function can be assessed, which cannot be measured using traditional pencil and paper measures. Major tests of attention such as simple and choice reaction time have always been automated, as have intensive vigilance tests like the continuous performance test and digit vigilance tasks. Further, computerized tests of verbal and object recognition permit, besides the assessment of the accuracy of recognition, the time actually taken to successfully retrieve the information from memory. This important aspect of memory has been overlooked by traditional tests which cannot make this assessment, but this aspect of memory declines markedly and independently of accuracy with normal aging, and is severely compromised in many debilitating diseases such as dementia (e.g., Simpson et al., 1991; Nicholl et al., 1994; Wesnes et al., 2002). Further in MCI, such slowed speed of retrieval of information is an early characteristic of the disease (Nicholl et al., 1995), which also can respond to pharmacological treatment (Newhouse et al., 2012). Automation also provides the same benefits for tests of the ability to retain information in working memory, as the role of working memory is to facilitate the performance of ongoing tasks; and clearly it is not just the ability to correctly retain the information that is important but also the time taken to decide correctly retrieve this information, something which cannot be assessed with traditional tests such as digit span. A further important benefit of assessing speed is that it permits "speed-accuracy trade-offs" to be identified, which helps to avoid misinterpretations of study findings.

GUIDELINES FOR ESTABLISHING COGNITION ENHANCEMENT

Our understanding of cognition enhancement is at an early stage, and there are few, if any, established criteria. For a compound to be established as an enhancer of one or more aspects of cognitive function, the following criteria have been recently proposed (Wesnes, 2010).

1. Improvements must be identified by well recognized and extensively validated tests of cognitive function.
2. Improvements should be to one or more major domains of cognitive function.
3. Improvements must be seen on core measures of task performance, and any suggestions of speed-accuracy trade-offs should be interpreted with caution.
4. Improvements in one cognitive domain should not occur at the cost to another.
5. Improvements should not be followed by rebound declines.
6. Improvements should be of magnitudes which are behaviorally and clinically relevant.
7. Improvements should not be subject to tachyphylaxis over the period for which the treatment is intended to be used.
8. Self-ratings are of interest, and may be used as supportive evidence, but are not sufficient in the absence of objective test results.

COGNITIVE FUNCTION AND NORMAL AGING

There is much debate about the declines in the quality of mental functioning which accompany aging. A traditional approach has been to compare young adults (e.g., 18–25 years) to the elderly (e.g., 65–80 years), and much research has shown that a variety of aspects of cognitive functioning are poorer in the elderly. One consistent criticism of this approach is that the elderly group grew up in a different era, which may have limited their subsequent abilities (for example, due to socioeconomic factors such as more limited educational abilities and poorer nutrition), and thus the differences may not simply have been due to aging. A research group based at the University of Virginia, the United States, led by Timothy Salthouse, has comprehensively investigated this area over the past few decades. The outcomes of this research program have been recently summarized (Salthouse, 2010). The approach of Salthouse and colleagues has been to assess thousands of healthy individuals across the age range on a variety of traditional neuropsychological tests and to evaluate the pattern of change by decade from early adulthood until the 1980s. The consistent finding has been for linear declines to be present in a range of measures of attention, information processing, reasoning, and various aspects of memory from the twenties onward. Using a variety of analytic techniques, the groups have established that despite common assumptions to the contrary, age-related declines in measures of cognitive functioning are relatively large, begin in early adulthood, are evident in several different types of cognitive abilities, and are not always accompanied by increases in between-person variability. This pattern has also been identified over the same age range using computerized tests of cognitive function, showing linear declines in 5 year cohorts to the speed and accuracy of various aspects of attention, working and episodic memory (Wesnes and Ward, 2000; Wesnes, 2003, 2006). This effect is illustrated in Figure 1.1

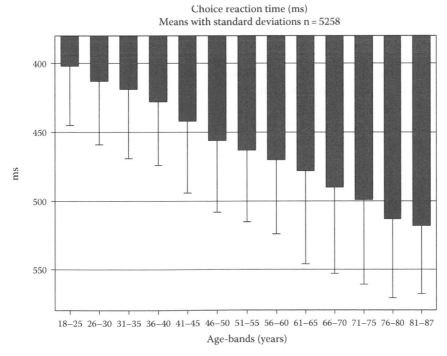

FIGURE 1.1   (**See color insert.**) Declines in normal aging on choice reaction time, a test of attention.

for choice reaction time. As can be seen, the declines are linear across the age range, starting in the late 1920s, which is entirely consistent with the work of Salthouse. Further, a decline of one standard deviation can be seen by early middle age, and by at least another by the 1960s. An important aspect of the latter findings is that the individuals tested had participated in clinical trials as healthy volunteers, and had thus undergone extensive medical screening. These individuals were thus free of major medical or psychiatric conditions, and such declines actually represent a best case for normal aging. The same tests have been administered to patients with a variety of conditions including hypertension, heart disease, fibromyalgia, ADHD, epilepsy, narcolepsy, chronic fatigue syndrome, schizophrenia, and multiple sclerosis. When each of these populations is compared to age-matched healthy controls, cognitive deficits of one or more standard deviations are seen on, for example, the ability to focus attention (Wesnes, 2006). This body of research therefore indicates that major aspects of cognitive function decline with normal aging, and that a variety of mental and physical illnesses will further exacerbate this deterioration.

In recognition of cognitive declines in normal aging, the U.S. National Institute of Mental Health (NIMH) set up a working group in 1986 to agree criteria for the condition of age-associated memory impairment (AAMI) (Crook et al., 1986). The aims of the criteria were to identify those elderly individuals (50 years and older) who were aware of memory loss that had occurred gradually, who scored at least one standard deviation below the normal score for that of the young on a widely recognized test of memory (e.g., the Benton visual retention test; the logical memory subtest of the Wechsler memory scale, etc.), who showed evidence of adequate intellectual functioning (using the vocabulary subtest of the Wechsler adult intelligence scale), and who showed no evidence of dementia (as assessed by a mini-mental status examination score of 24 or above). The exclusion criteria were designed to exclude those whose poor performance was not due to normal aging, for example, being secondary to disease or actually being dementia. A number of clinical trials subsequently evaluated the effect of various pharmacological and herbal treatments for AAMI, with some limited success (for review, see Wesnes and Ward, 2000). The Fourth Edition of the Diagnostic and Statistical Manual of Mental Disorders of the American Psychiatric Association (DSM-IV) identified age-associated cognitive decline (AACD) as a condition which may be a focus of clinical attention (diagnostic code 780.9). The advantage of AACD is that it extended the range of impairments from simply memory to cognitive functioning in general, thus encompassing attention, information processing, and a range of other aspects now known to deteriorate with aging. The definition was for a "decline in cognitive functioning consequent to the aging process that is within normal limits given the person's age. Individuals with this condition may report problems remembering names or appointments or may experience difficulty in solving complex problems. This category should be considered only after it has been determined that the cognitive impairment is not attributable to a specific mental disorder or neurological complaint" (p. 684, DSM-IV).

However, regulatory bodies have not accepted AAMI, AACD, or other similar conditions as legitimate conditions for drug registration, and much of the focus of drug development in the past decade has moved to the condition of MCI (Petersen and Morris, 2005). However, the criteria for an individual to be classified for this condition is to be 1.5 standard deviations poorer than age-matched controls on a recognized test of memory, which limits the condition to less than 10% of the population, which thus has no relevance for the majority of the population who are experiencing age-related cognitive decline. Further, despite some very large clinical trials of potential treatments for MCI, only occasional findings of enhancements have been identified in the condition (e.g., Newhouse et al., 2012). Part of the problem was that the endpoint of many trials was the rate of conversion to Alzheimer's disease, which required long-term trials with large samples of patients, and the fact that in many trials the expected rate of conversion did not occur in the placebo-treated groups.

NATURAL PRODUCTS AS COGNITION ENHANCERS

Amphetamine was the first synthetic compound shown to improve human cognitive function (e.g., Mackworth, 1965), and both caffeine and nicotine are widely recognized to improve aspects of cognitive function such as attention, even in non-habitual users (e.g., Wesnes and Warburton, 1984; Haskell et al., 2005). There is a commonly held opinion that normal individuals, especially the young, are operating at optimum levels and cannot be enhanced. However, there exist hundreds of studies that demonstrate acute improvements to attention and memory in healthy student populations with a wide variety of substances such as oxygen (Moss et al., 1998), chewing gum (Wilkinson et al., 2000), nicotine (Wesnes and Warburton, 1984), caffeine (Haskell et al., 2005), gingko biloba (Kennedy et al., 2000), amphetamine, and methamphetamine (Silber et al., 2006).

The reticence of regulatory bodies to accept age-related cognitive decline as a suitable condition for treatment is obviously at odds with current general medical practice which seeks to treat a huge variety of other age-related conditions, ranging from failing hearing and eyesight to hip replacement. In the absence of regulatory acceptance or any consistent pressure from advocate groups, individuals the world over are left alone to seek to attempt to preserve their cognitive abilities as they age through a variety of techniques including physical and mental exercise (e.g., brain training), as well as by taking "smart drugs."

An alternative approach available to individuals who wish to minimize age-related declines in mental efficiency is to seek various natural and nutritional substances which can be obtained "over the counter." Certainly, there has been a large research effort over recent decades to evaluate the effects of natural therapies upon cognitive functioning. While many naturally occurring plant extracts are commonly misconstrued to be "safe," the use in Eastern cultures over millennia of substances such as ginkgo biloba and ginseng has identified the general absence of side effects of such products. *Ginkgo biloba*, e.g., has been the subject of enduring worldwide research interest for the past four decades, and a large and generally consistent body of research identifying positive effects on cognitive function has been identified by various research groups in healthy young and elderly volunteers (e.g., D'Angelo et al., 1986; Brautigam et al., 1998; Kennedy et al., 2000) and mildly cognitively impaired elderly patients (e.g., Wesnes et al., 1987; Rai et al., 1991; Kleijnen and Knipschild, 1992). While large well-controlled trials have shown the ability of ginkgo to treat the cognitive deficits in patients with Alzheimer's disease and other dementias (e.g., LeBars et al., 1997; Napryeyenko and Borzenko, 2007), a very large trial has shown that the compound is not able to prevent the development of dementia (DeKoskey et al., 2008); though it has to be acknowledged that none of the registered treatments for the disease have been demonstrated to do this either. A follow-up publication on the DeKoskey study, however, showed that gingko did not prevent the rates of cognitive decline over a median 6 year period in individuals aged 72–96 years (Snitz et al., 2009). On balance, while ginkgo clearly does not prevent cognitive decline in elderly individuals, or prevent the onset of dementia, it does appear to have beneficial cognitive effects on younger populations, and also patients with dementia.

Other research programmes have evaluated the effects of a combination of standardized extracts of *Ginkgo biloba* and *Panax ginseng*, showing improvements to working and episodic memory with acute doses in volunteers (Kennedy et al., 2001, 2002), patients with neurasthenia (Wesnes et al., 1997), and middle-aged volunteers (Wesnes et al., 2000). In each of these four studies, statistically reliable improvements were seen to the ability to successfully hold and retrieve information in short-term (working) and long-term (episodic) memory. There were no improvements to attention, or to the speed with which the information could be retrieved from memory. In the Wesnes et al. (2000) study, 256 healthy volunteers with a mean age of 56 years (range 38–66) were tested in a 14 week randomized placebo-controlled double-blind study, and over the period of the study, an overall improvement in the ability to store and retrieve information in memory of 7.5% was identified. A subsequent analysis of these data showed that the magnitude of the improvement was sufficient to counteract the decline that would have occurred in the population compared to a younger population of 18–25 years. This is evidence that age-related cognitive declines can be reversed by natural substances, which can be purchased over the counter in pharmacies, and offers individuals the chance to self-medicate with relatively safe substances to maintain cognitive function into late middle age.

In addition to the above research, a wide range of other natural substances have been found to improve cognitive function in the young and elderly, including caffeine (e.g., Smit and Rogers, 2000; Haskell et al., 2005; Smith et al., 2005), pyroglutamic acid (Grioli et al., 1990), phosphatidylserine (Crook et al., 1991), guanfacine (McEntee et al., 1991), huperzine (Wang, 1994; Zangara et al., 2003), ginseng + vitamins (Neri et al., 1995; Wesnes et al., 2003), *Panax ginseng* (Kennedy et al., 2001b; 2007; Sunram-Lea et al., 2004), acetyl-L-carnitine (Salvioli and Neri M, 1994; Thal et al., 1995), *Bacopa monniera* (Maher et al., 2002; Stough et al., 2008), sage (Tildesley et al., 2003; 2005; Scholey et al., 2008), *Melissa officinalis* (Kennedy et al., 2002; 2003), alpha lipoic acid (Hager et al., 2001), guarana (Kennedy et al., 2004), essential oils and aromas (Moss et al., 2003, 2008), pycnogenol (Ryan et al., 2008), and thiamine (Haskell et al., 2008). Benefits have also been identified with breakfast cereals (Wesnes et al., 2003; Ingwersen et al., 2007), energy drinks (e.g., Scholey and Kennedy, 2004), and chewing gum (Wilkinson et al., 2002).

The level of evidence required in this field should not differ from any other field of clinical research. Therefore, randomized, double-blind, placebo-controlled trials must be employed, and cognitive test systems utilized, which are fit-for-purpose for the requirement of detecting enhancements to various aspects of cognitive function. Only properly characterized substances should be tested, and standardized extracts are clearly essential to allow replication in different laboratories. Safety is of crucial concern; only substances which have an established safety profile should be evaluated, and safety should be carefully monitored in any clinical trial in this field. One large well-conducted study has just been accepted for publication, which satisfies these various requirements. The trial evaluated the effects of docosahexaenoic acid (DHA) on cognitive function in 485 elderly people who fulfilled the DSM-IV criteria described earlier for AACD (Yurko-Mauroa et al., 2010). Six months of supplementation was found to produce statistically reliable improvements to memory. Though the effect size of the improvement was small (0.19), as with the Wesnes et al. (2000) trial, the computerized cognitive assessment system used in the study had a normative database; and using this database the authors were able to identify that the effect reflected a 7 year reduction in normal aging (3.4 years when compared to placebo), which may well be attractive to the population studied (mean age 70 years). An important aspect of this study was the careful monitoring of safety, the adverse events not being different between the placebo and active treated groups. Besides being conducted to the rigorous standards required in this field and carefully monitoring safety, an important aspect of the study for future research was the presentation of effect sizes as well as an assessment of the potential "cognitive age-reducing" effect of treatment.

CONCLUSIONS

Until national and international regulatory bodies recognize the cognitive declines in normal aging as a legitimate target for drug development, individuals who wish to reduce age-related declines in mental efficiency should follow recommended guidelines for optimal levels of physical exercise and diet, as well as partaking in mental exercise; which while not widely demonstrated to help protect against mental decline, is almost certainly not harmful. In addition, the availability of well-standardized nutritional and natural products with established safety profiles, which have been convincingly demonstrated to possess cognition-enhancing properties, will be attractive to many. A number of widely available substances are already showing such potential; and the developers of such products have the opportunity to structure their research programmes to fulfil the various criteria and standards required by regulatory groups, in order to bring products to the market, which can help counteract the cognitive declines that appear to occur in the majority of individuals as they age, without compromising health and well-being.

REFERENCES

Bannerman L (2010). Bring smart drugs out of the closet, experts urge Government. *The Times*, London, February 27, 2010.

Bostrom N and Sandberg A (2009). Cognitive enhancement: Methods, ethics, regulatory challenges. *Science and Engineering Ethics* 15: 311–341.

Brautigam MRH, Blommaert FA, Verleye G, Castermans J, Steur ENH, and Kleijnen J (1998). Treatment of age-related memory complaints with Ginkgo biloba extract: A randomised double blind placebo-controlled study. *Phytomedicine* 5: 425–434.

Budson AE and Price BH (2005). Memory dysfunction. *New England Journal of Medicine* 352: 692–699.

Cakic V (2009). Smart drugs for cognitive enhancement: Ethical and pragmatic considerations in the era of cosmetic neurology. *Journal of Medical Ethics* 35: 611–615

Crook TH, Bartus RT, Ferris SH, Whitehouse P, Cohen GD, and Gershon S (1986). Age-associated memory impairment: Proposed diagnostic criteria and measures of clinical change. Report of a National Institute of Mental Health Workgroup. *Developmental Neuropsychology* 5: 295–306.

Crook TH, Tinklenberg J, Yesavage J, Petrie W, Nunzi MG, and Massari DC (1991). Effects of phosphatidylserine in age-associated memory impairment. *Neurology* 41: 644–649.

D'Angelo L, Grimaldi R, Caravaggi M, Marcoli E, Perucca E, Lecchini S, Frigo GM, and Crema A (1986). A double-blind, placebo-controlled clinical study on the effect of a standardised ginseng extract on psychomotor performance in healthy volunteers. *Journal of Ethnopharmacology* 16(1): 15–22.

DeKosky ST, Williamson JD, Fitzpatrick AL et al. (2008). *Ginkgo biloba* for prevention of dementia: A randomized clinical trial. *JAMA* 300: 2253–2262.

Grioli S, Lomeo C, Quattropani MC, Spignoli G, and Villardita C (1990). Pyroglutamic acid improves the age associated memory impairment. *Fundamentals of Clinical Pharmacology* 4: 169–173.

Haase J, Halama P, and Horr R (1996). Effectiveness of brief infusions with *Ginkgo biloba* special extract EGb 761 in dementia of the vascular and Alzheimer type. *Zeitschrift für Gerontologie und Geriatrie* 29(4): 302–309.

Hager K, Marahrens A, Kenklies M, Riederer P, and Münch G (2001). Alpha-lipoic acid as a new treatment option for Alzheimer type dementia. *Archives of Gerontology and Geriatrics* 32: 275–282.

Haskell CF, Kennedy DO, Milne AL, Wesnes KA, and Scholey AB (2008). The effects of L-theanine, caffeine and their combination on cognition and mood. *Biological Psychology* 77: 113–122.

Haskell CF, Kennedy DO, Wesnes KW, and Scholey AB (2005). Cognitive and mood improvements of caffeine in habitual consumers and habitual non-consumers of caffeine. *Psychopharmacology* 179: 813–825.

Hopfenmüller W (1994). Proof of the therapeutical effectiveness of a *Ginkgo biloba* special extract: Meta-analysis of 11 clinical trials in aged patients with cerebral insufficiency. *Arzneimittelforschung* 44(9): 1005–1013.

Ingwersen J, Defeyter MA, Kennedy DO, Wesnes KA, and Scholey AB (2007). A low glycaemic index breakfast cereal preferentially prevents children's cognitive performance from declining throughout the morning. *Appetite* 49: 240–244.

Kanowski S, Herrmann WM, Stephan K, Wierich W, and Horr R (1996). Proof of efficacy of the *Ginkgo biloba* special extract EGb 761 in outpatients suffering from mild to moderate primary degenerative dementia of the Alzheimer type or multi-infarct dementia. *Pharmacopsychiatry* 29(2): 47–56.

Kennedy DO, Scholey AB, Tildesley NTJ, Perry EK, and Wesnes KA (2002). Modulation of mood and cognitive performance following acute administration of Melissa officinalis (lemon balm). *Pharmacology, Biochemistry and Behavior* 72: 953–964.

Kennedy DO, Scholey AB, and Wesnes KA (2000). The dose-dependent cognitive effects of acute administration of ginkgo biloba in healthy young volunteers. *Psychopharmacology* 151: 416–423.

Kennedy DO, Scholey AB, and Wesnes KA (2001a). Differential, dose dependent changes in cognitive performance following acute administration of a *Ginkgo biloba/Panax ginseng* combination to healthy young volunteers. *Nutritional Neuroscience* 4: 339–412.

Kennedy DO, Scholey AB, and Wesnes KA (2001b). Dose dependent changes in cognitive performance following acute administration of Ginseng to healthy young volunteers. *Nutritional Neuroscience* 4: 259–310.

Kennedy DO, Scholey AB, and Wesnes KA (2002). Modulation of cognition and mood following administration of single doses of *Ginkgo biloba*, ginseng and a ginkgo/Ginseng combination to healthy young adults. *Physiology and Behaviour* 75: 1–13.

Kennedy DO, Wake G, Savelev S, Tildesley NTJ, Perry EK, Wesnes KA, and Scholey AB (2003). Modulation of mood and cognitive performance following acute administration of single doses of melissa officinalis (Lemon Balm) with human CNS nicotinic and muscarinic receptor-binding Properties. *Neuropsychopharmacology* 28: 1871–1881.

Kleijnen J and Knipschild P (1992). *Ginkgo biloba* for cerebral insufficiency. *British Journal of Clinical Pharmacology* 34: 352–358.

Le Bars PL, Katz MM, Berman N, Itil TM, Freedman AM, and Schatzberg AF (1997). A placebo-controlled, double-blind, randomised trial of an extract of *Ginkgo biloba* for dementia. North American EGb study group. *Journal of the American Medical Association* 278(16): 1327–1332.

Mackworth JF (1965). The effect of amphetamine on the detectability of signals in a vigilance task. *Canadian Journal of Psychology* 19: 104–109.

Maher B (2008). Poll results: Look who's doping. *Nature* 452: 674–675.

Maher BFG, Stough C, Shelmerdine A, Wesnes KA, and Nathan PJ (2002). The acute effects of combined administration of ginkgo biloba and bacopa monneira on cognitive function in humans. *Human Psychopharmacology* 17: 163–164.

Maurer K, Ihl R, Dierks T, and Frolich L (1997). Clinical efficacy of *Ginkgo biloba* special extract EGb 761 in dementia of the Alzheimer type. *Journal of Psychiatric Research* 31(6): 645–655.

McEntee WJ, Crook TH, Jenkyn LR, Petrie W, Larrabee GJ, and Coffey DJ (1991). Treatment of age-associated memory impairment with guanfacine. *Psychopharmacology Bulletin* 27: 41–46.

Moss M, Cook J, Wesnes KA, and Duckett P (2003). Aromas of rosemary and lavender essential oils differentially affect cognition and mood in healthy adults. *International Journal of Neuroscience* 113: 15–38.

Moss M, Howarth R, Wilkinson L, and Wesnes KA (2006). Expectancy and the aroma of Roman chamomile influence mood and cognition in healthy volunteers. *International Journal of Aromatherapy* 16: 63–73.

Moss MC, Scholey AB, and Wesnes KA (1998). Oxygen administration selectively enhances cognitive performance in healthy young adults: A placebo-controlled double-blind crossover study. *Psychopharmacology* 138: 27–33.

Napryeyenko O and Borzenko I (2007). *Ginkgo biloba* special extract in dementia with neuropsychiatric features. A randomised, placebo-controlled, double-blind clinical trial. *Arzneimittelforschung* 57: 4–11.

Neri M, Andermacher E, Pradelli JM, and Salvioli G (1995). Influence of a double blind pharmacological trial on two domains of well-being in subjects with age associated memory impairment. *Archives of Gerontology and Geriatrics* 21: 241–252.

Newhouse P, Kellar K, Aisen P, White H, Wesnes K, Coderre E, Pfaff A, Wilkins H, Howard D, and Levin ED (2012). Nicotine treatment of mild cognitive impairment: A six-month double-blind pilot clinical trial. *Neurology* 84: 91–101.

Nicholl CG, Lynch S, Kelly CA, White L, Simpson L, Simpson PM, Wesnes KA, and Pitt BMN (1995). The Cognitive Drug Research computerised assessment system in the evaluation of early dementia—Is speed of the essence? *International Journal of Geriatric Psychiatry* 10: 199–206.

Oken BS, Storzbach DM, and Kaye JA (1998). The efficacy of *Ginkgo biloba* on cognitive function in Alzheimer's disease. *Archives of Neurology* 55(11): 1409–1415.

Petersen RC and Morris JC (2005). Mild cognitive impairment as a clinical entity and treatment target. *Archives of Neurology* 62: 1160–1163.

Rai GS, Shovlin C, and Wesnes K (1991). A double-blind, placebo controlled study of *Ginkgo biloba* extract ('Tanakan') in elderly out-patients with mild to moderate memory impairment. *Current Medical Research Opinion* 12: 350–355.

Ryan J, Croft K, Mori T, Wesnes KA, Spong J, and Stough C (2008). An examination of the effects of the antioxidant Pycnogenol® on cognitive performance, serum lipid profile, Endocrinological and oxidative stress biomarkers in an elderly population. *Journal of Psychopharmacology* 22: 553–562.

Sahakian B and Morein-Zamir S (2007). Professor's little helper. *Nature* 450: 1157–1159.

Salthouse TA (2010). *Major Issues in Cognitive Aging*. Oxford University Press, New York. ISBN13: 9780195372151.

Salvioli G and Neri M (1994). L-acetylcarnitine treatment of mental decline in the elderly. *Drugs Experimental and Clinical Research* 20:169–176.

Schilcher H (1988). *Ginkgo biloba* L. Untersuchungen zur Qualität, Wirkung, Wirksamkeit und Unbedenklichkeit. *Zeitschift für Phytotherapie* 9: 119–127.

Scholey AB and Kennedy DO (2004). Cognitive and physiological effects of an "energy drink": An evaluation of the whole drink and of glucose, caffeine and herbal flavouring fractions. *Psychopharmacology* 176: 320–330.

Scholey A, Tildesley N, Ballard C, Wesnes KA, Tasker A, Perry EK, and Kennedy DO (2008). An extract of *Salvia* (sage) with anticholinesterase properties improves memory and attention in healthy older volunteers. *Psychopharmacology* 198: 127–139.

Silber BY, Croft RJ, Papafotiou K, and Stough C (2006). The acute effects of d-amphetamine and methamphetamine on attention and psychomotor performance. *Psychopharmacology* 187: 154–169.

Simpson PM, Surmon DJ, Wesnes KA, and Wilcock GR (1991). The cognitive drug research computerised assessment system for demented patients: A validation study. *International Journal of Geriatric Psychiatry* 6: 95–102.

Smit HJ and Rogers PJ (2000). Effects of low doses of caffeine on cognitive performance, mood and thirst in low and higher caffeine consumers. *Psychopharmacology* 152: 167–173.

Smith A, Sutherland D, and Christopher G (2005). Effects of repeated doses of caffeine on mood and performance of alert and fatigued volunteers. *Journal of Psychopharmacology* 19: 620–626.

Snitz BE, O'Meara ES, and Carlson MC (2009). Ginkgo biloba for preventing cognitive decline in older adults. *JAMA* 302: 2663–2670.

Solfrizzi V, Panza F, Torres F, Mastroianni F, Del Parigi A, Venezia A, and Capurso A (1999). High monounsaturated fatty acids intake protects against age-related cognitive decline. *Neurology* 52: 1553–1569.

Stough C, Downey LA, Lloyd J, Silber B, Redman S, Hutchison C, Wesnes KA, and Nathan P (2008). Examining the nootropic effects of *Bacopa monniera* on human cognitive functioning: 90 day, double-blind, placebo-controlled randomized trial. *Phytotherapy Research* 22: 1629–1634.

Sünram-Lea SI, Birchall RJ, Wesnes KA, and Petrini O (2004). The effect of acute administration of 400MG of *Panax ginseng* on cognitive performance and mood in healthy young volunteers. *Current Topics in Nutraceutical Research* 3: 251–254.

Szalavitz M (2009). Popping smart pills: The case for cognitive enhancement. January 6, 2009. http://www.time.com/time/printout/0,8816,1869435,00.html

Thal LJ, Carta A, Clarke WR et al. (1996). A one year multicenter placebo-controlled study of acetyl-L-carnitine in patients with Alzheimer's disease. *Neurology* 47: 705–711.

Wang ZX, Ren QY, and Shen YC (1994). A double-blind control study of Huperzine A and Piracetam in patients with age-associated memory impairment and Alzheimer's disease. *Neuropsychopharmacology* 10: 763S.

Wesnes KA (2003). The Cognitive Drug Research computerised assessment system: Application to clinical trials. In: De Deyn P, Thiery E, and D'Hooge R (Eds.). *Memory: Basic Concepts, Disorders and Treatment*. Uitgeverij Acco, Leuven, Belgium, pp. 453–472.

Wesnes KA (2006). Cognitive function testing: The case for standardisation and automation. *Journal of the British Menopause Society* 12: 158–163.

Wesnes KA (2010). Cognition enhancement—Expanding opportunities in drug development: Wake up to the MATRICS. *International Clinical Trials* June 2010: 66–72.

Wesnes KA, Faleni RA, Hefting NR, Houben JJG, Jenkins E, Jonkman JHJ, Leonard J, Petrini O, and van Lier JJ (1997). The cognitive, subjective and physical effects of a *Ginkgo biloba/Panax ginseng* combination in healthy volunteers with neurasthenic complaints. *Psychopharmacology Bulletin* 33: 677–683.

Wesnes KA, Hildebrand K, and Mohr E (1999). Computerised cognitive assessment. In: Wilcock GW, Bucks RS, and Rockwood K (Eds.). *Diagnosis and Management of Dementia: A Manual for Memory Disorders Teams*. Oxford University Press, Oxford, U.K., pp.124–136.

Wesnes KA, Luthringer R, Ambrosetti L, Edgar C, and Petrini O (2003). The effects of a combination of Panax ginseng, vitamins and minerals on mental performance, mood and physical fatigue in nurses working night shifts: A double-blind, placebo controlled trial. *Current Topics in Nutraceutical Research* 1: 169–174.

Wesnes KA, McKeith IG, Ferrara R, Emre M, Del Ser T, Spano PF, Cicin-Sain A, Anand R, and Spiegel R (2002). Effects of rivastigmine on cognitive function in dementia with Lewy bodies: A randomised placebo-controlled international study using the cognitive drug research computerised assessment system. *Dementia and Geriatric Cognitive Disorders* 13: 183–192.

Wesnes KA, Pincock C, Richardson D, Helm G, and Hails S (2003). Breakfast reduces declines in attention and memory over the morning in schoolchildren. *Appetite* 41: 329–331.

Wesnes KA, Simmons D, Rook M, and Simpson PM (1987). A double blind placebo controlled trial of Tanakan in the treatment of idiopathic cognitive impairment in the elderly. *Human Psychopharmacology* 2: 159–171.

Wesnes K and Warburton DM (1984). Effects of scopolamine and nicotine on human rapid information processing performance. *Psychopharmacology* 82: 147–150.

Wesnes KA and Ward T (2000). Treatment of age-associated memory impairment. In: Qizilbash N, Schneider L, Chui H, Tariot P, Brodaty H, Kaye J, and Erkinjuntti T (Eds.). *Evidence-Based Dementia Practice: A Practical Guide to Diagnosis and Management (with Internet Updates)*. Blackwell Science Publications, Oxford, U.K., pp. 639–653.

Wilkinson L, Scholey A, and Wesnes KA (2002). Chewing gum selectively improves aspects of memory in healthy volunteers. *Appetite* 38: 1–2.

Yurko-Mauroa K, McCarthya D, Romb D et al. (2010). Beneficial effects of docosahexaenoic acid on cognition in age-related cognitive decline. *Alzheimer's & Dementia* 6: 456–464.

Zangara A (2003). The psychopharmacology of huperzine A: An alkaloid with cognitive enhancing and neuroprotective properties of interest in the treatment of Alzheimer's disease. *Pharmacology, Biochemistry and Behaviour* 75: 675–686.

2 Measuring and Interpreting the Efficacy of Nutraceutical Interventions for Age-Related Cognitive Decline

Andrew Pipingas and David Camfield

CONTENTS

AGE-RELATED COGNITIVE DECLINE

The world's population is aging rapidly, with the proportion of the population over 60 growing at a rate of around 2% per annum in the developed world (United Nations, 2009). In the most developed regions, 264 million people (21% of the population) were estimated to be 60 years and older in 2009, with this figure projected to increase to around 416 million (33% of the population) by the year 2050 (United Nations, 2009). A major societal health issue for an aging population is not only the greater incidence of neurodegenerative disorders such as Alzheimer's disease but also the impact of normal age-related cognitive decline. Up to 50% of adults aged 64 and over have reported difficulties with their memory (Reid and MacLullich, 2006). In response to the reality of an aging population, there has been increased research focus in recent years on the development of effective interventions that may ameliorate the declines in cognitive ability.

Age-related deficits in cognitive abilities have been consistently reported across a range of cognitive domains including processing speed, attention, episodic memory, spatial ability, and executive function (Craik, 1994; Hultsch et al., 2002; Park et al., 1996, 2002; Rabbitt and Lowe, 2000; Salthouse, 1996; Schaie, 1996; Verhaeghen and Cerella, 2002; Zelinski and Burnight, 1997). While an overall decline in processing speed may explain some of the age-related variance in cognitive ability (Salthouse, 1996), processing speed alone cannot explain why a range of neuropsychological measures still remain significantly related to age once processing speed is taken into account (Pipingas et al., 2010). Further, there is growing evidence documenting a more rapid decline for certain cognitive functions in comparison to others, a finding which suggests that factors in addition to processing speed are also involved in cognitive decline (Buckner, 2004).

Hedden and Gabrieli (2004) differentiate three categories of cognitive decline in normal aging: (1) lifelong declines, including processing speed, working memory, and encoding of information into episodic memory; (2) late-life declines, including well-practiced tasks and those that require previous knowledge such as vocabulary and semantic knowledge; and (3) life-long stability, including autobiographical memory, emotional processing, and implicit memory. Another common distinction is often made between crystallized abilities (e.g., vocabulary and general knowledge), which remain stable until later life versus fluid abilities (e.g., attention, executive function, and memory) that decline from middle adulthood until late old age (Gunstad et al., 2006).

The reason for disproportionate declines across cognitive domains is that certain brain structures are more heavily affected by these processes than others during aging (Buckner, 2004; Grieve et al., 2005; Hedden and Gabrieli, 2004). Cortical volume decreases in the frontostriatal system are most strongly correlated with age-related cognitive decline (Bugg et al., 2006; Hedden and Gabrieli, 2004; Kramer et al., 2006; Schretlen et al., 2000; West, 1996). It has been estimated that decreases in Prefrontal Cortex (PFC) volume occur at a rate of around 5% per decade after the age of 20, in contrast to relatively small volumetric declines in the hippocampus and medial temporal lobe structures, which occur at a rate of around 2%–3% per decade (Hedden and Gabrieli, 2004; Kramer et al., 2006). In the absence of neurodegenerative diseases such as Alzheimer's (AD), the medial temporal lobes are relatively spared by the aging process (Albert, 1997).

Age-related reductions in the brain's gray matter are due to a number of factors including neuron apoptosis, neuron shrinkage, and lowered numbers of synapses; whereas reductions in white matter may be attributed in part to large age-related decreases in the length of myelinated axons (Fjell and Walhovd, 2010). During aging the brain suffers accumulative damage due to a number of cellular processes including reactive oxygen species formation (Halliwell, 1992), chronic inflammation (Sarkar and Fisher, 2006), redox metal accumulation (Connor et al., 1995), and homocysteine accumulation (Kruman et al., 2000). In addition to direct cellular damage, the brain is also indirectly impaired by insults to the cardiovascular system (Pase et al., 2010).

While the unfortunate decline in cognitive ability is ubiquitous, it is also evident that a great deal of variability exists in both the rate and the extent of cognitive decline experienced by individuals as they age (Shammi et al., 1998; Wilson et al., 2002). While some of the variance may be explained by genetic factors (e.g., Hariri et al., 2003; Price and Sisodia, 1998), there is also a great deal of research highlighting the importance of diet and lifestyle during aging. Chronic nutraceutical interventions hold great promise in ameliorating age-related cognitive decline because they simultaneously target multiple cellular mechanisms of cognitive decline. Many natural substances already identified through in vivo as well as clinical studies have been found to have potent anti-oxidant and anti-inflammatory properties as well as being of benefit to the cardiovascular system (Ghosh and Scheepens, 2009; Head, 2009; Kidd, 1999). In order to be able to accurately assess and interpret the clinical efficacy of these natural substances, it is recommended in the following review that highly accurate and specific cognitive tests are needed, together with the creation of normative databases that may be used to interpret clinical data in terms of years of cognitive function recovered.

TESTS OF COGNITIVE FUNCTION IN THE ELDERLY

Traditional ways of measuring cognitive decline in the elderly have involved clinical neuropsychological tests, tests that were often designed for the diagnosis of dementia. Commonly used dementia assessment scales include the mini mental state exam (MMSE; Folstein et al., 1975), the clinical dementia rating scale (Morris, 1997) and the cognitive subtest of the Alzheimer's disease assessment scale (ADAS-cog; Mohs et al., 1983). Such scales involve structured interviews to determine the presence of dementia, and if present then the severity of dementia symptoms. While these scales may be useful in the diagnosis of dementia, they lack the sensitivity to be able to assess cognitive decline in the normal population, and hence a strong ceiling effect would be expected. Another limitation of these clinical assessment scales is that they primarily assess global cognitive function, as opposed to specific cognitive domains that may be disproportionately affected by the aging process. Further, these tests often rely on pen and paper recording, which lacks the measurement precision associated with modern computerized testing.

For these reasons it is recommended that computerized tests that target specific cognitive abilities and have a high degree of sensitivity to fluctuations in cognitive function be used for testing the efficacy of interventions for age-related cognitive decline, rather than the more traditional dementia assessment scales. The cognitive drug research (CDR; Wesnes et al., 1999) neuropsychological assessment battery has previously been found to be a particularly sensitive measure for the detection of changes to cognitive function associated with chronic nutraceutical and dietary interventions (Ryan et al., 2008; Stough et al., 2008; Wesnes et al., 2000). The computerized mental performance assessment system (COMPASS; Scholey et al., 2010), which was developed at Northumbria University, United Kingdom, has also been found to be a sensitive measure in nutraceutical intervention trials, as has the Cambridge neuropsychological test automated battery (CANTAB; Cambridge Cognition, Cambridge, United Kingdom). Using large normative samples, the CANTAB has been found to be sensitive enough to detect declines in cognitive ability associated with normal aging as well as mild cognitive impairment preceding dementia (Égerházi et al., 2007; Robbins et al., 1994).

More recently our laboratory has developed a neuropsychological assessment battery designed specifically for the assessment of age-related cognitive decline, the Swinburne University computerized cognitive aging battery (SUCCAB; Pipingas et al., 2008, 2010). The SUCCAB is a computerized test battery consisting of nine tasks designed to capture the range of cognitive functions that decline with age: immediate/delayed word recall, simple reaction time, choice reaction time, immediate/delayed recognition, visual vigilance, *n*-back working memory, Stroop color-word, spatial working memory, and contextual memory (Pipingas et al., 2010). In preliminary studies from our laboratory, the tasks contained in the SUCCAB have been found to be highly sensitive to age-related cognitive decline (Pipingas et al., 2008).

COMPARING THE EFFECTS OF NUTRACEUTICAL INTERVENTIONS

When reporting the results of timed cognitive tasks times, it is informative to not only state the results of parametric statistical tests and p-values but also report the average millisecond improvement that is observed in the treatment group. Further, if the mean difference in reaction times has also been found to change in the placebo group, then the mean change in the treatment group above and beyond the change observed in the placebo group is the most informative metric.

An example of reported millisecond improvements in SUCCAB tasks that were found to be associated with a chronic intervention is provided by Pipingas et al. (2008). In a randomized, placebo-controlled trial, Pipingas et al. (2008) investigated the cognitive effects associated with 5 weeks supplementation with the *Pinus radiata* bark extract Enzogenol® in 42 males aged 50–65 years. Significant differences between the treatment and placebo groups were found for SUCCAB spatial working memory (SWM) and immediate recognition memory. The average reduction in reaction time for the SWM task in the Enzogenol group was found to be 65 ms, while reaction times for the control group were unchanged. Similarly, for SUCCAB immediate recognition memory, the average reduction in reaction time for the Enzogenol group was found to be 60 ms, while the reaction time in the control group increased by 7 ms (Pipingas et al., 2008).

The same approach can also be used for computerized cognitive tests that are scored according to accuracy (percent correct). An example of improvements to SWM accuracy associated with a nutraceutical intervention is provided by Stough et al. (2008). In a 90 day randomized placebo-controlled trial, Stough et al. (2008) investigated the cognitive effects of *Bacopa monniera* in 62 participants aged 18–60 years. Significant differences between the treatment and placebo groups were found for change in CDR SWM accuracy over the 3 month period. The average improvement in accuracy for the SWM task in the *Bacopa monniera* group was found to be 5.44%, while the average improvement in the control group was found to be 2.3%. If we make the assumption that the average improvement in accuracy for the control group is a measure of improvement due to practice effects, then we can see that there is still a 3.14% improvement observed in the treatment group above and beyond this value.

INTERPRETATION OF COGNITIVE CHANGE IN TERMS OF COGNITIVE AGING

When baseline normative data are collected in regard to the results of computerized cognitive tests across a wide age range, the regression coefficient of age can be compared with the treatment effect when a nutraceutical intervention is applied.

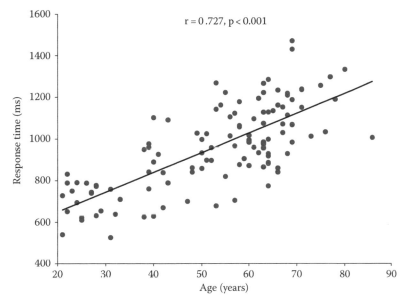

FIGURE 2.1   SUCCAB spatial working memory reaction time as a function of age. (Data from Pipingas, A. et al., *Curr. Top. Nutraceutical Res.*, 8(2–3), 79, 2010.)

This approach was recently used in our laboratory, whereby the SUCCAB cognitive battery was administered to 120 participants between the ages of 21 and 86 years (Pipingas et al., 2010). Significant correlations between both accuracy and reaction time measures were found across a wide range of cognitive domains including word recall, recognition memory, SWM, contextual memory, simple and choice reaction time, visual vigilance, n-back working memory, and Stroop reaction times (Pipingas et al., 2010). Regression analysis was used to predict reaction time and accuracy measures as a function of age for each cognitive domain from the SUCCAB. SWM ability was found to display the greatest degree of age-related decline amongst all cognitive measures, followed by contextual memory and immediate recognition tasks. SWM reaction time and accuracy as a function of age, together with Pearson's correlation coefficient, are displayed in Figures 2.1 and 2.2, respectively.

Here it can be seen that there is a steady linear increase in reaction time and a decrease in task accuracy from the age of 20 years onward. While further normative data for the SUCCAB is required in order to predict this relationship more accurately, this preliminary study nevertheless illustrates the strong relationship between age and cognitive decline.

An illustrative example is the previously mentioned intervention study by Pipingas et al. (2008) using *P. radiata* bark extract in elderly adults. Average SWM reaction time for the treatment group decreased from 1018 ms at baseline to 953 ms post-treatment. By using the regression equation for SWM reaction time as a function of age established in the SUCCAB normative study (Pipingas et al., 2010), the 65 ms improvement in reaction time can be interpreted as a SWM cognitive age-recovery of approximately 6.5 years (Figure 2.3).

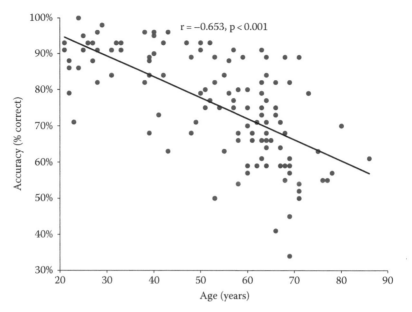

FIGURE 2.2 SUCCAB spatial working memory accuracy as a function of age. (Data from Pipingas, A. et al., *Curr. Top. Nutraceutical Res.*, 8(2–3), 79, 2010.)

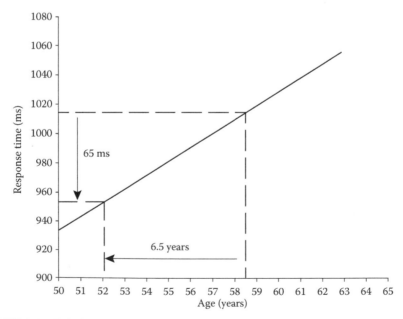

FIGURE 2.3 Relating improvements in reaction time to years recovered in cognitive age. Example from Pipingas et al. (2008) using *Pinus radiata* bark extract. (From Pipingas, A. et al., *Phytother. Res.*, 22(9), 1168, 2008.)

Another example of this approach was a study by Durga et al. (2007), which investigated the effects of 3 year folic acid supplementation on cognition. By comparing the regression coefficient of age with the treatment effect, the authors were able to calculate that 3 year supplementation conferred an individual the performance of someone 4.7 years younger for memory, 1.7 years younger for sensorimotor speed, 2.1 years younger for information processing speed, and 1.5 years younger for global cognitive function (Durga et al., 2007). Such interpretations of the data place cognitive change in a more meaningful context, changes that can be directly interpreted in terms of the amelioration of cognitive decline.

RECOMMENDATIONS FOR FUTURE STUDIES

In summary, in order to accurately assess and interpret the efficacy of nutraceutical interventions for the amelioration of age-related cognitive decline, a number of guidelines are recommended:

- Computerized cognitive test batteries be used in preference to traditional pen and paper neuropsychological tasks. These batteries should be difficult enough to not produce ceiling effects in younger age groups, and should be specific enough to cover different cognitive domains in separate sub-tests.
- In order to properly interpret the results of cognitive tests, it is advantageous to establish a large (ongoing) normative database from participants of all ages across the lifespan. By doing this, the relationship between age and reaction time and accuracy measures can be accurately established for the population.
- In addition to the conventional reporting of parametric statistics and p-values, it is also recommended that these findings be placed in a larger context, whereby the improvement for the treatment group (above and beyond improvements observed in the placebo group) be interpreted in terms of years recovered in cognitive ability.

REFERENCES

Albert, M. S. (1997). The ageing brain: Normal and abnormal memory. *Philosophical Transactions of the Royal Society B: Biological Sciences, 352*(1362), 1703–1709.

Buckner, R. L. (2004). Memory and executive function in aging and ad: Multiple factors that cause decline and reserve factors that compensate. *Neuron, 44*(1), 195–208.

Bugg, J. M., Zook, N. A., DeLosh, E. L., Davalos, D. B., and Davis, H. P. (2006). Age differences in fluid intelligence: Contributions of general slowing and frontal decline. *Brain and Cognition, 62*, 9–16.

Connor, J. R., Snyder, B. S., Arosio, P., Loeffler, D. A., and Lewitt, P. (1995). A quantitative analysis of isoferritins in select regions of aged, Parkinsonian, and Alzheimer's diseased brains. *Journal of Neurochemistry, 65*(2), 717–724.

Craik, F. I. M. (1994). Memory changes in normal aging. *Current Directions in Psychological Science, 3*(5), 155–158.

Durga, J., van Boxtel, M. P., Schouten, E. G., Kok, F. J., Jolles, J., Katan, M. B., and Verhoef, P. (2007). Effect of 3-year folic acid supplementation on cognitive function in older adults in the FACIT trial: A randomised, double blind, controlled trial. *Lancet, 369*(9557), 208–216.

Égerházi, A., Berecz, R., Bartók, E., and Degrell, I. (2007). Automated neuropsychological test battery (CANTAB) in mild cognitive impairment and in Alzheimer's disease. *Progress in Neuro-Psychopharmacology and Biological Psychiatry,* 31(3), 746–751.

Fjell, A. M. and Walhovd, K. B. (2010). Structural brain changes in aging: Courses, causes and cognitive consequences. *Reviews in the Neurosciences,* 21(3), 187–221.

Folstein, M. F., Folstein, S. E., and McHugh, P. R. (1975). 'Mini mental state'. A practical method for grading the cognitive state of patients for the clinician. *Journal of Psychiatric Research,* 12(3), 189–198.

Ghosh, D. and Scheepens, A. (2009). Vascular action of polyphenols. *Molecular Nutrition and Food Research,* 53(3), 322–331.

Grieve, S. M., Clark, C. R., Williams, L. M., Peduto, A. J., and Gordon, E. (2005). Preservation of limbic and paralimbic structures in aging. *Human Brain Mapping,* 25(4), 391–401.

Gunstad, J., Paul, R. H., Brickman, A. M., Cohen, R. A., Arns, M., Roe, D., Lawrence, J. J., and Gordon, E. (2006). Patterns of cognitive performance in middle-aged and older adults: A cluster analytic examination. *Journal of Geriatric Psychiatry and Neurology,* 19(2), 59–64.

Halliwell, B. (1992). Reactive oxygen species and the central nervous system. *Journal of Neurochemistry,* 59(5), 1609–1623.

Hariri, A. R., Goldberg, T. E., Mattay, V. S., Kolachana, B. S., Callicott, J. H., Egan, M. F., and Weinberger, D. R. (2003). Brain-derived neurotrophic factor val[66]met polymorphism affects human memory-related hippocampal activity and predicts memory performance. *Journal of Neuroscience,* 23(17), 6690–6694.

Head, E. (2009). Oxidative damage and cognitive dysfunction: Antioxidant treatments to promote healthy brain aging. *Neurochemical Research,* 34(4), 670–678.

Hedden, T. and Gabrieli, J. D. E. (2004). Insights into the ageing mind: A view from cognitive neuroscience. *Nature Reviews Neuroscience,* 5(2), 87–96.

Hultsch, D. F., MacDonald, S. W. S., and Dixon, R. A. (2002). Variability in reaction time performance of younger and older adults. *Journals of Gerontology—Series B Psychological Sciences and Social Sciences,* 57(2), 101–115.

Kidd, P. M. (1999). A review of nutrients and botanicals in the integrative management of cognitive dysfunction. *Alternative Medicine Review,* 4(3), 144–161.

Kramer, A. F., Fabiani, M., and Colcombe, S. J. (2006). Contributions of cognitive neuroscience to the understanding of behaviour and aging. In J. E. Birren and K. W. Schaie (Eds.), *Handbook of the Psychology of Aging* (6th edn., pp. 57–83). Burlington, MA: Elsevier Academic Press.

Kruman, I. I., Culmsee, C., Chan, S. L., Kruman, Y., Guo, Z., Penix, L., and Mattson, M. P. (2000). Homocysteine elicits a DNA damage response in neurons that promotes apoptosis and hypersensitivity to excitotoxicity. *Journal of Neuroscience,* 20(18), 6920–6926.

Mohs, R. C., Rosen, W. G., and Davis, K. L. (1983). The alzheimer's disease assesment scale: An instrument for assessing treatment efficacy. *Psychopharmacology Bulletin,* 19, 448–450.

Morris, J. C. (1997). Clinical dementia rating: A reliable and valid diagnostic and staging measure for dementia of the Alzheimer type. *International Psychogeriatrics,* 9(Suppl. 1), 173–176.

Park, D. C., Lautenschlager, G., Hedden, T., Davidson, N. S., Smith, A. D., and Smith, P. K. (2002). Models of visuospatial and verbal memory across the adult life span. *Psychology and Aging,* 17(2), 299–320.

Park, D. C., Lautenschlager, G., Smith, A. D., Earles, J. L., Frieske, D., Zwahr, M., and Gaines, C. L. (1996). Mediators of long-term memory performance across the life span. *Psychology and Aging,* 11(4), 621–637.

Pase, M. P., Pipingas, A., Kras, M., Nolidin, K., Gibbs, A. L., Wesnes, K. A., Scholey, A. B., and Stough, C. (2010). Healthy middle-aged individuals are vulnerable to cognitive deficits as a result of increased arterial stiffness. *Journal of Hypertension,* 28(8), 1724–1729.

Pipingas, A., Harris, E., Tournier, E., King, R., Kras, M., and Stough, C. K. (2010). Assessing the efficacy of nutraceutical interventions on cognitive functioning in the elderly. *Current Topics in Nutraceutical Research,* 8(2–3), 79–87.

Pipingas, A., Silberstein, R. B., Vitetta, L., Van Rooy, C., Harris, E. V., Young, J. M., Frampton, C. M., Sali, A., and Nastasi, J. (2008). Improved cognitive performance after dietary supplementation with a *Pinus radiata* bark extract formulation. *Phytotherapy Research,* 22(9), 1168–1174.

Price, D. L. and Sisodia, S. S. (1998). Mutant genes in familial Alzheimer's disease and transgenic models. *Annual Review of Neuroscience,* 21, 479–505.

Rabbitt, P. and Lowe, C. (2000). Patterns of cognitive ageing. *Psychological Research,* 63, 308–316.

Reid, L. M. and MacLullich, A. M. J. (2006). Subjective memory complaints and cognitive impairment in older people. *Dementia and Geriatric Cognitive Disorders,* 22(5–6), 471–485.

Robbins, T. W., James, M., Owen, A. M., Sahakian, B. J., McInnes, L., and Rabbitt, P. (1994). Cambridge neuropsychological test automated battery (CANTAB): A factor analytic study of a large sample of normal elderly volunteers. *Dementia,* 5(5), 266–281.

Ryan, J., Croft, K., Mori, T., Wesnes, K., Spong, J., Downey, L., Kure, C., Lloyd, J., and Stough, C. (2008). An examination of the effects of the antioxidant Pycnogenol on cognitive performance, serum lipid profile, endocrinological and oxidative stress biomarkers in an elderly population. *Journal of Psychopharmacology,* 22(5), 553–562.

Salthouse, T. A. (1996). The processing-speed theory of adult age differences in cognition. *Psychological Review,* 103 (3), 403–428.

Sarkar, D. and Fisher, P. B. (2006). Molecular mechanisms of aging-associated inflammation. *Cancer Letters,* 236(1), 13–23.

Schaie, K. W. (1996). Intellectual development in adulthood: The seattle longitudinal study. *Intellectual Development in Adulthood: The Seattle Longitudinal Study.* Cambridge University Press, New York, NY.

Scholey, A., Ossoukhova, A., Owen, L., Ibarra, A., Pipingas, A., He, K., Roller, M., and Stough, C. (2010). Effects of American ginseng (*Panax quinquefolius*) on neurocognitive function: An acute, randomised, double-blind, placebo-controlled, crossover study. *Psychopharmacology,* 212(3), 345–356.

Schretlen, D., Pearlson, G. D., Anthony, J. C., Aylward, E. H., Augustine, A. M., Davis, A., and Barta, P. (2000). Elucidating the contributions of processing speed, executive ability, and frontal lobe volume to normal age-related differences in fluid intelligence. *Journal of International Neuropsychological Society,* 6, 52–61.

Shammi, P., Bosman, E., and Stuss, D. T. (1998). Aging and variability in performance. *Aging, Neuropsychology, and Cognition,* 5(1), 1–13.

Stough, C., Downey, L. A., Lloyd, J., Silber, B., Redman, S., Hutchison, C., Wesnes, K., and Nathan, P. J. (2008). Examining the nootropic effects of a special extract of Bacopa monniera on human cognitive functioning: 90 Day double-blind placebo-controlled randomized trial. *Phytotherapy Research,* 22(12), 1629–1634.

United Nations. (2009). *World Population Prospects: The 2008 Revision, Highlights, Working Paper No. ESA/P/WP.210.* Department of Economic and Social Affairs Population Division, New York.

Verhaeghen, P. and Cerella, J. (2002). Aging, executive control, and attention: A review of meta-analyses. *Neuroscience and Biobehavioural Reviews,* 26, 849–857.

Wesnes, K. A., Ward, T., Ayre, G., and Pincock, C. (1999). Validity and utility of the cognitive drug research (CDR) computerised assessment system: A review following fifteen years of usage. *European Neuropsychopharmacology,* 9(Suppl. 5), S368.

Wesnes, K. A., Ward, T., McGinty, A., and Petrini, O. (2000). The memory enhancing effects of a *Ginkgo biloba/Panax ginseng* combination in healthy middle-aged volunteers. *Psychopharmacology,* 152(4), 353–361.

West, R. L. (1996). An application of prefrontal cortex function theory to cognitive aging. *Psychological Bulletin,* 120(2), 272–292.

Wilson, R. S., Beckett, L. A., Barnes, L. L., Schneider, J. A., Bach, J., Evans, D. A., and Bennett, D. A. (2002). Individual differences in rates of change in cognitive abilities of older persons. *Psychology and Aging,* 17(2), 179–193.

Zelinski, E. M. and Burnight, K. P. (1997). Sixteen-year longitudinal and time lag changes in memory and cognition in older adults. *Psychology and Aging,* 12(3), 503–513.

Part II

Vitamins and Nutrients

3 Actions of N-Acetylcysteine in the Central Nervous System

Implications for the Treatment of Neurodegenerative and Neuropsychiatric Disorders

David Camfield

CONTENTS

PHARMACOKINETICS

N-Acetylcysteine (NAC) is the N-acetyl derivative of cysteine, and is less reactive, less toxic, and less susceptible to oxidation than cysteine, as well as being more soluble in water. For these reasons it is a better source of cysteine than the parenteral administration of cysteine itself [1]. NAC is rapidly absorbed, with time to peak plasma levels (t_{max}) being 1.4 ± 0.7 h following oral administration. The average elimination half-life ($t_{1/2}$) has been reported to be 2.5 ± 0.6 h [2]. The bioavailability of NAC increases according to the dose, with the peak serum level being on average 16 µmol/L after 600 mg and 35 µmol/L after 1200 mg [3]. When taken orally NAC is readily taken up in the stomach and gut and sent to the liver where it is converted almost entirely to cysteine and used for glutathione (GSH) synthesis [4]. Cysteine that is not converted to GSH is capable of crossing the blood–brain barrier by means of sodium-dependent transport systems [5].

NAC IN THE TREATMENT OF DEMENTIA

CYSTEINE AVAILABILITY AND GLUTATHIONE PRODUCTION

The endogenous tripeptide GSH is the most abundant low-molecular-weight thiol in human cells and plays a central role in antioxidant defense from ROS [6] as well as protection against toxic compounds [7]. GSH is synthesized in tissue from the amino acids L-cysteine, L-glutamic acid, and glycine, where the availability of cysteine is generally the rate-limiting factor in its production [8]. Across a number of studies, supplementation using NAC has been found to be an effective way of increasing intracellular GSH levels, in clinical cases of deficiency [9–11] as well as amongst healthy volunteers [12]. Due to its effectiveness in raising GSH levels and protecting the human body from oxidative stress and toxins, NAC supplementation has been investigated as a treatment for a wide number of conditions including paracetamol intoxication, HIV, cancer, radiocontrast-induced nephropathy, and chronic obstructive pulmonary disease [6]. There is evidence to suggest that in Alzheimer's disease (AD), GSH levels are decreased in both cortical areas and the hippocampus [13]. For this reason NAC may play a neuroprotective role by restoring GSH levels to a normal state.

PRECLINICAL STUDIES

In vitro research by Chen and colleagues [14] revealed that pretreatment of cortical neurons with NAC protected mitochondrial function and membrane integrity under conditions of oxidative stress. Similarly, Olivieri and colleagues [15] found a neuroprotective effect for NAC in neuroblastoma cells exposed to oxidative stress. Pretreatment with NAC resulted in a reduction in oxidative stress resulting from exposure to amyloid-β proteins, as well as a reduction in phospho-tau levels. Research by Martinez and colleagues [16] revealed that aged mice fed NAC for 23 weeks performed better on a passive avoidance memory test than age-matched controls. Furthermore, lipid peroxide and protein carbonyl contents of the synaptic mitochondria were found to be significantly decreased in the NAC-supplemented animals compared to the controls.

Mice deficient in apolipoprotein E undergo increased oxidative damage to brain tissue and cognitive decline when maintained on a folate-free diet. Tchantchou et al. [17] found that dietary supplementation with NAC (1 g/kg diet) alleviated oxidative damage and cognitive decline, and restored GSH synthase and GSH levels to those of normal mice. There is also evidence to suggest that NAC supplementation may bring about a reduction in amyloid-β formation. In an animal model of AD using 12-months-old SAMP8 mice with an overexpression of the amyloid precursor protein, Farr et al. [18] found chronic administration of NAC to bring about significantly improved memory performance on the T-maze avoidance paradigm and lever press appetitive task. In an animal model of AD using TgCRND8 mice, Tucker et al. [19] found chronic treatment of NAC for 3 months to result in a significant reduction of amyloid-β in cortex. Similarly, research by Fu et al. [20] revealed that mice with amyloid-β peptide intracerebroventricularly injected performed significantly better in behavioral tests of memory and learning when pretreated with NAC in comparison to those without pretreatment. NAC pretreatment was also found to significantly reverse reductions in GSH and ACh.

CLINICAL STUDIES

While considerable experimental evidence exists for the neuroprotective role of NAC, there is currently a scarcity of clinical studies examining its efficacy in the treatment and prevention of dementia. One such study, a double-blind clinical trial of NAC in patients with probable AD was conducted by Adair and colleagues [21]. Forty-three patients were randomized to either placebo or 50 mg/kg/day NAC for 6 months and tested at baseline as well as at 3 and 6 months on MMSE as well as a cognitive battery. NAC supplementation was not found to be associated with significant differences in MMSE scores compared to placebo at either 3 or 6 months; however, patients receiving NAC showed significantly better performance on the letter fluency task compared to placebo, as well as a trend toward improvement in performance on the Wechsler memory scale immediate figure recall test. Further, ANOVA using a composite measure of cognitive tests favored NAC treatment at both 3 and 6 months.

In a 12 month open-label study of the efficacy of a nutraceutical and vitamin formulation in the treatment of early-stage AD, Chan et al. [22] administered 600 mg of NAC daily as part of a larger formulation of substances including ALCAR, alpha-tocopherol, B6, folate, and S-adenosyl methionine to 14 community-dwelling individuals. Participants were found to be significantly improved on the dementia rating scale (DRS) at both 6 and 12 months, with an overall improvement of 31%. However, a limitation of this study was that no placebo group was used for comparison, although the authors claim that the efficacy of their nutraceutical formulation exceeded that of historical placebos in previous studies of mild-to-moderate AD. In a follow-up study by the same group [23], the efficacy of the same nutraceutical formulation containing NAC was tested in a group of 12 nursing home residents with moderate to late-stage AD over a 9 month period. This time participants were randomized to either treatment or placebo. The nutraceutical formulation was found to delay cognitive decline as measured by the DRS for approximately 6 months, whereas in the placebo group a similar rate of decline was observed at only 3 months. While it is difficult to differentiate the efficacy of NAC from the other substances included in the formulation, these studies provide preliminary evidence for the efficacy of NAC in improving symptom severity in early-stage AD, and delaying the onset of decline in moderate to late-stage AD.

NAC IN THE TREATMENT OF NEUROPSYCHIATRIC DISORDERS

CYSTINE–GLUTAMATE ANTIPORTER

Group II metabotropic glutamate receptors (mGluR2/3) are located presynaptically on neurons in a large number of brain regions including the cortex, amygdala, hippocampus, and striatum [24], and play an important role in the regulation of the synaptic release of glutamate [25]. Stimulation of mGluR2/3 receptors by extracellular glutamate has an inhibitory effect on the synaptic release of glutamate [26]. Extracellular levels of glutamate are maintained primarily by means of the cystine–glutamate antiporter [27]. This Na^+-independent antiporter is bound to plasma membranes and is found ubiquitously throughout the body, while being located predominantly on glial cells in the human brain [28]. Cystine is the disulfide derivative of cysteine, consisting of two oxidized cysteine residues. When extracellular levels of cystine are increased in the brain, the antiporters on glial cells exchange extracellular cystine for intracellular glutamate. This leads to the stimulation of mGluR2/3 receptors and inhibition of synaptic glutamate release. For this reason, cysteine prodrugs have the ability to reduce the synaptic release of glutamate, with important implications for the treatment of psychiatric disorders.

INHIBITION OF GLUTAMATE RELEASE IN OBSESSIVE COMPULSIVE DISORDER

A number of magnetic resonance spectroscopy (MRS) studies of obsessive compulsive disorder (OCD) have revealed abnormal glutamate transmission in brain regions associated with cortico-striatal-thalamo-cortical (CSTC) neurocircuitry. Glx, a composite measure of glutamate, glutamine, homocarnosine, and GABA, has been found to be elevated in the caudate in OCD patients and to normalize again following SSRI treatment [29–33]. This finding is consistent with the metabolic hyperactivity in CSTC circuits, which is a known hallmark of OCD [34]. In contrast, Glx levels have been found to be decreased in the anterior cingulate [35], a finding that parallels the inverse relationship between anterior cingulate and basal ganglia volume in OCD patients [36]. Further evidence of elevated glutamate levels associated with OCD comes from a study by Chakrabarty et al. [37], who reported increased levels of glutamate in the CSF of drug-naive OCD patients.

A number of studies have investigated the effects of glutamate-modulating drugs in the treatment of OCD spectrum disorders. In an open-label study using Riluzole, a pharmacological agent which reduces synaptic glutamate release, Pittenger et al. [38–39] reported a significant decrease in symptoms in 13 treatment-resistant OCD patients over a 12 week period. However, it has been found that not all anti-glutamatergic agents have been found to be effective, with topiramate (Topamax) being found to exacerbate OCD symptoms and Lamotrigine found to be ineffective [36]. There have also been mixed results found to date for the efficacy of memantine in the treatment of OCD [40]. This is most likely due to differences in the mechanism of action associated with each of these varied compounds.

Due to the effects of inhibiting synaptic glutamate release through glial cystine–glutamate exchange, NAC also been investigated as a possible treatment for OCD. In a case study of a 58 year-old woman with SRI-refractory OCD, Lafleur et al. [41] reported that NAC augmentation of fluvoxamine resulted in a marked reduction in

OCD symptoms (Y-BOCS), and a clinically significant improvement in OCD symptoms. The NAC dose used in this study was titrated up from 1200 mg PO daily to 3000 mg daily over a 6 week period, and then maintained at this dosage level for a further 7 weeks. It is interesting to note that a reduction of 8 points on the Y-BOCS scale was noticed after only 1 week of treatment, which is indicative of rapid onset of treatment effects in comparison to conventional SSRI treatments for OCD, which may take several weeks for effects to become noticeable [42]. The possibility is also raised that there are acute effects associated with NAC use, whereby a patient with OCD may be able to use NAC on as-needed basis as an augmentation strategy for days when their symptoms are worse than usual. Two clinical trials are currently underway to test the efficacy of NAC in the treatment of OCD. Costa and colleagues from the University of Sao Paulo are conducting a 16-week intervention study using 3000mg/day NAC as an adjunctive treatment in OCD (NCT01555970) while Pittenger and colleagues from Yale University are conducting a 12-week study using 2400mg/day NAC for children aged 8–17 years (NCT01172275). It is hoped that these studies will provide important data as to the efficacy of NAC as treatment strategy for OCD.

A disorder related to OCD, which is classified as part of the OCD spectrum disorders is trichotillomania, characterized by repetitive hair pulling. Grant et al. [43] conducted a double-blind trial to assess the efficacy of NAC (1200–2400 mg/day) in 50 participants with trichotillomania over a 12 week period. Patients in the NAC treatment group were found to have significantly greater reductions in hair-pulling symptoms in comparison to placebo. Significant improvements were observed from 9 weeks of treatment onwards. Fifty-six percent of patients were found to be "much or very much improved" from the NAC treatment group in comparison to only 16% assigned to the placebo group. Another OCD spectrum disorder that NAC use has been investigated as a potential treatment strategy is compulsive nail-biting. Berk et al. [44] present three case studies where patients with a life-long history of compulsive nail-biting were found to benefit from NAC treatment. In the first case study, a 46 years old woman is reported to stop nail-biting altogether over a 7 months period using a dosage of 1000 mg NAC BID. In the second case study, a 44 years old woman is reported to stop nail biting after 4 months of treatment with NAC 1000 mg BID, and to have not recommenced on a 2 month follow-up. In the third case study, a 46 years-old patient was not reported to stop nail-biting all together, but noticed a reduction in this behavior after 28 weeks after starting NAC treatment. In addition to trichotillomania and nail-biting, NAC has also been reported to be effective in the reduction of skin-picking behavior [45].

A number of concerns will need to be addressed in assessing the suitability of NAC as a viable treatment option for OCD and related disorders, beyond a demonstration of efficacy. Considering the high degree of comorbidity of depression with OCD [46], it is necessary in further research to investigate the effects of NAC on mood. Although preclinical evidence to date is promising, it suggests that agonists acting on the mGluR2/3 receptors may dampen responses to stress and have a potential antidepressant effect [47–48], and the case study by Lafleur et al. [41] also reported a decrease in depression in their patient as measured by the HAM-D over the course of the trial. It may be important to monitor possible acute side effects of cognitive slowing that may result from over-regulation of glutamatergic tone with high-dose NAC use, as has been occasionally reported in relation to Riluzole [49].

NORMALIZATION OF EXTRACELLULAR GLUTAMATE IN SUBSTANCE ABUSE

Increased glutamate transmission in the nucleus accumbens has been found to be a mediator of drug-seeking behavior, while in the case of repeated use of drugs of abuse such as cocaine, a reduction in basal levels of extracellular glutamate in the nucleus accumbens are also observed [50,51]. Alterations in cystine–glutamate exchange and metabotropic glutamate receptor activity has also been found to regulate vesicular release of dopamine, another central neurotransmitter in reward-related behavior [26,52]. Due to its effects in inhibiting the synaptic release of glutamate in the CNS, NAC has been investigated for use in the treatment of substance abuse. Preclinical research by Baker et al. [52] revealed that systemic administration of NAC to cocaine-treated rats restored extracellular glutamate levels in the nucleus accumbens in vivo. Further, due to its effects on stimulating cystine–glutamate exchange, NAC was found to block cocaine-primed reinstatement of drug-taking behavior. In rats withdrawn from cocaine use, there is a change in the ability to create synaptic plasticity, which is related to alterations in prefrontal glutamatergic innervation of the nucleus accumbens core. Moussawi et al. [53] reported that the administration of NAC to cocaine-treated rats reversed the deficit in synaptic plasticity by indirect stimulation of mGlu2/3 and mGlu5 receptors, responsible for long-term potentiation and long-term depression, respectively.

In a pilot study investigating the effects of NAC on craving in 15 cocaine-dependent humans, LaRowe et al. [54] reported that 600 mg NAC administered at 12 h intervals over a 3 day period resulted in a significant reduction in the desire to use cocaine, interest in cocaine and cue viewing time, in the presence of cocaine-related cues. An open-label dose-ranging study of NAC in the treatment of cocaine dependence in humans was conducted by Mardikian et al. [55]. Twenty-three treatment-seeking cocaine-dependent patients were assigned to either NAC 1200, 2400, or 3600 mg/day over a 4 week trial. Sixteen of the patients completed the trial, and the majority of these either stopped using cocaine or significantly reduced their intake by the end of the trial. The higher doses of 2400 and 3600 mg/day were found to be more effective in treating cocaine-dependence, with higher retention rates in comparison to the lower dose of 1200 mg/day.

In human research using other drugs of abuse, similar results have been reported. In an open-label study investigating the use of NAC in cannabis addiction, Gray et al. [56] reported that 1200 mg NAC twice daily resulted in significant reductions in marijuana craving amongst 24 cannabis-dependent participants, as well as a trend-level reduction in marijuana usage, over a 4 week period. Knackstedt et al. [57] conducted a study to investigate the effect of nicotine on cystine–glutamate exchange in the nucleus accumbens and the efficacy of NAC in the treatment of nicotine addiction, using both animal and human data. Over a 21 day period, rats self-administered nicotine intravenously and 12 h following the last nicotine dose the brains were removed and immunoblotting was conducted in order to investigate changes in the catalytic subunit of the cystine–glutamate exchanger (xCT) or the glial glutamate transporter (GLT-1) in the nucleus accumbens, the ventral tegmental area (VTA), the amygdala, and the PFC. Decreased expression of the xCT was observed in the nucleus accumbens and the VTA, and decreased GLT-1 expression was observed in the nucleus accumbens. In the second part of the study, 29 nicotine-dependent human subjects were administered

2400 mg NAC/daily versus placebo for 4 weeks in a double-blind design. Smokers treated with NAC were found to report a greater reduction in the number of cigarettes smoked over the 4 week period in comparison to placebo, with a significant time × treatment group interaction when controlling for alcohol consumption.

Preclinical studies have demonstrated that levels of glutamate in the nucleus accumbens mediate reward-seeking behaviors in general [58], not only addictive behaviors related to pharmacological agents. A pilot study by Grant et al. [59] investigated the efficacy of NAC in the treatment of pathological gambling. Twenty-seven pathological gamblers were administered NAC over an 8 week period in an open-label design, starting with an initial dose of 600 mg/day that was titrated up over the first 4 weeks until a noticeable clinical improvement was seen, with a maximum possible dose being 1800 mg/day. The Yale-Brown obsessive compulsive scale modified for pathological gambling (PG-YBOCS) was used as the primary endpoint. PG-YBOCS scores were found to be significantly decreased by the end of the 8 weeks, with a mean effective NAC dose of 1476.9 ± 311.13 mg/day. Sixteen participants were classified as responders, defined by a 30% or greater reduction in PG-YBOCS score. Of these, 13 participants entered a double-blind follow-up phase, where they were randomized to either continue receiving their maximum dose from the open-label phase versus placebo over a 6 week period. At the end of the double-blind phase, 83.3% of the NAC group still met responder criteria in comparison to only 28.6% of those assigned to placebo. Although the first of its kind, this well-designed pilot study provides preliminary data in support of the efficacy of NAC in the treatment of pathological gambling.

A longer-term study by Bernardo et al. [60] investigating the effect of 2 g/day NAC on the use of alcohol, tobacco, and caffeine use in patients with bipolar disorder failed to find efficacy for NAC. Seventy-five participants were randomized to NAC or placebo over a 6 month period, with no significant changes in substance use observed over the length of the trial, with the exception of reduced caffeine intake in the NAC group at week 2. However, it is important to note that patients were selected for the study on the basis of clinical criteria for bipolar disorder, rather than a primary substance abuse disorder. For this reason, there were low rates of substance use in the cohort, which detracted from the statistical power necessary to determine a treatment effect.

INCREASING GLUTATHIONE PRODUCTION IN SCHIZOPHRENIA AND BIPOLAR DISORDER

CSF levels of GSH have been found to be decreased by 27% in drug-naive schizophrenia patients, while MRS has revealed that levels in the medial PFC are reduced by as much as 52% [61]. Decreased levels of GSH have also been reported in the caudate region in schizophrenia patients, as revealed by post-mortem assay [62]. There is evidence to suggest that decreased levels of GSH in schizophrenia are due to genetic polymorphisms in the genes responsible for GSH synthesis [63,64]. Due to the efficacy of NAC in boosting GSH levels in the CNS, it has been investigated for possible clinical benefits in the treatment of schizophrenia. Berk et al. [65] administered NAC 2000 mg/day versus placebo over a 6 month period to 140 patients with chronic schizophrenia, as augmentation to their regular antipsychotic medication. Patients receiving NAC were found to have a significant reduction in negative symptoms of schizophrenia as measured by the positive and negative symptoms scale (PNSS) as well as a reduction in clinical global impression of symptom severity (CGI-S) and CGI-improvement. These findings are corroborated by the research of Lavoie et al. [66], which has demonstrated that chronic NAC use at 2000 mg/day over 60 days improves mismatch negativity, a measure of NMDA receptor function, in schizophrenia patients. While the reason why restoring GSH levels and reducing oxidative stress in the brain brings about a clinical improvement in the negative symptoms of schizophrenia remains to be elucidated, these findings provide encouraging preliminary evidence for the efficacy of NAC as an augmentation strategy in treating this disorder.

Alterations in GSH metabolism have also been described as a feature of bipolar disorder as well as schizophrenia [67–69]. By applying the same rationale to bipolar disorder, Berk et al. [70] investigated whether boosting GSH levels through NAC supplementation would improve depressive symptoms in this disorder. Using a randomized controlled study design 75 individuals with bipolar disorder were administered NAC 2000 mg/day versus placebo over a 6 month period. NAC treatment was found to be associated with a significant improvement on the Montgomery Asberg depression rating scale (MASRS) after 20 weeks. The authors hypothesized that the clinical improvement could be attributed to the restoration of oxidative imbalances that are perturbed in bipolar disorder.

NAC SAFETY AND TOLERABILITY

Oral doses of NAC up to 8000 mg/day have not been known to cause clinically significant adverse reactions [10], and in a review of over 46 placebo-controlled trials, with NAC administered orally to a total of 4000 people, no significant adverse effects from NAC treatment were observed [4]. In relation to high oral doses of NAC (around 10,000 mg) typically used in cases of acetaminophen overdose, a review by Miller and Rumack [71] reported that mild symptoms such as headache, lethargy, fever, or skin rash occur in around 1%–5% of patients, while more moderate symptoms such as increased blood pressure, chest pain, hypertension, rectal bleeding, and respiratory distress occur in less than 1% of patients. One potential cause for concern over NAC supplementation was raised in a study by Palmer et al. [72] where rats receiving high-dose NAC in vivo for 3 weeks developed pulmonary arterial hypertension (PAH). The authors linked the finding of PAH to the conversion of NAC to S-nitroso-N-acetylcysteine (SNOAC) and a resultant hypoxia-mimetic effect. However, it is important to note that the rats were continuously exposed to a dose per weight roughly 40 times higher than the dose typically used in human studies. Good manufacturing practice (GMP) is important for NAC to ensure minimal oxidization to its dimeric form (di-NAC). Di-NAC is pharmacologically active at very low concentrations, and has immunological effects opposite to that of NAC [73]. For this reason it is important that any NAC obtained for chronic usage is from a trusted source.

SUMMARY

NAC is a substance with the potential to treat a diverse range of neuropathologies (see Table 3.1 for a summary of clinical research). In whole NAC is well tolerated, with a low incidence of adverse events in the dose ranges typically required for

TABLE 3.1

Clinical Studies Investigating the CNS Effects of NAC in Humans

Indication	N	Dosage (Daily)	Duration	Study Design	Outcomes	References
AD—probable	43	50 mg/kg	6 Months	RCT	Improved letter fluency	[21]
AD—early stage	14	600 mg	12 Months	Open label	Improvement on DRS	[22]
AD—mild to moderate	12	600 mg	9 Months	RCT	Delayed decline in DRS by 3 months	[23]
OCD	1	1200– 3000 mg	13 Weeks	Case study	23 point Y-BOCS decrease	[41]
Trichotillomania	50	1200– 2400 mg	12 Weeks	RCT	Significant reduction in hair pulling (56% patients)	[43]
Nail-biting	3	2000 mg	6 Months	Case studies	Significant reduction in symptoms	[44]
Cocaine dependence	15	1200 mg	3 Days	RCT crossover	Reduction in interest, desire to use, cue viewing time.	[54,74]
Cocaine dependence	23	1200, 2400, or 3600 mg	4 Weeks	Open label dosing	Majority of completers significantly reduced usage or stopped using	[55]
Cannabis dependence	24	2400 mg	4 Weeks	Open label	Significant reduction in craving, reduction in usage	[56]
Nicotine dependence	29	2400 mg	4 Weeks	RCT	Reduction in cigarettes smoked	[57]
Pathological gambling	27	600– 1800 mg	8 Weeks	Open label	16 responded, >30% reduction in PG-YBOCS	[59]
Pathological gambling	13	1477 mg avg	6 Weeks	RCT	83% in NAC group continued to meet responder criteria	[59]
Alcohol, tobacco and caffeine use	75	2000 mg	6 Months	RCT	No significant change in usage (insufficient power)	[60]

(continued)

TABLE 3.1 (continued)
Clinical Studies Investigating the CNS Effects of NAC in Humans

Indication	N	Dosage (Daily)	Duration	Study Design	Outcomes	References
Schizophrenia	140	2000 mg	6 Months	RCT	Significant improvement in PANSS and CGI-severity and CGI-improvement	[65]
Depression in bipolar disorder	75	2000 mg	6 Months	RCT	Significant improvement on MADRS	[70]

AD, Alzheimer's disease; CGI, clinical global impression; DRS, dementia rating scale; MADRS, Montgomery Asberg depression rating scale; OCD, obsessive compulsive disorder; PANSS, positive and negative symptoms scale; PG-YBOCS, Yale-Brown obsessive compulsive scale modified for pathological gambling; RCT, double-blind randomized placebo-controlled trial; Y-BOCS, Yale-Brown obsessive compulsive scale.

clinical effects. As a highly effective cysteine prodrug, NAC can both significantly boost endogenous GSH production and influence the synaptic release of glutamate due to its effects on cystine–glutamate exchange. For these reasons, NAC can be used as a means of ameliorating symptoms in wide range of disorders of the CNS including neurodegenerative disorders as well as obsessive-compulsive spectrum disorders, substance abuse disorders, behavioral addictions, schizophrenia, and bipolar disorder. Further large-scale trials of NAC are warranted in order to better establish clinically effective dosage ranges and treatment schedules for these varied neurodegenerative and neuropsychiatric conditions.

REFERENCES

1. Bonanomi, L. and A. Gazzaniga, Toxicology, pharmacokinetics and metabolism of acetylcysteine. *European Journal of Respiratory Disease*, 1980. **111**: 45–51.
2. Pendyala, L. and P.J. Creaven, Pharmacokinetic and pharmacodynamic studies of N-acetylcysteine, a potential chemopreventive agent during a Phase I trial. *Cancer Epidemiology Biomarkers and Prevention*, 1995. **4**(3): 245–251.
3. Allegra, L. et al., Human neutrophil oxidative bursts and their in vitro modulation by different N-acetylcysteine concentrations. *Arzneimittel-Forschung/Drug Research*, 2002. **52**(9): 669–676.
4. Atkuri, K.R., J.J. Mantovani, and L.A. Herzenberg, N-Acetylcysteine-a safe antidote for cysteine/glutathione deficiency. *Current Opinion in Pharmacology*, 2007. **7**(4): 355–359.
5. Smith, Q.R., Transport of glutamate and other amino acids at the blood-brain barrier. *Journal of Nutrition*, 2000. **130**(4 Suppl.): 1016S-1022S.
6. Aitio, M.L., N-acetylcysteine—Passe-partout or much ado about nothing? *British Journal of Clinical Pharmacology*, 2006. **61**(1): 5–15.
7. Townsend, D.M., K.D. Tew, and H. Tapiero, The importance of glutathione in human disease. *Biomedicine and Pharmacotherapy*, 2003. **57**(3): 145–155.

8. Wu, G. et al., Glutathione metabolism and its implications for health. *Journal of Nutrition*, 2004. **134**(3): 489–492.

9. Skrzydlewska, E. and R. Farbiszewski, Protective effect of N-acetylcysteine on reduced glutathione, reduced glutathione-related enzymes and lipid peroxidation in methanol intoxication. *Drug and Alcohol Dependence*, 1999. **57**(1): 61–67.

10. De Rosa, S.C. et al., N-acetylcysteine replenishes glutathione in HIV infection. *European Journal of Clinical Investigation*, 2000. **30**(10): 915–929.

11. Bridgeman, M.M.E. et al., Cysteine and glutathione concentrations in plasma and bronchoalveolar lavage fluid after treatment with N-acetylcysteine. *Thorax*, 1991. **46**(1): 39–42.

12. Roes, E.M. et al., Effects of oral N-acetylcysteine on plasma homocysteine and whole blood glutathione levels in healthy, non-pregnant women. *Clinical Chemistry and Laboratory Medicine*, 2002. **40**(5): 496–498.

13. Adams Jr, J.D. et al., Alzheimer's and Parkinson's disease: Brain levels of glutathione, glutathione disulfide, and vitamin E. *Molecular and Chemical Neuropathology*, 1991. **14**(3): 213–226.

14. Chen, G.J. et al., Transient hypoxia causes Alzheimer-type molecular and biochemical abnormalities in cortical neurons: Potential strategies for neuroprotection. *Journal of Alzheimer's Disease*, 2003. **5**(3): 209–228.

15. Olivieri, G. et al., N-acetyl-L-cysteine protects SHSY5Y neuroblastoma cells from oxidative stress and cell cytotoxicity: Effects on Beta-amyloid secretion and tau phosphorylation. *Journal of Neurochemistry*, 2001. **76**(1): 224–233.

16. Martinez, M., A.I. Hernández, and N. Martinez, N-Acetylcysteine delays age-associated memory impairment in mice: Role in synaptic mitochondria. *Brain Research*, 2000. **855**(1): 100–106.

17. Tchantchou, F. et al., N-acteyl cysteine alleviates oxidative damage to central nervous system of ApoE-deficient mice following folate and vitamin E-deficiency. *Journal of Alzheimer's Disease*, 2005. **7**(2): 135–138.

18. Farr, S.A. et al., The antioxidants alpha lipoic acid and N-acetylcysteine reverse memory impairment and brain oxidative stress in aged SAMP8 mice. *Journal of Neurochemistry*, 2003. **84**(5): 1173–1183.

19. Tucker, S. et al., Pilot study of the reducing effect on amyloidosis in vivo by three FDA pre-approved drugs via the Alzheimer's APP 5'untranslated region. *Current Alzheimer Research*, 2005. **2**(2): 249–254.

20. Fu, A.L., Z.H. Dong, and M.J. Sun, Protective effect of N-acetyl-l-cysteine on amyloid Beta-peptide-induced learning and memory deficits in mice. *Brain Research*, 2006. **1109**(1): 201–206.

21. Adair, J.C., J.E. Knoefel, and N. Morgan, Controlled trial of N-acetylcysteine for patients with probable Alzheimer's disease. *Neurology*, 2001. **57**(8): 1515–1517.

22. Chan, A. et al., Efficacy of a vitamin/nutriceutical formulation for early-stage Alzheimer's disease: A 1-year, open-label pilot study with an 16-month caregiver extension. *American Journal of Alzheimer's Disease and other Dementias*, 2009. **23**(6): 571–585.

23. Remington, R. et al., Efficacy of a vitamin/nutriceutical formulation for moderate-stage to later-stage alzheimer's disease: A placebo-controlled pilot study. *American Journal of Alzheimer's Disease and other Dementias*, 2009. **24**(1): 27–33.

24. Wright, R.A. et al., [3H]LY341495 binding to group II metabotropic glutamate receptors in rat brain. *Journal of Pharmacology and Experimental Therapeutics*, 2001. **298**(2): 453–460.

25. Schoepp, D.D., Unveiling the functions of presynaptic metabotropic glutamate receptors in the central nervous system. *Journal of Pharmacology and Experimental Therapeutics*, 2001. **299**(1): 12–20.

26. Moran, M.M. et al., Cystine/glutamate exchange regulates metabotropic glutamate receptor presynaptic inhibition of excitatory transmission and vulnerability to cocaine seeking. *Journal of Neuroscience*, 2005. **25**(27): 6389–6393.
27. Baker, D.A. et al., The origin and neuronal function of in vivo nonsynaptic glutamate. *Journal of Neuroscience*, 2002. **22**(20): 9134–9141.
28. Pow, D.V., Visualising the activity of the cystine-glutamate antiporter in glial cells using antibodies to aminoadipic acid, a selectively transported substrate. *GLIA*, 2001. **34**(1): 27–38.
29. Moore, G.J. et al., Case study: Caudate glutamatergic changes with paroxetine therapy for pediatric obsessive-compulsive disorder. *Journal of the American Academy of Child and Adolescent Psychiatry*, 1998. **37**(6): 663–667.
30. Rosenberg, D.R. et al., Increased medial thalamic choline in pediatric obsessive-compulsive disorder as detected by quantitative in vivo spectroscopic imaging. *Journal of Child Neurology*, 2001. **16**(9): 636–641.
31. Rosenberg, D.R. et al., Decrease in caudate glutamatergic concentrations in pediatric obsessive-compulsive disorder patients taking paroxetine. *Journal of the American Academy of Child and Adolescent Psychiatry*, 2000. **39**(9): 1096–1103.
32. Rosenberg, D.R., S.N. MacMillan, and G.J. Moore, Brain anatomy and chemistry may predict treatment response in paediatric obsessive-compulsive disorder. *International Journal of Neuropsychopharmacology*, 2001. **4**(2): 179–190.
33. Bolton, J. et al., Case study: Caudate glutamatergic changes with paroxetine persist after medication discontinuation in pediatric OCD. *Journal of the American Academy of Child and Adolescent Psychiatry*, 2001. **40**(8): 903–906.
34. Saxena, S. et al., Neuroimaging and frontal-subcortical circuitry in obsessive-compulsive disorder. *British Journal of Psychiatry. Supplement*, 1998(35): 26–37.
35. Rosenberg, D.R. et al., Reduced anterior cingulate glutamatergic concentrations in childhood OCD and major depression versus healthy controls. *Journal of the American Academy of Child and Adolescent Psychiatry*, 2004. **43**(9): 1146–1153.
36. Pittenger, C., J.H. Krystal, and V. Coric, Glutamate-modulating drugs as novel pharmacotherapeutic agents in the treatment of obsessive-compulsive disorder. *NeuroRx*, 2006. **3**(1): 69–81.
37. Chakrabarty, K. et al., Glutamatergic dysfunction in OCD. *Neuropsychopharmacology*, 2005. **30**(9): 1735–1740.
38. Coric, V. et al., Beneficial effects of the antiglutamatergic agent riluzole in a patient diagnosed with obsessive-compulsive disorder and major depressive disorder. *Psychopharmacology*, 2003. **167**(2): 219–220.
39. Coric, V. et al., Riluzole augmentation in treatment-resistant obsessive-compulsive disorder: An open-label trial. *Biological Psychiatry*, 2005. **58**(5): 424–428.
40. Pasquini, M. and I. Berardelli, Anxiety levels and related pharmacological drug treatment: A memorandum for the third millennium. *Annali dell'Istituto Superiore di Sanita*, 2009. **45**(2): 193–204.
41. Lafleur, D.L. et al., N-acetylcysteine augmentation in serotonin reuptake inhibitor refractory obsessive-compulsive disorder. *Psychopharmacology*, 2006. **184**(2): 254–256.
42. Fineberg, N.A. and T.M. Gale, Evidence-based pharmacotherapy of obsessive-compulsive disorder. *International Journal of Neuropsychopharmacology*, 2005. **8**(1): 107–129.
43. Grant, J.E., B.L. Odlaug, and W.K. Suck, N-acetylcysteine, a glutamate modulator, in the treatment of trichotillomania: A double-blind, placebo-controlled study. *Archives of General Psychiatry*, 2009. **66**(7): 756–763.
44. Berk, M. et al., Nail-biting stuff? The effect of N-acetyl cysteine on nail-biting. *CNS Spectrums*, 2009. **14**(7): 357–360.
45. Odlaug, B.L. and J.E. Grant, N-acetyl cysteine in the treatment of grooming disorders. *Journal of Clinical Psychopharmacology*, 2007. **27**(2): 227–229.

46. Pigott, T.A. et al., Obsessive compulsive disorder: Comorbid conditions. *Journal of Clinical Psychiatry*, 1994. **55**(10 Suppl.): 15–32.
47. Cartmell, J. and D.D. Schoepp, Regulation of neurotransmitter release by metabotropic glutamate receptors. *Journal of Neurochemistry*, 2000. **75**(3): 889–907.
48. Schoepp, D.D. et al., LY354740, an mGlu2/3 receptor agonist as a novel approach to treat anxiety/stress. *Stress*, 2003. **6**(3): 189–197.
49. Pittenger, C. et al., Riluzole in the treatment of mood and anxiety disorders. *CNS Drugs*, 2008. **22**(9): 761–786.
50. Cornish, J.L. and P.W. Kalivas, Cocaine sensitization and craving: Differing roles for dopamine and glutamate in the nucleus accumbens. *Journal of Addictive Diseases*, 2001. **20**(3): 43–54.
51. Cornish, J.L. and P.W. Kalivas, Glutamate transmission in the nucleus accumbens mediates relapse in cocaine addiction. *Journal of Neuroscience*, 2000. **20**(15): RC 89.
52. Baker, D.A. et al., Neuroadaptations in cystine-glutamate exchange underlie cocaine relapse. *Nature Neuroscience*, 2003. **6**(7): 743–749.
53. Moussawi, K. et al., N-Acetylcysteine reverses cocaine-induced metaplasticity. *Nature Neuroscience*, 2009. **12**(2): 182–189.
54. Larowe, S.D. et al., Is cocaine desire reduced by N-acetylcysteine? *American Journal of Psychiatry*, 2007. **164**(7): 1115–1117.
55. Mardikian, P.N. et al., An open-label trial of N-acetylcysteine for the treatment of cocaine dependence: A pilot study. *Progress in Neuro-Psychopharmacology and Biological Psychiatry*, 2007. **31**(2): 389–394.
56. Gray, K.M. et al., N-Acetylcysteine (NAC) in young marijuana users: An open-label pilot study. *American Journal on Addictions*, 2010. **19**(2): 187–189.
57. Knackstedt, L.A. et al., The role of cystine-glutamate exchange in nicotine dependence in rats and humans. *Biological Psychiatry*, 2009. **65**(10): 841–845.
58. Kalivas, P.W. and N.D. Volkow, The neural basis of addiction: A pathology of motivation and choice. *American Journal of Psychiatry*, 2005. **162**(8): 1403–1413.
59. Grant, J.E., S.W. Kim, and B.L. Odlaug, N-acetyl cysteine, a glutamate-modulating agent, in the treatment of pathological gambling: A pilot study. *Biological Psychiatry*, 2007. **62**(6): 652–657.
60. Bernardo, M. et al., Effects of N-acetylcysteine on substance use in bipolar disorder: A randomised placebo-controlled clinical trial. *Acta Neuropsychiatrica*, 2009. **21**(6): 285–291.
61. Do, K.Q. et al., Schizophrenia: Glutathione deficit in cerebrospinal fluid and prefrontal cortex in vivo. *European Journal of Neuroscience*, 2000. **12**(10): 3721–3728.
62. Yao, J.K., S. Leonard, and R. Reddy, Altered glutathione redox state in schizophrenia. *Disease Markers*, 2006. **22**(1–2): 83–93.
63. Gysin, R. et al., Impaired glutathione synthesis in schizophrenia: Convergent genetic and functional evidence. *Proceedings of the National Academy of Sciences of the United States of America*, 2007. **104**(42): 16621–16626.
64. Tosic, M. et al., Schizophrenia and oxidative stress: Glutamate cysteine ligase modifier as a susceptibility gene. *American Journal of Human Genetics*, 2006. **79**(3): 586–592.
65. Berk, M. et al., N-acetyl cysteine as a glutathione precursor for schizophrenia-A double-blind, randomized, placebo-controlled trial. *Biological Psychiatry*, 2008. **64**(5): 361–368.
66. Lavoie, S. et al., Glutathione precursor, N-acetyl-cysteine, improves mismatch negativity in schizophrenia patients. *Neuropsychopharmacology*, 2008. **33**(9): 2187–2199.
67. Kuloglu, M. et al., Lipid peroxidation and antioxidant enzyme levels in patients with schizophrenia and bipolar disorder. *Cell Biochemistry and Function*, 2002. **20**(2): 171–175.
68. Abdalla, D.S.P. et al., Activities of superoxide dismutase and glutathione peroxidase in schizophrenic and manic-depressive patients. *Clinical Chemistry*, 1986. **32**(5): 805–807.

69. Andreazza, A.C. et al., Serum S100B and antioxidant enzymes in bipolar patients. *Journal of Psychiatric Research*, 2007. **41**(6): 523–529.

70. Berk, M. et al., N-acetyl cysteine for depressive symptoms in bipolar disorder-A double-blind randomized placebo-controlled trial. *Biological Psychiatry*, 2008. **64**(6): 468–475.

71. Miller, L.F. and B.H. Rumack, Clinical safety of high oral doses of acetylcysteine. *Seminars in Oncology*, 1983. **10**(Suppl. 1): 76–85.

72. Palmer, L.A. et al., S-Nitrosothiols signal hypoxia-mimetic vascular pathology. *Journal of Clinical Investigation*, 2007. **117**(9): 2592–2601.

73. Särnstrand, B. et al., N,N'-diacetyl-L-cystine—The disulfide dimer of N-acetylcysteine - Is a potent modulator of contact sensitivity/delayed type hypersensitivity reactions in rodents. *Journal of Pharmacology and Experimental Therapeutics*, 1999. **288**(3): 1174–1184.

74. LaRowe, S.D. et al., Safety and tolerability of N-acetylcysteine in cocaine-dependent individuals. *American Journal on Addictions*, 2006. **15**(1): 105–110.

4 Neurocognition and Micronutrients in the Elderly

Helen Macpherson

CONTENTS

INTRODUCTION

The concept of healthy cognitive aging is gaining importance as rapid population aging is taking place in the Western world. As a greater percentage of the population moves toward old age, their medical and primary care needs will be exacerbated, resulting in increased financial pressure on the health-care system. Furthermore, an aging population will result in an increased prevalence of age-related neurodegenerative disorders such as dementia. In 2005, it was estimated that 24 million people worldwide were living with dementia, and it is predicted that this quantity will double every 20 years to reach 81 million by the year 2040 (Ferri et al., 2005). Age constitutes the major risk factor for Alzheimer's disease (AD), and it is estimated that in developed countries such as the United States, approximately 13% of individuals over the age of 65 and 48% of persons over the age of 85 are affected by this disease (Thies and Bleiler, 2011). The cost of caring for those with dementia constitutes a significant financial burden on tax payers and emotional burden on family members. With longer life expectancy anticipated, there are many benefits to maintaining both physical and cognitive well-being, including enhanced quality of life and a lower risk of dementia onset.

Cognitive deterioration occurs across the life span and is a feature not only of AD, but also the normal aging process. In many cases the experience of cognitive decline may be a precursor to the development of AD (Goedert and Spillantini, 2006). Consequently, to delay the onset or rate of dementia, it may be necessary to target interventions to individuals prior to the appearance of cognitive decline. Currently there is a scientific interest in the potential of health and lifestyle interventions to improve cognitive function or slow the rate of decline in the elderly. There is evidence from randomized controlled trials to indicate that aerobic exercise programs (Baker et al., 2010), mental training (Valenzuela and Sachdev, 2009), and dietary supplements consisting of omega-3 fatty acids (Yurko-Mauro et al., 2010), herbal preparations (Mix and Crews, 2002; Pipingas et al., 2008), or vitamins (Durga et al., 2007) may be capable of improving cognitive function and potentially serve as interventions against cognitive decline.

Observations from prospective and epidemiological studies have demonstrated that maintaining adequate vitamin and nutritional status may be particularly important for cognitive function in the elderly. For instance, vitamin depletion has been shown to precede cognitive decline (Kado et al., 2005), and both intake of specific nutrients and circulating levels of vitamins in the blood have been correlated with cognitive function in healthy elderly (Perrig et al., 1997; Maxwell et al., 2005).

Vitamins and micronutrients are chemicals in the diet that do not belong to the major categories of fats, proteins, or carbohydrates and cannot be synthesized by the body in large enough quantities for normal requirements (Huskisson et al., 2007). Vitamins and mineral micronutrients must be introduced through food intake and this can become a problem for the elderly who are vulnerable to vitamin and mineral deficiencies due to decreased appetite caused by lower energy needs (Nieuwenhuizen et al., 2010) and reduced absorption in the gut (Baik and Russell, 1999). To date research in this area has focused on several important questions. First, can dietary interventions using micronutrients slow the rate of cognitive decline in those with dementia, and do they possess the potential to prevent the onset of dementia? Second, and of greater relevance to this chapter, can nutritional interventions with selected micronutrients improve cognitive performance or slow the rate of cognitive decline in healthy elderly? This chapter will address the latter question, with a focus on relationships between micronutrients and neurocognition in elderly humans. Predominantly, our discussion will center on recent studies that have investigated the association between cognition in healthy elderly and specific B group vitamins, antioxidant vitamins, and multivitamins.

AGE-RELATED NEUROCOGNITIVE CHANGES

Even in healthy elderly, changes in cognition occur across the life span, with memory decline widely held to be one of the hallmarks of advanced age. Decline occurs to processes that rely on complex, controlled, goal-oriented behavior, including performance monitoring, the generation of future goals, and the ability to adjust behavior in response to feedback (Budson and Price, 2005). These cognitive operations are referred to as executive function and are particularly important for effective working memory and episodic memory performance. In comparison, other more declarative forms of memory including vocabulary and verbal IQ remain relatively intact in older adults (Christensen, 2001). Crystallized intelligence, described as knowledge gained from cultural influences and experience tends to increase until age 60, rather than decrease with age (Jones and Conrad, 1933). Beyond the age of 70, crystallized abilities decline, albeit to a lesser degree than fluid intelligence (Christensen et al., 1994).

Disruptions in the frontal–striatal system and medial temporal cortical regions are thought to contribute to decline in executive function over the life span (Buckner, 2004). Evidence from the field of neuroimaging has corroborated these findings, demonstrating age-related changes in frontal neural activity during episodic memory, semantic memory, and working memory activation tasks (Cabeza, 2002; Phillips and Andrés, 2010). Loss of brain volume can occur as neurons become more susceptible to the effects of mitochondrial dysfunction, changes in metabolic rate, oxidative stress, and neuroinflammation (Floyd and Hensley, 2002). This decrease in brain volume exerts effects upon cognitive function, with age-related atrophy shown to be a predictor of cognitive decline on IQ measures (Rabbit et al., 2008). Volumetric imaging studies indicate that the frontal white matter is preferentially vulnerable to aging (Raz et al., 2005; Firbank et al., 2007). Small infarcts induce anterior white-matter deterioration (Pugh and Lipsitz, 2002), which in turn may decrease the efficiency of frontally mediated executive processes (Buckner, 2005).

Longitudinal studies have shown that cognitive change in healthy elderly is not a unitary process and that advancing age is accompanied by greater interindividual variance, particularly in the domains of memory (Rabbitt et al., 2004) and cognitive speed (Christensen, 2001; Wilson et al., 2004). The diversity of these trajectories indicates that cognitive aging does not occur at the same rate for all individuals and some will experience greater decline than others. Individual differences stemming from genetic risk factors (Alexander et al., 2007), education (Ardila et al., 2000), cardiovascular function (Beeri et al., 2009), physical activity (Larson et al., 2006), depression (Depp and Jeste, 2006), and nutritional status (Moreiras et al., 2007) represent a few of the more widely researched predictors of cognitive decline in the elderly. In some cases, poorer memory performers may be in the prodromal stage of an age-related neurodegenerative disease such as AD as many of the same cognitive domains to be affected by the normal aging process are exacerbated in AD (Collie and Maruff, 2000). AD is the most common form of dementia in the elderly and is characterized by the DSM-IV as a marked decline from previous functioning in short-term memory and a severe disruption to language, planning, or visual processing (American Psychiatric Association, 2000). Risk factors for AD include advanced age, genetic factors, vascular disease, hypercholesterolemia, hypertension, atherosclerosis, coronary heart disease, smoking, diabetes, and obesity (Carr et al., 1997; Duron and Hanon, 2008; Lange-Asschenfeldt and Kojda, 2008).

Computed tomography (CT) and magnetic resonance imaging (MRI) have shown that the neuropathological process of AD includes enlargement of ventricles due to substantial loss of brain tissue, nerve cells, synapse, and dendrites caused by the presence of neurofibrillary tangles and beta amyloid plaques (Rusinek et al., 1991; Petrella et al., 2003; Kidd, 2008). Pathologically, AD is characterized by beta amyloid plaques (Aβ) and neurofibrillary tangles caused by the deposition of abnormal proteins. Neutritic or senile plaques are extracellular deposits of Aβ, whereas tangles are intracellular aggregates formed by a hyperphosphorylated form of the microtubule-associated protein tau (Blennow et al., 2006). The pathogenic mechanism of AD has been suggested to be an imbalance between the production and clearance of Aβ in the brain, leading to neuronal degeneration and dementia (Hardy and Higgins, 1992). Damage initially occurs to the large cortical neurons in the temporal lobe and later in the association areas (Braak and Braak, 1991; Norfray and Provenzale, 2004).

Cholinesterase inhibitors are the most commonly used treatment for AD and are used to prevent the breakdown of acetylcholine (ACh), a neurotransmitter involved in learning, memory, and attention at the synaptic junctions (Lleo, 2007). By increasing the brain synaptic availability of acetylcholine, remaining neurons are able to function more effectively, but cognitive benefits fade as deterioration worsens and cholinesterase inhibitors are unable to reverse the process of this disease (Kidd, 2008).

Mild cognitive impairment (MCI) is a term used to define individuals who have experienced a decline in cognition that is greater than expected for their age and education level but is not severe enough to meet the criteria for dementia (De Mendonca et al., 2004). In many cases, MCI represents a preclinical stage of AD, particularly in those with the amnestic form of MCI who have been observed to develop dementia at a rate of approximately 10%–15% per year as compared with healthy controls who convert to dementia at a rate of 1%–2% per annum (Petersen, 2007). Trials of cholinesterase inhibitors in patients with MCI have shown only limited cognitive improvements and a high rate of adverse events suggesting that they are not suitable to serve as interventions in the dementia process (Birks and Flicker, 2006; Allain et al., 2007; Sobów and Kłoszewska, 2007). Research has shown that low levels of B vitamins (Clarke et al., 1998) and antioxidants (Rinaldi et al., 2003; Mecocci, 2004) exist in cases of MCI and AD indicating that low vitamin status may be associated with the neuropathological process of AD and possibly precede this age-related condition. Consequently, it has been posited that individuals at high risk of developing cognitive decline or dementia may experience benefits from dietary interventions consisting of vitamin supplements (Jelic and Winblad, 2003). Further evidence reviewed in the subsequent sections of this chapter may indicate that vitamins including B vitamins and antioxidants are also important for the maintenance of cognition in healthy elderly.

ROLE OF VITAMINS IN THE CENTRAL NERVOUS SYSTEM

B VITAMINS AND HOMOCYSTEINE

The B group vitamins belong to a group of micronutrients essential for healthy functioning of the brain and nervous system, and are required for energy metabolism. The vitamin B complex includes B_1 (thiamine), which plays a role in the metabolism of carbohydrates, B_{12} is a cofactor for two enzymes, methionine synthase and L-methylmalonyl-CoA mutase, and B_6 (pyridoxine) is required for the synthesis of multiple neurotransmitters including adrenaline, serotonin, dopamine, GABA, and tyramine. Folic acid, also referred to as folate, is involved in the metabolism of amino acids, the synthesis of nucleic acids, and formation of blood cells and nerve tissue. Vitamin B_2 (riboflavin) is required for the conversion of B_6 and folic acid into their coenzyme forms (Huskisson et al., 2007).

Vitamins, B_{12}, B_6, and folate have long been associated with cognitive functioning and are vital for maintaining brain and cardiovascular health (McCaddon, 2006). These members of the vitamin B complex lower concentrations of homocysteine, a sulfur-containing amino acid. Concentrations of homocysteine increase with age (Joosten et al., 1996) and have been associated with conditions of aging including dementia and cardiovascular disease (Selhub, 2006; Solfrizzi et al., 2006). Folate and vitamin B_{12} are necessary for the methylation of homocysteine to methionine and B_6 is required for homocysteine to be metabolized to cysteine (see Figure 4.1). When intake of folate or vitamin B_{12} is insufficient, disruptions to the methylation cycle

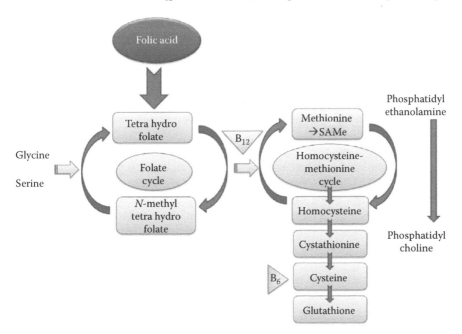

FIGURE 4.1 Folate and homocysteine–methionine cycle. (Adapted from Malouf, R. and Grimley Evans, J., *Cochrane Database Syst. Rev.*, 4, 2008, DOI: 10.1002/14651858. CD004514.pub2.)

occur, causing an intracellular accumulation of homocysteine, which may be toxic to neurons or lead to amyloid and tau protein accumulation (Obeid and Herrmann, 2006). Hyperhomocysteinemia may also induce gene expression in the brain and interact with specific neural targets including cellular receptors, intracellular proteins, and molecules of nitric oxide, leading to neuropathology (McCaddon, 2006).

Vitamin B deficiency may exert more direct effects on cognitive function via the role of these vitamins in the production of neurotransmitters in the brain. In order to maintain normal levels of homocysteine, vitamin B_6, B_{12}, and folic acid are required in the methylation of homocysteine to methionine. The amino acid methionine is essential to one-carbon metabolism, a series of biological processes crucial for DNA synthesis, DNA repair, and various methylation reactions (Mattson and Shea, 2003). In the central nervous system, methionine is required in the synthesis of S-adenosylmethionine (SAM), the sole donor for methylation reactions in the brain. The neurotransmitters dopamine, norepinephrine, and serotonin, as well as proteins, phospholipids, DNA, and myelin, are the products of these reactions (Selhub et al., 2000). Additionally, SAM is essential for maintenance of choline in the CNS as well as the production of acetylcholine and the antioxidant glutathione (Tchantchou et al., 2008). In accordance with the hypomethylation hypothesis, some loss of cognitive function in the elderly may be the end result of a lower production of SAM and consequently neurotransmitters due to B_{12} and folate deficiency (Calvaresi and Bryan, 2001).

A second hypothesis has proposed that impaired neurocognitive function in the elderly may stem from the injurious effects of elevated levels of homocysteine (Calvaresi and Bryan, 2001). In the brain, elevated homocysteine increases oxidative stress and DNA damage, triggers apoptosis, and imparts excitotoxic effects (Sachdev, 2005). In vitro, homocysteine disrupts neuronal homeostasis by multiple routes including N-methly-D-aspartate (NMDA) channel activation leading to excessive calcium influx and glutamate excitotoxicity (Ho et al., 2002). Hyperhomocysteinemia may also induce gene expression and interact with specific targets including cellular receptors, intracellular proteins, and molecules of nitric oxide, leading to neuropathology (McCaddon, 2006).

Elevated homocysteine also exerts a detrimental effect on the cardiovascular system via pro-coagulant actions toward platelets and vascular endothelium (De Koning et al., 2003). Vascular damage may be related to reduced bioavailability of endothelial nitric oxide, which is a powerful vasodilator (Obeid and Herrmann, 2006). Furthermore, homocysteine-induced damage to endothelial cells promotes arteriosclerosis, which is in turn associated with poorer cognitive function (Breteler et al., 1994; Knopman et al., 2001; Seshadri, 2006). Collectively these findings suggest homocysteine may affect cognitive function in the elderly, via direct effects on the brain or through indirect mechanisms operating on the cardiovascular system.

ANTIOXIDANTS AND OXIDATIVE STRESS

It is thought that free radicals and oxidative stress contribute to the neuronal changes responsible for cognitive decline in normal and pathological aging. Oxidative stress represents a disturbance in the equilibrium status of prooxidant reactions involving oxygen free radicals and those involving antioxidants (Valko et al., 2007). It is the maintenance of this prooxidant/antioxidant balance via redox homeostasis that is vital for healthy cellular function. Vitamin A, C, E, selenium, and coenzyme Q10 possess potent antioxidant actions and protect neural tissue from aggression by free radicals (Bourre, 2006). Beta-carotene is a precursor to vitamin A and is a powerful antioxidant. Vitamin C is known to interact synergistically with B complex vitamins and is essential for the metabolism and utilization of folic acid (Huskisson et al., 2007). Vitamin C levels are particularly high in the brain and this antioxidant is also necessary for the production of some neurotransmitters and the transformation of dopamine into noradrenalin (Bourre, 2006). Vitamin E is a lipid soluble chain-breaking antioxidant, which exercises neuroprotective effects against reactive oxygen species injuries (Cantuti-Castelvetri et al., 2000). Vitamin E interacts synergistically with selenium possibly increasing the antioxidant capacity of this vitamin (Bourre, 2006). Coenzyme Q10 is also a lipid soluble molecule that acts as an antioxidant and coenzyme for mitochondrial enzymes (Boreková et al., 2008).

According to the free radical hypothesis of aging, detrimental age-related changes take place in the brain as the result of an inability to cope with oxidative stress that occurs throughout the life span (Beckman and Ames, 1998). Oxidative stress can be defined as an excessive bioavailability of reactive oxygen species (ROS) caused by an imbalance between production of ROS and destruction of ROS by antioxidants (Kregel and Zhang, 2007). ROS are produced in the mitochondria in aerobic cells and cause damage to mitochondrial components and initiate degradative processes including damage to lipids, proteins, and DNA (Floyd and Carney, 1992; Cadenas and Davies, 2000). In particular, the brain is vulnerable to the effects of oxidative stress as it possesses reduced free radical scavenging ability and requires large quantities of oxygen (Floyd and Carney, 1992; Cantuti-Castelvetri et al., 2000). In the body, oxidative stress can lead to lipid peroxidation, a deleterious process that modifies the fluidity and permeability of neuronal membranes leading to an alteration of cellular functioning and damaged membrane bound receptors and enzymes (Mariani et al., 2005). Vitamin E exerts antioxidant activity in cell membranes and can inhibit lipid peroxidation (Isaac et al., 2008). Vitamins C and E have been demonstrated to decrease oxidative DNA damage markers (Boothby and Doering, 2005). Markers of oxidative stress were reduced, including high sensitivity C reactive protein, LDL oxidation F2-isoprostanes, and monocyte superoxide anion concentrations following supplementation with vitamin E for a period of 2 years (Devaraj et al., 2007). Subsequently, neuroprotective effects of antioxidants against oxidative damage in the brain may represent a mechanism that can enhance cognition or delay the rate of cognitive decline in the elderly.

In older adults, low intake and peripheral levels of antioxidants appear to be associated with greater risk of vascular disease (Jialal and Devaraj, 2003). Oxidative stress has been further implicated in the pathophysiological process of cardiovascular disease and stroke (Mariani et al., 2005). Throughout the progression of cardiovascular disease, low density lipoprotein (LDL) accumulates in the sub-endothelial space in arteries where it becomes oxidized leading to foam cell formation, endothelial dysfunction and injury, and generation of atherosclerotic lesions (Diaz et al., 1997). Cellular antioxidants protect against the cytotoxic effects of oxidized LDL, with vitamin C suggested to affect atherogenesis, with a high intake of this antioxidant protective against cerebrovascular disease (Gale et al., 1996).

OTHER MICRONUTRIENTS

Calcium, magnesium, and zinc are essential for optimum neural functioning, although their relationship to cognitive processes in the elderly has not been researched as extensively as the B vitamins or antioxidants. To summarize the roles of these minerals in the central nervous system, magnesium is essential for enzymes requiring vitamin B_1 as a cofactor, and is needed for the synthesis and action of ATP (Bourre, 2006). Neurotransmission is regulated by calcium, and zinc is required as a structural component of many proteins, hormones, hormone receptors, and neuropeptides (Huskisson et al., 2007). Zinc deficiencies are relatively common and can impair neuropsychological function even at a mild to moderate level of deficiency (Sandstead, 2000). Zinc increases antioxidant activity, and supplementation has been demonstrated to increase the activity of zinc-dependent antioxidant enzymes in healthy elderly subjects (Mariani et al., 2008).

Vitamin D is a steroid hormone that maintains levels of phosphorus, calcium, and bone mineralization. More pertinent to cognition, vitamin D may also play a biological role in neurocognitive function as receptors are located in regions of the brain important for planning, processing, and forming new memories (Buell and Dawson-Hughes, 2008). Although research into the importance of vitamin D for cognition in normal and pathological aging is only beginning to gain momentum, vitamin D has been suggested to exert protective effects against cardiovascular and cerebrovascular disease, peripheral artery disease inflammation, and to promote neuronal health (Buell and Dawson-Hughes, 2008; Buell et al., 2010). Low sunlight exposure, age-related decreases in cutaneous synthesis, and diets low in vitamin D contribute to the high prevalence of vitamin D inadequacy in the elderly (Holick, 2006).

VITAMIN STATUS AND COGNITION IN THE ELDERLY

Before turning to the literature examining the nature of the relationship between vitamin status and cognitive function, this chapter will cover a brief description of the measures used to assess cognition in the elderly. Most commonly, within large-scale epidemiological studies, cognitive performance has been measured in one of two ways. The first way of assessing cognitive performance has involved investigating individual cognitive domains using standardized cognitive or neuropsychological measures. The second way has relied on global cognitive measures such as the mini mental state examination (MMSE; Folstein et al., 1975) or the 3MS modified version of the MMSE. These assessments have been used to provide an overall estimate of an individual's level of functioning, without reference to any specific cognitive domain. Scores derived from these measures can then be correlated with reported dietary intake of vitamins or levels of vitamins taken from the serum or plasma. Other global measures of cognition have comprised a composite performance score consisting of an averaged or summed score across a range of individual tests from different cognitive domains. Most commonly global measures have been administered together with, at minimum, a verbal memory and verbal fluency assessment of executive function. As a separate entity from cognitive performance, cognitive decline has also been investigated using these instruments, not withstanding that decline can only be measured longitudinally.

B VITAMINS

Vitamin B_{12} depletion is common in old age (Wolters et al., 2003). The effects of atrophic gastritis are thought to contribute to insufficient B_{12} absorption in the elderly, rather than a lack of the micronutrient in the diet per se (Selhub et al., 2000). Observations of psychiatric symptoms associated with B_{12} deficiency have been documented since 1849 and over the past 50 years low B_{12} status has been related to memory impairment, personality change, and psychosis (McCaddon, 2006), many of which overlap with dementia (American Psychiatric Association, 2000). More recent evidence has shown that folate and homocysteine are equally important for mental function. Interestingly relationships between MCI and low serum folate levels (Quadri et al., 2004) and elevated plasma total homocysteine have been reported (Quadri et al., 2005). Based on these findings, the authors of these studies proposed that low folate and elevated homocysteine may predate dementia onset and that maintenance of micronutrient levels through dietary supplementation may aid in dementia prevention. While this is a promising assertion, it is still under debate as to whether deficient B_{12} and folate may contribute to the neurodegenerative process associated with AD, or whether deficiencies may reflect a consequence of the disease process (Seshadri, 2006).

Numerous studies have also investigated the association between B vitamins, homocysteine and cognitive function in healthy elderly. In the cross-sectional, Maine-Syracuse study of 812 young through to elderly subjects, a relationship was identified between vitamin B_6 and multiple cognitive domains encompassing visual–spatial organization, working memory, scanning-tracking, and abstract reasoning (Elias et al., 2006). Outcomes from the Third National Health and Nutrition Examination Survey revealed that individuals aged over 60 years, with elevated homocysteine accompanied by low folate levels, demonstrated poorer story recall than those with normal levels of homocysteine (Morris et al., 2001). Findings from the Singapore Longitudinal Ageing study, which investigated 451 high functioning Chinese elders, revealed that higher levels of folate were associated with better scores on a verbal learning instrument (Feng et al., 2006). A study by Riggs et al. (1996) revealed that middle aged to elderly men from the Boston Veterans Affairs Normative Aging study who had higher concentrations of plasma homocysteine and low concentrations of B_{12} and folate also displayed poorer spatial copying skills. Follow-up analysis from this study revealed that plasma folate became the strongest predictor of spatial copying ability, independent of homocysteine, 3 years later (Tucker et al., 2005).

Other trials have examined the relationship between B vitamins or homocysteine and cognition using a longitudinal design. Findings from the MacArthur Studies of Successful Aging showed that low folate levels in individuals aged in their 70s were predictive of cognitive decline 7 years later (Kado et al. 2005). It was found by Nurk et al. (2005) that elevated homocysteine at baseline predicted memory deficit after

6 years in those aged 65–67 who took part in the Hordaland Homocysteine study. Although a historical connection between B_{12} and cognition has been established, several epidemiological studies have failed to find an association between B_{12} concentration and cognitive status (Kado et al., 2005; Mooijaart et al., 2005; Nurk et al., 2005; Feng et al., 2006). Alternatively, measurement of methylmalonic acid (MMA), a product of amino acid metabolism, may provide a more useful diagnostic tool for B_{12} deficiency (Moretti et al., 2004). Results from the Oxford Healthy Aging Project showed that when holotranscobalamin (holoTC), the biologically active fraction of vitamin B_{12}, and MMA were used as measures of vitamin B_{12} status, each were associated with a more rapid cognitive decline on the MMSE over 10 years (Clarke et al., 2007). When considered together, outcomes from these longitudinal studies point to a more causal role of B_{12} and folate deficiency in cognitive decline.

ANTIOXIDANT VITAMINS

Antioxidant vitamin status in the blood has been associated with cognitive function in the elderly. The Third National Health and Nutrition Examination Survey investigated blood vitamin levels taken from a multiethnic sample of 4809 elderly residents of the United States (Perkins et al., 1999). Findings from this trial revealed that higher serum levels of vitamin E were associated with better memory recall of a three sentence story. Interestingly, no correlations between vitamin A, beta-carotene, selenium, or vitamin C with memory performance were identified. By contrast, in a sample of Swiss elderly aged 64–95 years, both past and current levels of ascorbic acid and beta-carotene were associated with free recall, recognition, and vocabulary performance (Perrig et al., 1997). In the Cognitive Change in Women study, a comprehensive neuropsychological test battery was utilized to assess the domains of memory, executive function, language, attention, and visual function in 526 women aged 60 or above with cognitive impairment (Dunn et al., 2007). Baseline results from this study revealed that low serum alpha-tocopherol status was cross sectionally associated with increased odds ratio of memory and mixed cognitive impairments. In contrast, previous vitamin E supplement intake was not associated with any type of cognitive impairment.

Intake of antioxidant vitamins from the diet has also been linked to cognitive function in older adults. The Chicago Health and Aging Project investigated the dietary habits of 2889 community residents, aged 65–102 using a food frequency questionnaire (Morris et al., 2002). The results of this study revealed that higher intake of vitamin E from the diet or from supplements was associated with a slower rate of cognitive decline over 3 years as measured by a combined cognitive score derived from tests of immediate and delayed story recall, a measure of perceptual speed and the MMSE. In contrast, findings from the Rotterdam study revealed that intake of beta-carotene, and not vitamin C or E, was associated with performance on the MMSE in elderly aged 55–95 years (Warsama Jama et al., 1996).

Use of vitamin E and C supplements by elderly men has been linked to better cognitive function at follow up 3–5 years later (Masaki et al., 2000). Results from the Cache County study revealed that participants aged 65 years or older who had a lower a intake of vitamin C, vitamin E, and carotene also had a greater acceleration of the rate of cognitive decline on the 3MS over 7 years compared to those with a higher use. The conclusions from this study were that higher antioxidant dietary vitamin C, vitamin E, and carotene may delay cognitive decline in the elderly (Wengreen et al., 2007). Separate analyses from the same trial demonstrated that those with the APOE e4 allele using vitamin C, E, or multivitamin supplements in combination with nonsteroidal anti-inflammatory drugs showed less cognitive decline over an 8 year period than those with the E4 allele who were taking antioxidant supplements (Fotuhi et al., 2008). It was suggested from these results that those at a higher genetic risk of developing AD may benefit from use of anti-inflammatories and antioxidant supplementation. Thus, the use of combined vitamin C and E supplements appears to influence cognition in the elderly. Long-term consumption (>10 years) of vitamin C and E supplements was associated with better performance on cognitive tests including verbal fluency, digit span backward, and a telephone administered version of the MMSE, in a large sample of community-dwelling elderly women who participated in the Nurse's Health Study (Grodstein, 2003). Women taking both supplements displayed equivalent cognitive function to individuals 2 years younger. Similarly, elderly subjects from the Canadian Study of Health and Ageing using combined vitamin C and E and supplements were less likely to experience significant cognitive decline on the 3MS after 5 years (Maxwell et al., 2005). Collectively, results from these studies suggest that chronic use of antioxidant supplements may help slow the rate of cognitive decline in the elderly.

ZINC

Few studies have investigated the relationship between zinc and cognition in older adults. In older adults recruited from European countries including Italy, Greece, Germany, France, and Poland, plasma zinc status was correlated with global cognitive functioning as measured by the MMSE (Marecllini et al., 2006). The observation that inhabitants from countries with diets rich in zinc exhibited superior cognitive function suggests that zinc supplementation may have beneficial effects on cognition in elderly who are deficient.

VITAMIN D

In a large sample of middle aged to elderly men free from dementia, levels of vitamin D have been associated with digit symbol substitution performance (Lee et al., 2009). Findings from the Nutrition and Memory in Elders study revealed a positive correlation between levels of vitamin D and measures of executive function and attention processing speed in over 1000 elderly subjects (Buell et al., 2009). As this relationship remained robust after adjustment for a number of factors including homocysteine, apoE4 allele, plasma B vitamins, and multivitamin use, it is possible that vitamin D may also represent an important predictor of cognitive function in the elderly. In a sample of elderly French community-dwelling women, dietary intake of vitamin D, as estimated from a food frequency questionnaire, has been associated with performance on a global cognitive measure (Annweiler et al., 2010), indicating intake of vitamin D may also be important for cognition.

In summary, the findings from population studies have demonstrated that in the elderly cognitive function is related to dietary intake, long-term supplement use, and circulating levels of individual vitamins. The results from these studies indicate that dietary intake of B vitamins, antioxidants, and selected minerals may comprise important predictors of cognitive function, particularly as individuals enter the later stages of the life span.

VITAMIN STATUS AND BRAIN STRUCTURAL CORRELATES IN THE ELDERLY

When considering the neurocognitive effects of dietary factors, it may be useful to examine the association between vitamin status and brain structural parameters. Most commonly structural MRI has been used for this purpose, with more comprehensive examinations undertaken at post mortem. In a fascinating study, blood nutrient levels were correlated with brain pathology in elderly nuns who lived in the same convent, ate at the same kitchen, and consequently had comparable environmental factors and overall lifestyle (Snowdon et al., 2000). Blood was collected and analyzed for nutrients, lipids, and nutrient markers. Following the death of 30 elderly nuns, a neuropathologist examined the brains for signs of atrophy, AD lesions (neurofibrillary tangles, senile plaques, and neutritic plaques), and atherosclerosis in the major arteries at the base of the brain. The results demonstrated that serum folate levels were negatively related to atrophy of the neocortex, particularly in those with a significant number of AD lesions in the neocortex. It was proposed that the association between low folate levels and cortical atrophy may not be entirely due to the deleterious effects of vascular disease, as folate was negatively correlated with subgroups of participants with minimal signs of vascular neuropathology such as arteriosclerosis and brain infarcts. Similarly, De Lau et al. (2009) have posited that neuropathology related to low B_{12} levels may arise from other nonvascular factors. The larger-scale, population-based, Rotterdam Scan study provided evidence that poorer vitamin B_{12} status in the normal range was significantly associated with greater severity of white-matter lesions, in particular periventricular white-matter lesions in healthy elderly. However, as B_{12} levels were not related to cerebral infarcts, these researchers hypothesized that the association between B_{12} levels and white-matter lesions may due to effects on myelin integrity in the brain, rather than through vascular mechanisms alone.

Vitamin B_{12} levels in healthy community-dwelling elderly have also been associated with brain volume loss over a 5 year period (Vogiatzoglou et al., 2008). The results of this study revealed that the decrease in brain volume was greater among those with lower vitamin B_{12} and holoTC levels. Specifically, those in the bottom tertile for B_{12} (<308 pmol/L) at baseline experienced the greatest rate of brain-volume loss. Based on these findings, the authors concluded that plasma vitamin B_{12} status may provide an early marker of brain atrophy and consequently may represent a potentially modifiable risk factor for cognitive decline in the elderly. Findings from a recent randomized trial indicated that 2 years B vitamin supplementation was capable of slowing the rate of brain atrophy (Smith et al., 2010). In this study elderly with MCI were assigned to a treatment of combined vitamins B_6, B_{12}, and folate or a placebo. Volumetric MRI scans revealed that the vitamin treatment reduced the rate of atrophy by approximately 30%, and biochemical measures indicated this was accompanied by a reduction in homocysteine. These findings indicate that reducing homocysteine via B vitamin supplementation may exert protective effects on brain structural parameters.

There is evidence that concentrations of homocysteine and B vitamins may be related to cortical volume in specific brain regions. Increasing plasma homocysteine levels have been associated with atrophy to cortical and hippocampal regions, but not the amygdala, in healthy elderly (Den Heijer et al., 2003) and thinner hippocampal width in seniors above 80 years of age (Williams et al., 2002). In a more recent investigation by Firbank et al. (2010), white-matter atrophy rate and hippocampal atrophy rate over 2 years was correlated with homocysteine levels independent of B_{12} levels in hypertensive individuals aged 70–89 years. In contrast, homocysteine was not related to rate of gray-matter atrophy. These findings may point to an effect of homocysteine on selected brain structures.

Erickson et al. (2008) have shown that greater intake of vitamins B_6 and B_{12} is related to gray-matter volume in specific cortical regions. In this study, participants aged between 59 and 79 years underwent MRI and a 3 day food diary was collected to determine dietary vitamin intake. Gray-matter volume in the left and right parietal regions was associated with B_{12}, whereas volume of the anterior and posterior cingulate, left parietal lobe, and superior frontal gyrus was related to B_6. Based on these findings, it was proposed that cognitive processes that rely on these regions could be most susceptible to vitamin B deficiency. To extrapolate from these findings, it may also be cognitive functions that rely on these cortical structures that demonstrate the best response to vitamin supplementation.

EVIDENCE FROM RANDOMIZED CONTROLLED TRIALS

The literature presented in this chapter has demonstrated that a relationship exists between cognitive function and vitamin status in the elderly, and further evidence has indicated that vitamins may exert effects on brain structure in older adults. While these findings demonstrate the importance of maintaining adequate B vitamin and antioxidant levels for brain health and cognitive function throughout old age, they do not provide direct evidence that supplementing the diet with vitamins will lead to enhanced cognitive function. Instead, trials using randomized, double blind, placebo-controlled methodology are necessary to address this question. The following section of this chapter will review findings from trials that have investigated the effects of B vitamins, antioxidants, zinc, and multivitamins on cognitive function in the elderly, using such methodology.

B VITAMINS

Promising results for B vitamins and cognitive function have been obtained from the Folic Acid and Carotid Intimamedia Thickness (FACIT) trial, which included over 800 adults aged 50–60 years (Durga et al., 2007). Participants in this trial were randomly allocated to either 800 μg folic acid, or a placebo each day for a period of 3 years. The results revealed that the folic acid treatment led to an improvement in information processing speed that was not evident for the placebo treatment. Benefits were also found for a composite score that included measures of information processing speed and memory.

Other less positive outcomes have been reported. Eussen et al. (2006) failed to identify any effects of B_{12} or combined B_{12} and folate supplementation on cognition in older persons with mild vitamin deficiency aged 70 or above. In this trial, participants allocated to either 1000 μg B_{12} alone or 1000 μg B_{12} combined with 400 μg folate daily for 24 weeks did not show any improvements on a battery of neuropsychological tests. Comparably a 4 month study into the effects of 4 months combined vitamin B_{16}, B_{12}, and folate supplementation in community-dwelling elderly did not identify any treatment-related improvements to neuropsychological task performance (Lewerin et al., 2005). In a cross-sectional study of young, middle aged, and older women, mixed results were uncovered regarding B_6, B_{12}, and folate supplementation (Bryan et al., 2002). While memory performance improved for the oldest group who received folate, participants supplemented with B_{12} or folate performed worse on a verbal fluency task than those who received B_6 or the placebo. The short treatment period coupled with the low dose of B_{12} used in this study (15 μg) may account for these results.

On the basis of the findings of Bryan et al. (2002) that the oldest participants demonstrated cognitive improvements with folate supplementation, it could be argued that the treatment effects of B vitamins may be specific to certain subgroups amongst the elderly, such as the very old or those with cognitive impairment. Support for this premise has been obtained from a study of patients with depleted vitamin B_{12} levels at baseline (Eastley et al., 2000). In this trial, 3 months intervention improved verbal fluency for individuals with cognitive impairment, while the same benefits were not observed for those with established dementia. Several smaller trials have demonstrated minor improvements to memory in individuals at risk of cognitive decline due to low vitamin status. For example, Deijen et al. (1992) identified improvements to long-term memory in vitamin B_6 deficient men, aged in their 70s following 3 months vitamin B_6 supplementation. It should be noted, however, that this effect could be partially attributed to a decrease in performance of the placebo group over this time period. Memory improvements were also observed in a small trial of elderly aged 70–90 years with low folate at baseline when treated for deficiency for a period of 2 months (Fioravanti et al., 1997). However, a trial in patients with ischemic vascular disease did not reveal any cognitive improvements to a letter digit coding test or a telephone interview of cognitive status following 12 months treatment with either folate and vitamin B_{12} or vitamin B_6 and riboflavin (Stott et al., 2005).

A Cochrane review conducted by Malouf and Grimley Evans (2008) indicated that there was not sufficient evidence to conclude that folate administered in isolation or combined with vitamin B_{12} improved cognition in the elderly. Based on a systematic review of randomized controlled trials, Balk et al. (2007) concluded that there was little beneficial effect of B_{12}, B_6, or folate supplementation on cognitive function and that there is still a need for longer, larger powered studies to accurately assess whether vitamin B vitamin supplementation is effective to slow cognitive decline or improve cognitive performance in the elderly. On the basis of their review, Balk et al. (2007) further articulated that only cognitive tests that adequately differentiate individual cognitive domains should be used to evaluate the efficacy of B vitamin supplementation.

In summary, evidence from intervention studies suggests that supplementation with vitamin B_{12} or folate may provide only limited improvements to processing speed and memory in the elderly.

ANTIOXIDANT VITAMINS

The effects of selected antioxidant supplementation on cognitive measures in elderly adults have been investigated in several large-scale placebo-controlled studies. Within some of these studies the primary aim was not to measure cognitive performance changes with supplementation and in some studies no cognitive data was available prior to supplementation. Participants in the Age-Related Eye Disease study were randomly assigned to receive daily antioxidants (vitamin C, 500 mg; vitamin E, 400 IU; beta-carotene, 15 mg), zinc and copper (zinc, 80 mg; cupric oxide, 2 mg), antioxidants plus zinc and copper, or placebo. After approximately 7 years, there was no difference between the treatment groups on any of the cognitive tests (Yaffe et al., 2004). A cognitive testing component including assessment of general cognition, verbal memory, and category fluency was added to the Physician's Health study, where elderly men supplemented their diet with 50 mg beta-carotene or a placebo on alternate days (Grodstein et al., 2007). There was no impact of treatment with beta-carotene for 3 years or less on cognitive performance, whereas treatment duration of at least 15 years provided significant benefits for verbal memory, cognitive status, and a composite of these measures. These findings indicate that long-term interventions, implemented at early stages of brain aging may provide cognitive benefits. Using the same outcome measures, subjects in the Women's Health study received vitamin E supplementation (600 IU) on alternate days or placebo (Kang et al., 2006). The results of this study showed no effect of the treatment on a global composite score after $9\frac{1}{2}$ years and it was concluded that long-term use of vitamin E supplements did not provide cognitive benefits among generally healthy older women.

In brief, there appears to be some advantage to long-term beta-carotene supplementation in the elderly; however, the mainly telephone-administered cognitive tests used in these trials may not be sensitive to the neurocognitive effects of antioxidant supplementation.

ZINC

Fewer studies have utilized RCT methodology to explicitly evaluate the cognitive effects of trace minerals such as zinc. In a trial of healthy adults aged 55–87 years, subjects received either 15 or 30 mg of zinc per day (Maylor et al., 2006). Treatment effects were assessed at 3 and 6 months using measures of visual memory, working memory, attention, and reaction time from the Cambridge Automated Neuropsychological Test Battery. At the 3 month testing period, spatial working memory was improved at both dosages; however, a detrimental effect of the lower dose was observed on an attentional measure, indicating that the beneficial treatment effects of zinc may be limited to specific cognitive domains.

MULTIVITAMINS

Multivitamins are a combination formula of the B vitamins and antioxidant vitamins described previously in this chapter, in addition to minerals such as calcium, magnesium, zinc, and iron. Dietary supplementation with multivitamins is relatively common in the elderly, with data from the U.S. National Health and Nutrition Examination Survey showing that 63% of individuals aged 60 years or above had used a dietary supplement in the past month and 40% had specifically used a multivitamin supplement (Radimer et al., 2004). In adults above the age of 75 years, this figure appears to be somewhat higher, with another study reporting that over 59% of this demographic supplemented their diet with multivitamins (Nahin et al., 2006). While these supplements are widely used in the older population, there are not a large number of studies that have used randomized controlled designs to investigate the effects of multivitamin supplementation on cognitive function in the elderly.

The findings from several RCTs investigating the effects of multivitamins in the elderly have not uncovered cognitive benefits. For instance, Cockle et al. (2000) did not identify improvements to reaction time, short-term memory, or recognition memory, in seniors following up to 24 weeks of multivitamin supplementation. Similarly, there were no improvements on measures of pattern recognition, IQ or symbol search after 24 weeks of multivitamin supplementation in women aged 60 and above (Wolters et al., 2005). In a 12 month trial conducted by McNeill et al. (2007), there were no alterations in verbal fluency or digit span performance in individuals aged 65 years or older. Once the age groups were separated into younger and older subjects, there was a small beneficial effect of the multivitamin treatment on verbal fluency for subjects over 75 years of age.

More recently, positive results have been obtained from a trial that investigated a combined multivitamin and herbal supplement. Specifically, benefits to verbal fluency and recall were identified in elderly aged 50–75 years after 4 months treatment with a complex antioxidant blend consisting of 34 vitamins, minerals, amino acids, lipids, and herbal extracts, all with antioxidant properties (Summers et al., 2010). Intake of herbs contributes significantly to intake of plant antioxidants (Dragland et al., 2003) and may contribute to synergistic benefits when in the presence of other antioxidant vitamins (Cantuti-Castelvetri et al., 2000). Subsequently, it is conceivable that combined multivitamin and herbal supplements may exert greater effects on cognition than formulas consisting of vitamins alone.

Several other smaller trials have examined the cognitive effects of multivitamin supplementation. In a study of only 20 subjects, there were no treatment related improvements to the MMSE following 12 months supplementation with a multivitamin or placebo in elderly with cognitive impairments (Baker et al., 1999). A recently published 3 month study of 47 young adults compared the cognitive effects of supplementation with a multivitamin, a poly-herbal formula, and a placebo. Despite small, unequal treatment group sizes, there appeared to be some memory enhancements associated with the multivitamin and poly-herbal treatments (Shah and Goyal, 2010). Replication of this study with larger, equal group sizes may be required to confirm these results. In a trial of young to elderly subjects, improved performance on a digit memory task and the trails making test have been documented after time periods of 2 weeks and 3 months treatment with a dietary supplement consisting of folic acid, B_{12}, vitamin E, S-adenosylmethionine, N-acetylcysteine, and acetyl-L-carnitine (Chan et al., 2009). Interestingly, in an open label extension, memory benefits were diminished when the treatment was withdrawn for 3 months and reemerged when treatment was reinstated for a further 3 months. As only a subset of elderly subjects were shown to respond to the treatment, and only a fraction of the initial sample remained at the 12 month conclusion of this study, replication of this study may be necessary.

Investigations in younger subject groups have also identified cognitive improvements following multivitamin supplementation. After 12 months supplementation with a multivitamin, benefits to speed of attention has been observed, although improvements were restricted to female participants (Benton et al., 1995). The results from other trials indicate that the cognitive benefits of multivitamin supplementation in young adults may be limited to measures that require high levels of cognitive demand. In a 9 week study of young to middle aged females, multivitamin-related cognitive enhancements were observed on a computerized multitasking framework consisting of mathematical, stroop, numerical processing, and memory search tasks (Haskell et al., 2010). Similarly cognitive performance has been augmented on a highly demanding serial subtraction task following 33 days multivitamin supplementation in men of a comparable age range (Kennedy et al., 2010). When considered together, these findings indicate that cognitive benefits of multivitamin supplementation may not only be restricted to the elderly.

LIMITATIONS AND FUTURE DIRECTIONS

Despite relatively promising findings that have linked blood levels of vitamins and long-term antioxidant use to better cognitive function in the elderly, the translation of this evidence to randomized controlled investigations into the effects of dietary vitamin supplementation on cognition has been less positive. A meta-analysis of randomized controlled trials has concluded that B vitamins, antioxidants, and multivitamins do not exert clinically important effects on global cognition in the elderly (Jia et al., 2008). Due to methodological differences pertaining to the specific ingredients, vitamin concentrations, and cognitive instruments, it is difficult to formulate cross-study comparisons of the efficacy of vitamin use for this purpose. In order to obtain clear evidence that dietary supplementation with micronutrients can improve cognitive performance or slow the rate of cognitive decline in the elderly, there needs to be greater forethought and justification of the cognitive domains under investigation in randomized controlled trials.

For instance, global measures such as the MMSE do not refer to any specific cognitive domains and provide only an overall estimate of an individual's level of functioning. The use of such measures may result in difficulty distinguishing the effects of dietary interventions on memory or other more general cognitive processes (Benton et al., 2005). Instruments such as the MMSE have been primarily designed to diagnose dementia and possess limited ability to detect cognitive change in healthy individuals due to ceiling effects (Reisberg, 2007). In addition, such global instruments have demonstrated limited ability to detect treatment effects in healthy individuals (Summers et al., 1990). It has been argued that nutritional effects are likely to be subtle and that global measures may not be sensitive enough to capture improvements in performance in healthy individuals (Bryan et al., 2002; MacReady et al., 2011). The same criticism may apply to neuropsychological or IQ measures that were not developed with the intention of facilitating the measurement of small nutraceutical benefits.

Haskell et al. (2008) have argued that the outcome measures previously employed in randomized controlled trials to investigate the effects of multivitamins on cognitive function in children have relied on measures of IQ rather than true cognitive performance measures. These authors proposed that measures of attention and nonverbal IQ would be more sensitive to the effects of multivitamins, due to the potential modulating effect on neurotransmitters involved in these processes. However, in the elderly, the onset of other age-related neural and cognitive changes that are not applicable to children may complicate this interpretation.

In the elderly, several factors may influence the cognitive domains that respond to dietary supplementation with vitamins or minerals. First, it can be postulated that the domains most vulnerable to the deleterious effects of aging will respond best to intervention. For example, chronic folic acid supplementation has been shown to improve performance on tests that measure information processing speed and working memory (Durga et al., 2007), domains that are known to decline with age (Babcock and Salthouse, 1990; Salthouse, 1990, 1996). Recent studies have also shown that supplementation with powerful antioxidants known as flavonoids can enhance mental functions on working memory–related indices (Pipingas et al., 2008; Ryan et al., 2008).

Second, it could be hypothesized that the cognitive processes mediated by cortical structures most vulnerable to vitamin deficiency in the elderly may show the greatest improvement with supplementation. Gray-matter volume in the parietal regions has been correlated with vitamin B_{12} intake (Erickson et al., 2008), and elevated homocysteine has been shown to selectively compromise the hippocampus (Williams et al., 2002; Den Heijer et al., 2003). Subsequently the cognitive domains supported by these regions, such as memory, may prove to be more vulnerable to vitamin depletion than others and may even respond better to dietary supplementation. Randomized controlled trials integrating both structural and functional brain imaging techniques may have the potential to address this research question.

Within the elderly population, there is a need to identify subgroups of people who would experience the most cognitive benefits from dietary supplementation with vitamins. For example, the results of a study conducted by Bunce et al. (2005) showed that possession of the APOE e4 allele, in combination with low B vitamin levels, impeded retrieval memory performance in demanding face recognition conditions. These findings were independent of dementia up to 6 years later, indicating that these individuals were not in a preclinical phase of AD. Based on the results of this study, the authors proposed that in healthy elderly, genetic risk factors for AD such as the APOE e4 allele may predispose some brain structures and processes to be more vulnerable to the deleterious effects of low B_{12} levels. In addition to genetic factors, elderly experiencing ongoing illness, stress, poor appetite, and vision impairments are also at risk of vitamin and micronutrient deficiencies (Payette et al., 1995).

CONCLUSION

The evidence presented in this chapter summarizes recent research findings that link vitamin status to cognition in the elderly. A healthy brain and cardiovascular system are essential for cognition across the life span and adequate nutrient intake can help maintain optimal neurophysiological function. There is a biological role of the B vitamins and antioxidants in neurotransmitter production, and these vitamins exert neuroprotective and cardioprotective mechanisms (Cantuti-Castelvetri et al., 2000; Floyd and Hensley, 2002; Ferrari, 2004; Bourre, 2006). Low levels of vitamin B_{12} and folate and elevated homocysteine have been demonstrated to predate cognitive decline in the elderly and are related to factors that may influence cognition such as brain pathology and poor cardiovascular health. Further evidence demonstrates that the use of antioxidant supplements may help slow the rate of cognitive decline in the elderly. The findings from randomized controlled trials are less promising but do demonstrate improvements to processing speed and memory associated with vitamin B_{12} and folate and combined multivitamin and herbal formulas in the elderly. Larger, longer duration studies may be required to fully assess the ability of nutritional interventions with B vitamins, antioxidant vitamins, and multivitamins to improve cognitive performance or slow the rate of cognitive decline in healthy elderly. Functional imaging studies utilizing MRI or electrophysiological techniques such as electroencephalograph (EEG) may be beneficial in future trials to provide greater insights into neurocognitive effects of vitamin supplementation.

REFERENCES

Alexander, D. M., Williams, L. M., Gatt, J. M., Dobson-Stone, C., Kuan, S. A., Todd, E. G., Schofield, P. R., Cooper, N. J., and Gordon, E. (2007). The contribution of apolipoprotein E alleles on cognitive performance and dynamic neural activity over six decades. *Biological Psychology*, 75(3), 229–238.

Allain, H., Bentue-Ferrer, D., and Akwa, Y. (2007). Treatment of the mild cognitive impairment (MCI). *Human Psychopharmacology*, 22(4), 189–197.

American Psychiatric Association (2000). *Diagnostic Criteria from DSM-IV-TR*. Washington, DC: American Psychiatric Association.

Annweiler, C., Schott, A. M., Rolland, Y., Blain, H., Herrmann, F. R., and Beauchet, O. (2010). Dietary intake of vitamin D and cognition in older women: A large population-based study. *Neurology*, 75(20), 1810–1816.

Ardila, A., Ostrosky-Solis, F., Rosselli, M., and Gómez, C. (2000). Age-related cognitive decline during normal aging: The complex effect of education. *Archives of Clinical Neuropsychology*, 15(6), 495–513.

Babcock, R. L. and Salthouse, T. A. (1990). Effects of increased processing demands on age differences in working memory. *Psychology and Aging*, 5(3), 421–428.

Baik, H. W. and Russell, R. M. (1999). Vitamin B12 deficiency in the elderly. *Annual Review of Nutrition*, 19, 357–377.

Baker, H., De Angelis, B., Baker, E. R., Frank, O., and Jaslow, S. P. (1999). Lack of effect of 1 year intake of a high-dose vitamin and mineral supplement on cognitive function of elderly women. *Gerontology*, 45(4), 195–199.

Baker, L. D., Frank, L. L., Foster-Schubert, K., Green, P. S., Wilkinson, C. W., McTiernan, A., Plymate, S. R. et al. (2010). Effects of aerobic exercise on mild cognitive impairment: A controlled trial. *Archives of Neurology*, 67(1), 71–79.

Balk, E. M., Raman, G., Tatsioni, A., Chung, M., Lau, J., and Rosenberg, I. H. (2007). Vitamin B6, B12, and folic acid supplementation and cognitive function: A systematic review of randomized trials. *Archives of Internal Medicine*, 167(1), 21–30.

Beckman, K. B. and Ames, B. N. (1998). The free radical theory of aging matures. *Physiological Reviews*, 78(2), 547–581.

Beeri, M. S., Ravona-Springer, R., Silverman, J. M., and Haroutunian, V. (2009). The effects of cardiovascular risk factors on cognitive compromise. *Dialogues in Clinical Neuroscience*, 11(2), 201–212.

Benton, D., Fordy, J., and Haller, J. (1995). The impact of long-term vitamin supplementation on cognitive functioning. *Psychopharmacology*, 117(3), 298–305.

Benton, D., Kallus, K. W., and Schmitt, J. A. J. (2005). How should we measure nutrition-induced improvements in memory? *European Journal of Nutrition*, 44(8), 485–498.

Birks, J. and Flicker, L. (2006). Donepezil for mild cognitive impairment. *Cochrane Database of Systematic Reviews* (Online), 3, CD006104.

Blennow, K., de Leon, M. J., and Zetterberg, H. (2006). Alzheimer's disease. *Lancet*, 368(9533), 387–403.

Boothby, L. A. and Doering, P. L. (2005). Vitamin C vitamin E for Alzheimer's disease. *Annals of Pharmacotherapy*, 39, 2073–2080.

Boreková, M., Hojerová, J., Koprda, V., and Bauerová, K. (2008). Nourishing and health benefits of coenzyme Q10—A review. *Czech Journal of Food Sciences*, 26(4), 229–241.

Bourre, J. M. (2006). Effects of nutrients (in food) on the structure and function of the nervous system: Update on dietary requirements for brain. Part 1: Micronutrients. *Journal of Nutrition, Health and Aging*, 10(5), 377–385.

Braak, H. and Braak, E. (1991). Neuropathological stageing of Alzheimer-related changes. *Acta Neuropathologica*, 82(4), 239–259.

Breteler, M. M. B., Claus, J. J., Grobbee, D. E., and Hofman, A. (1994). Cardiovascular disease and distribution of cognitive function in elderly people: The Rotterdam study. *British Medical Journal*, 308(6944), 1604–1608.

Bryan, J., Calvaresi, E., and Hughes, D. (2002). Short-term folate, vitamin B-12 or vitamin B-6 supplementation slightly affects memory performance but not mood in women of various ages. *Journal of Nutrition*, 132(6), 1345–1356.

Buckner, R. L. (2004). Memory and executive function in aging and AD: Multiple factors that cause decline and reserve factors that compensate. *Neuron*, 44, 195–208.

Buckner, R. L. (2005). Three principles for cognitive aging research: Multiple causes and sequalae, variance in expression and response, and the need for integrative theory. In R. Cabeza, L. Myberg, and D. Park (Eds.) *Cognitive Neuroscience of Aging: Linking Cognitive and Cerebral Aging*, New York, Oxford University Press.

Budson, A. E. and Price, B. H. (2005). Memory dysfunction. *New England Journal of Medicine*, 352(7), 692–699.

Buell, J. S. and Dawson-Hughes, B. (2008). Vitamin D and neurocognitive dysfunction: Preventing "D"ecline? *Molecular Aspects of Medicine*, 29, 415–422.

Buell, J. S., Dawson-Hughes, B., Scott, T. M., Weiner, D. E., Dallal, G. E., Qui, W. Q., Bergethon, P. et al. (2010). 25-Hydroxyvitamin D, dementia, and cerebrovascular pathology in elders receiving home services. *Neurology*, 74(1), 18–26.

Buell, J. S., Scott, T. M., Dawson-Hughes, B., Dallal, G. E., Rosenberg, I. H., Folstein, M. F., and Tucker, K. L. (2009). Vitamin D is associated with cognitive function in elders receiving home health services. *Journals of Gerontology—Series A Biological Sciences and Medical Sciences*, 64(8), 888–895.

Bunce, D., Kivipelto, M., and Wahlin, Å. (2005). Apolipoprotein E, B vitamins, and cognitive function in older adults. *Journals of Gerontology—Series B Psychological Sciences and Social Sciences*, 60(1), P41–P48.

Cabeza, R. (2002). Hemispheric asymmetry reduction in older adults: The HAROLD model. *Psychology and Aging*, 17(1), 85–100.

Cadenas, E. and Davies, K. J. A. (2000). Mitochondrial free radical generation, oxidative stress, and aging. *Free Radical Biology and Medicine*, 29(3–4), 222–230.

Calvaresi, E. and Bryan, J. (2001). B vitamins, cognition, and aging: A review. *Journals of Gerontology—Series B Psychological Sciences and Social Sciences*, 56(6), P327–P339.

Cantuti-Castelvetri, I., Shukitt-Hale, B., and Joseph, J. A. (2000). Neurobehavioral aspects of antioxidants in aging. *International Journal of Developmental Neuroscience*, 18(4–5), 367–381.

Carr, D. B., Goate, A., Phil, D., and Morris, J. C. (1997). Current concepts in the pathogenesis of Alzheimer's disease. *American Journal of Medicine*, 103(3A), 3S–10S.

Chan, A., Remington, R., Kotyla, E., Lepore, A., Zemianek, J., and Shea, T. B. (2009). A vitamin/nutriceutical formulation improves memory and cognitive performance in community-dwelling adults without dementia. *Journal of Nutrition, Health and Aging*, 14, 1–7.

Christensen, H. (2001). What cognitive changes can be expected with normal ageing? *Australian and New Zealand Journal of Psychiatry*, 35(6), 768–775.

Christensen, H., Mackinnon, A., Jorm, A. F., Henderson, A. S., Scott, L. R., and Korten, A. E. (1994). Age differences and interindividual variation in cognition in community-dwelling elderly. *Psychology and Aging*, 9(3), 381–390.

Clarke, R., Birks, J., Nexo, E., Ueland, P. M., Schneede, J., Scott, J., Molloy, A., and Evans, J. G. (2007). Low vitamin B-12 status and risk of cognitive decline in older adults. *American Journal of Clinical Nutrition*, 86(5), 1384–1391.

Clarke, R., Smith, A. D., Jobst, K. A., Refsum, H., Sutton, L., and Ueland, P. M. (1998). Folate, vitamin B12, and serum total homocysteine levels in confirmed Alzheimer disease. *Archives of Neurology*, 55(11), 1449–1455.

Cockle, S. M., Haller, J., Kimber, S., Dawe, R. A., and Hindmarch, I. (2000). The influence of multivitamins on cognitive function and mood in the elderly. *Aging and Mental Health*, 4(4), 339–353.

Collie, A. and Maruff, P. (2000). The neuropsychology of preclinical Azheimer's disease and mild cognitive impairment. *Neuroscience and Biobehavioral Reviews*, 24, 365–374.

De Koning, L. A. B., Werstuck, G. H., Zhou, J., and Austin, R. C. (2003). Hyperhomocysteinemia and its role in the development of atherosclerosis. *Clinical Biochemistry*, 36(6), 431–441.

De Lau, L. M. L., Smith, A. D., Refsum, H., Johnston, C., and Breteler, M. M. B. (2009). Plasma vitamin B12 status and cerebral white-matter lesions. *Journal of Neurology, Neurosurgery and Psychiatry*, 80(2), 149–157.

De Mendonca, A., Guerreiro, M., Ribeiro, F., Mendes, T., and Garcia, C. (2004). Mild cognitive impairment: Focus on diagnosis. *Journal of Molecular Neuroscience*, 23(1–2), 143–147.

Deijen, J. B., Van der Beek, E. J., Orlebeke, J. F., and Van den Berg, H. (1992). Vitamin B-6 supplementation in elderly men: Effects on mood, memory, performance and mental effort. *Psychopharmacology*, 109(4), 489–496.

Den Heijer, T., Vermeer, S. E., Clarke, R., Oudkerk, M., Koudstaal, P. J., Hofman, A., and Breteler, M. M. B. (2003). Homocysteine and brain atrophy on MRI of non-demented elderly. *Brain*, 126(1), 170–175.

Depp, C. A. and Jeste, D. V. (2006). Definitions and predictors of successful aging: A comprehensive review of larger quantitative studies. *American Journal of Geriatric Psychiatry*, 14(1), 6–20.

Devaraj, S., Tang, R., Adams-Huet, B., Harris, A., Seenivasan, T., de Lemos, J. A., and Jialal, A. (2007). Effect of high-dose a-tocopherol supplementation on biomarkers of oxidative stress and inflammation and carotid atherosclerosis in patients with coronary artery disease. *American Journal of Clinical Nutrition*, 86, 1392–1398.

Diaz, M. N., Frei, B., Vita, J. A., and Keaney Jr, J. F. (1997). Antioxidants and atherosclerotic heart disease. *New England Journal of Medicine*, 337(6), 408–416.

Dragland, S., Senoo, H., Wake, K., Holte, K., and Blomhoff, R. (2003). Several culinary and medicinal herbs are important sources of dietary antioxidants. *Journal of Nutrition*, 133(5), 1286–1290.

Dunn, J. E., Weintraub, S., Stoddard, A. M., and Banks, S. (2007). Serum α-tocopherol, concurrent and past vitamin E intake, and mild cognitive impairment. *Neurology*, 68(9), 670–676.

Durga, J., van Boxtel, M. P., Schouten, E. G., Kok, F. J., Jolles, J., Katan, M. B., and Verhoef, P. (2007). Effect of 3-year folic acid supplementation on cognitive function in older adults in the FACIT trial: A randomised, double blind, controlled trial. *Lancet*, 369(9557), 208–216.

Duron, E. and Hanon, O. (2008). Vascular risk factors, cognitive decline, and dementia. *Vascular Health and Risk Management*, 4(2), 363–381.

Eastley, R., Wilcock, G. K., and Bucks, R. S. (2000). Vitamin B12 deficiency in dementia and cognitive impairment: The effects of treatment on neuropsychological function. *International Journal of Geriatric Psychiatry*, 15(3), 226–233.

Elias, M. F., Robbins, M. A., Budge, M. M., Elias, P. K., Brennan, S. L., Johnston, C., Nagy, Z., and Bates, C. J. (2006). Homocysteine, folate, and vitamins B6 and B12 blood levels in relation to cognitive performance: The Maine-Syracuse study. *Psychosomatic Medicine*, 68, 547–554.

Erickson, K. I., Suever, B. L., Prakash, R. S., Colcombe, S. J., McAuley, E., and Kramer, A. F. (2008). Greater intake of vitamins B6 and B12 spares gray matter in healthy elderly: A voxel-based morphometry study. *Brain Research*, 1199(C), 20–26.

Eussen, S. J., De Groot, L. C., Joosten, L. W., Bloo, R. J., Clarke, R., Ueland, P. M., Schneede, J. et al. (2006). Effect of oral vitamin B-12 with or without folic acid on cognitive function in older people with mild vitamin B-12 deficiency: A randomized, placebo-controlled trial. *American Journal of Clinical Nutrition*, 84(2), 361–370.

Feng, L., Ng, T. P., Chuah, L., Niti, M., and Kua, E. H. (2006). Homocysteine, folate, and vitamin B-12 and cognitive performance in older Chinese adults: Findings from the Singapore Longitudinal Ageing Study. *American Journal of Clinical Nutrition*, 84(6), 1506–1512.

Ferrari, C. K. B. (2004). Functional foods, herbs and nutraceuticals: Towards biochemical mechanisms of healthy aging. *Biogerontology*, 5(5), 275–289.

Ferri, C. P., Prince, M., Brayne, C., Brodaty, H., Fratiglioni, L., Ganguli, M., Hall, K. et al. (2005). Global prevalence of dementia: A Delphi consensus study. *Lancet*, 366(9503), 2112–2117.

Fioravanti, M., Ferrario, E., Massaia, M., Cappa, G., Rivolta, G., Grossi, E., and Buckley, A. E. (1997). Low folate levels in the cognitive decline of elderly patients and the efficacy of folate as a treatment for improving memory deficits. *Archives of Gerontology and Geriatrics*, 26(1), 1–13.

Firbank, M. J., Narayan, S. K., Saxby, B. K., Ford, G. A., and O'Brien, J. T. (2010). Homocysteine is associated with hippocampal and white matter atrophy in older subjects with mild hypertension. *International Psychogeriatrics*, 22(5), 804–811.

Firbank, M. J., Wiseman, R. M., Burton, E. J., Saxby, B. K., O'Brien, J. T., and Ford, G. A. (2007). Brain atrophy and white matter hyperintensity change in older adults and relationship to blood pressure: Brain atrophy, WMH change and blood pressure. *Journal of Neurology*, 254(6), 713–721.

Floyd, R. A. and Carney, J. M. (1992). Free radical damage to protein and DNA: Mechanisms involved and relevant observations on brain undergoing oxidative stress. *Annals of Neurology*, 32(Suppl.), S22–S27.

Floyd, R. A. and Hensley, K. (2002). Oxidative stress in brain aging: Implications for therapeutics of neurodegenerative diseases. *Neurobiology of Aging*, 23(5), 795–807.

Folstein, M. F., Folstein, S. E., and McHugh, P. R. (1975). "Mini-mental state." A practical method for grading the cognitive state of patients for the clinician. *Journal of Psychiatric Research*, 12, 189–198.

Fotuhi, M., Zandi, P. P., Hayden, K. M., Khachaturian, A. S., Szekely, C. A., Wengreen, H., Munger, R. G. et al. (2008). Better cognitive performance in elderly taking antioxidant vitamins E and C supplements in combination with nonsteroidal anti-inflammatory drugs: The Cache County Study. *Alzheimer's and Dementia*, 4(3), 223–227.

Gale, C. R., Martyn, C. N., and Cooper, C. (1996). Cognitive impairment and mortality in a cohort of elderly people. *British Medical Journal*, 312(7031), 607–611.

Goedert, M. and Spillantini, M. G. (2006). A century of Alzheimer's disease. *Science*, 314(5800), 777–781.

Grodstein, F., Chen, J., and Willett, W. C. (2003). High-dose antioxidant supplements and cognitive function in community-dwelling elderly women. *The American Journal of Clinical Nutrition*, 77(4), 975–984.

Grodstein, F., Kang, J. H., Glynn, R. J., Cook, N. R., and Gaziano, J. M. (2007). A randomized trial of beta carotene supplementation and cognitive function in men: The physicians' health study II. *Archives of Internal Medicine*, 167(20), 2184–2190.

Hardy, J. A. and Higgins, G. A. (1992). Alzheimer's disease: The amyloid cascade hypothesis. *Science*, 256(5054), 184–185.

Haskell, C. F., Robertson, B., Jones, E., Forster, J., Jones, R., Wilde, A., Maggini, S. et al. (2010). Effects of a multi-vitamin/mineral supplement on cognitive function and fatigue during extended multi-tasking. *Human Psychopharmacology*, 25(6), 448–461.

Haskell, C. F., Scholey, A. B., Jackson, P. A., Elliott, J. M., Defeyter, M. A., Greer, J., Robertson, B. C., Buchanan, T., Tiplady, B., and Kennedy, D. O. (2008). Cognitive and mood effects in healthy children during 12 weeks' supplementation with multi-vitamin/minerals. *British Journal of Nutrition*, 100(5), 1086–1096.

Ho, P. I., Ortiz, D., Rogers, E., and Shea, T. B. (2002). Multiple aspects of homocysteine neurotoxicity: Glutamate excitotoxicity, kinase hyperactivation and DNA damage. *Journal of Neuroscience Research*, 70(5), 694–702.

Holick, M. F. (2006). High prevalence of vitamin D inadequacy and implications for health. *Mayo Clinic Proceedings*, 81(3), 353–373.

Huskisson, E., Maggini, S., and Ruf, M. (2007). The influence of micronutrients on cognitive function and performance. *Journal of International Medical Research*, 35(1), 1–19.

Isaac, M. G. E. K. N., Quinn, R., and Tabet, N. (2008). Vitamin E for Alzheimer's disease and mild cognitive impairment. *Cochrane Database of Systematic Reviews*, (3), CD002854.

Jelic, V. and Winblad, B. (2003). Treatment of mild cognitive impairment: Rationale, present and future strategies. *Acta Neurologica Scandinavica*, 107(Suppl 179), 83–93.

Jia, X., McNeill, G., and Avenell, A. (2008). Does taking vitamin, mineral and fatty acid supplements prevent cognitive decline? A systematic review of randomized controlled trials. *Journal of Human Nutrition and Dietetics*, 21(4), 317–336.

Jialal, I. and Devaraj, S. (2003). Antioxidants and atherosclerosis: Don't throw out the baby with the bath water. *Circulation*, 107(7), 926–928.

Jones, H. E. and Conrad, H. S. (1933). The growth and decline of intelligence: A study of a homogeneous group between the ages of ten and sixty. *Genetic Psychology Monographs*, 13, 223–298.

Joosten, E., Lesaffre, E., and Riezler, R. (1996). Are different reference intervals for methylmalonic acid and total homocysteine necessary in elderly people? *European Journal of Haematology*, 57(3), 222–226.

Kado, D. M., Karlamangla, A. S., Huang, M. H., Troen, A., Rowe, J. W., Selhub, J., and Seeman, T. E. (2005). Homocysteine versus the vitamins folate, B6, and B12 as predictors of cognitive function and decline in older high-functioning adults: MacArthur Studies of Successful Aging. *American Journal of Medicine*, 118(2), 161–167.

Kang, J. H., Cook, N., Manson, J., Buring, J. E., and Grodstein, F. (2006). A Randomized trial of vitamin E supplementation and cognitive function in women. *Archives of Internal Medicine*, 166(22), 2462–2468.

Kennedy, D. O., Veasey, R., Watson, A., Dodd, F., Jones, E., Maggini, S., and Haskell, C. F. (2010). Effects of high-dose B vitamin complex with vitamin C and minerals on subjective mood and performance in healthy males. *Psychopharmacology*, 211(1), 1–14.

Kidd, P. M. (2008). Alzheimer's disease, amnestic mild cognitive impairment, and age-associated memory impairment: Current understanding and progress toward integrative prevention. *Alternative Medicine Review*, 13(2), 85–115.

Knopman, D., Boland, L. L., Mosley, T., Howard, G., Liao, D., Szklo, M., McGovern, P., and Folsom, A. R. (2001). Cardiovascular risk factors and cognitive decline in middle-aged adults. *Neurology*, 56(1), 42–48.

Kregel, K. C. and Zhang, H. J. (2007). An integrated view of oxidative stress in aging: Basic mechanisms, functional effects, and pathological considerations. *American Journal of Physiology—Regulatory Integrative and Comparative Physiology*, 292(1), R18–R36.

Lange-Asschenfeldt, C. and Kojda, G. (2008). Alzheimer's disease, cerebrovascular dysfunction and the benefits of exercise: From vessels to neurons. *Experimental Gerontology*, 43(6), 499–504.

Larson, E. B., Wang, L., Bowen, J. D., McCormick, W. C., Teri, L., Crane, P., and Kukull, W. (2006). Exercise is associated with reduced risk for incident dementia among persons 65 years of age and older. *Annals of Internal Medicine*, 144(2), 73–81.

Lee, D. M., Tajar, A., Ulubaev, A., Pendleton, N., O'Neill, T. W., O'Connor, D. B., Bartfai, G. et al. (2009). Association between 25-hydroxyvitamin D levels and cognitive performance in middle-aged and older European men. *Journal of Neurology, Neurosurgery and Psychiatry*, 80(7), 722–729.

Lewerin, C., Matousek, M., Steen, G., Johansson, B., Steen, B., and Nilsson-Ehle, H. (2005). Significant correlations of plasma homocysteine and serum methylmalonic acid with movement and cognitive performance in elderly subjects but no improvement from short-term vitamin therapy: A placebo-controlled randomized study. *American Journal of Clinical Nutrition*, 81(5), 1155–1162.

Lleo, A. (2007). Current therapeutic options for Alzheimer's disease. *Current Genomics*, 8(8), 550–558.

MacReady, A. L., Butler, L. T., Kennedy, O. B., Ellis, J. A., Williams, C. M., and Spencer, J. P. E. (2011). Cognitive tests used in chronic adult human randomised controlled trial micronutrient and phytochemical intervention studies. *Nutrition Research Reviews*, 23(2), 200–229.

Malouf, R. and Grimley Evans, J. (2008). Folic acid with or without vitamin B12 for the prevention and treatment of healthy elderly and demented people. *Cochrane Database of Systematic Reviews*, Vol. 4, DOI: 10.1002/14651858.CD004514.pub2.

Marcellini, F., Giuli, C., Papa, R., Gagliardi, C., Dedoussis, G., Herbein, G., Fulop, T. et al. (2006). Zinc status, psychological and nutritional assessment in old people recruited in five European countries: Zincage study. *Biogerontology*, 7(5–6), 339–345.

Mariani, E., Mangialasche, F., Feliziani, F. T., Cecchetti, R., Malavolta, M., Bastiani, P., Baglioni, M. et al. (2008). Effects of zinc supplementation on antioxidant enzyme activities in healthy old subjects. *Experimental Gerontology*, 43(5), 445–451.

Mariani, E., Polidori, M. C., Cherubini, A., and Mecocci, P. (2005). Oxidative stress in brain aging, neurodegenerative and vascular diseases: An overview. *Journal of Chromatography B: Analytical Technologies in the Biomedical and Life Sciences*, 827(1), 65–75.

Masaki, K. H., Losonczy, K. G., Izmirlian, G., Foley, D. J., Ross, G. W., Petrovitch, H., Havlik, R., and White, L. R. (2000). Association of vitamin E and C supplement use with cognitive function and dementia in elderly men. *Neurology*, 54(6), 1265–1272.

Mattson, M. P. and Shea, T. B. (2003). Folate and homocysteine metabolism in neural plasticity and neurodegenerative disorders. *Trends in Neurosciences*, 26(3), 137–146.

Maxwell, C. J., Hicks, M. S., Hogan, D. B., Basran, J., and Ebly, E. M. (2005). Supplemental use of antioxidant vitamins and subsequent risk of cognitive decline and dementia. *Dementia and Geriatric Cognitive Disorders*, 20(1), 45–51.

Maylor, E. A., Simpson, E. E. A., Secker, D. L., Meunier, N., Andriollo-Sanchez, M., Polito, A., Stewart-Knox, B., McConville, C., O'Connor, J. M., and Coudray, C. (2006). Effects of zinc supplementation on cognitive function in healthy middle-aged and older adults: The ZENITH study. *British Journal of Nutrition*, 96(4), 752–760.

McCaddon, A. (2006). Homocysteine and cognition—A historical perspective. *Journal of Alzheimer's Disease*, 9(4), 361–380.

McNeill, G., Avenell, A., Campbell, M. K., Cook, J. A., Hannaford, P. C., Kilonzo, M. M., Milne, A. C. et al. (2007). Effect of multivitamin and multimineral supplementation on cognitive function in men and women aged 65 years and over: A randomised controlled trial. *Nutrition Journal*, 6, 10.

Mecocci, P. (2004). Oxidative stress in mild cognitive impairment and Alzheimer disease: A continuum. *Journal of Alzheimer's Disease*, 6, 159–163.

Mix, J. A. and Crews, W. D. (2002). A double-blind, placebo-controlled, randomized trial of *Ginkgo biloba* extract EGb 761® in a sample of cognitively intact older adults: Neuropsychological findings. *Human Psychopharmacology*, 17(6), 267–277.

Mooijaart, S. P., Gussekloo, J., Frölich, M., Jolles, J., Stott, D. J., Westendorp, R. G. J., and De Craen, A. J. M. (2005). Homocysteine, vitamin B-12, and folic acid and the risk of cognitive decline in old age: The Leiden 85-Plus Study. *American Journal of Clinical Nutrition*, 82(4), 866–871.

Moreiras, T. O., Cuadrado Vives, C., Del Pozo De La Calle, S., and Rodríguez Sangrador, M. (2007). *Determinantes nutricionales de un envejecimiento sano: Proyecto HALE* (Nutritional determinants of healthy ageing. HALE project) *(Healthy Ageing: Longitudinal Study in Europe)*, 19(1), 30–36.

Moretti, R., Torre, P., Antonello, R. M., Cattaruzza, T., Cazzato, G., and Bava, A. (2004). Vitamin B12 and folate depletion in cognition: A review. *Neurology, India*, 52(3), 310–318.

Morris, M. C., Evans, D. A., Bienias, J. L., Tangney, C. C., and Wilson, R. S. (2002). Vitamin E and cognitive decline in older persons. *Archives of Neurology*, 59(7), 1125–1132.

Morris, M. S., Jacques, P. F., Rosenberg, I. H., and Selhub, J. (2001). Hyperhomocysteinemia associated with poor recall in the third national health and nutrition examination survey. *American Journal of Clinical Nutrition*, 73(5), 927–933.

Nahin, R. L., Fitzpatrick, A. L., Williamson, J. D., Burke, G. L., DeKosky, S. T., and Furberg, C. (2006). Use of herbal medicine and other dietary supplements in community-dwelling older people: Baseline data from the Ginkgo Evaluation of Memory study. *Journal of the American Geriatrics Society*, 54(11), 1725–1735.

Nieuwenhuizen, W. F., Weenen, H., Rigby, P., and Hetherington, M. M. (2010). Older adults and patients in need of nutritional support: Review of current treatment options and factors influencing nutritional intake. *Clinical Nutrition*, 29(2), 160–169.

Norfray, J. F. and Provenzale, J. M. (2004). Alzheimer's disease: Neuropathological findings and recent advances in imaging. *AJR*, 182, 2–13.

Nurk, E., Refsum, H., Tell, G. S., Engedal, K., Vollset, S. E., Ueland, P. M., Nygaard, H. A., and Smith, A. D. (2005). Plasma total homocysteine and memory in the elderly: The Hordaland homocysteine study. *Annals of Neurology*, 58(6), 847–857.

Obeid, R. and Herrmann, W. (2006). Mechanisms of homocysteine neurotoxicity in neurodegenerative diseases with special reference to dementia. *FEBS Letters*, 580(13), 2994–3005.

Payette, H., Gray-Donald, K., Cyr, R., and Boutier, V. (1995). Predictors of dietary intake in a functionally dependent elderly population in the community. *American Journal of Public Health*, 85(5), 677–683.

Perkins, A. J., Hendrie, H. C., Callahan, C. M., Gao, S., Unverzagt, F. W., Xu, Y., Hall, K. S. et al. (1999). Association of antioxidants with memory in a multiethnic elderly sample using the Third National Health and Nutrition Examination survey. *American Journal of Epidemiology*, 150(1), 37–44.

Perrig, W. J., Perrig, P., and Stähelin, H. B. (1997). The relation between antioxidants and memory performance in the old and very old. *Journal of the American Geriatrics Society*, 45(6), 718–724.

Petersen, R. C. (2007). Mild cognitive impairment: Current research and clinical implications. *Seminars in Neurology*, 27(1), 22–31.

Petrella, J. R., Coleman, R. E., and Doraiswamy, P. M. (2003). Neuroimaging and early diagnosis of Alzheimer disease: A look to the future. *Radiology*, 226(2), 315–336.

Phillips, L. H. and Andrés, P. (2010). The cognitive neuroscience of aging: New findings on compensation and connectivity. *Cortex*, 46(4), 421–424.

Pipingas, A., Silberstein, R. B., Vitetta, L., Van Rooy, C., Harris, E. V., Young, J. M., Frampton, C. M. et al. (2008). Improved cognitive performance after dietary supplementation with a *Pinus radiata* bark extract formulation. *Phytotherapy Research*, 22(9), 1168–1174.

Pugh, K. G. and Lipsitz, L. A. (2002). The microvascular frontal-subcortical syndrome of aging. *Neurobiology of Aging*, 23(3), 421–431.

Quadri, P., Fragiacomo, C., Pezzati, R., Zanda, E., Forloni, G., Tettamanti, M., and Lucca, U. (2004). Homocysteine, folate, and vitamin B-12 in mild cognitive impairment, Alzheimer disease, and vascular dementia. *American Journal of Clinical Nutrition*, 80(1), 114–122.

Quadri, P., Fragiacomo, C., Pezzati, R., Zanda, E., Tettamanti, M., and Lucca, U. (2005). Homocysteine and B vitamins in mild cognitive impairment and dementia. *Clinical Chemistry and Laboratory Medicine*, 43(10), 1096–1100.

Rabbit, P., Ibrahim, S., Lunn, M., Scott, M., Thacker, N., Hutchinson, C., Horan, M., Pendleton, N., and Jackson, A. (2008). Age-associated losses of brain volume predict longitudinal cognitive declines over 8 to 20 years. *Neuropsychology*, 22(1), 3–9.

Rabbitt, P. M. A., McInnes, L., Diggle, P., Holland, F., Bent, N., Abson, V., Pendleton, N., and Horan, M. (2004). The University of Manchester Longitudinal Study of Cognition in Normal Healthy Old Age, 1983 through 2003, Psychology Press. *Aging, Neuropsychology, and Cognition*, 11, 245–279.

Radimer, K., Bindewald, B., Hughes, J., Ervin, B., Swanson, C., and Picciano, M. F. (2004). Dietary supplement use by US adults: Data from the National Health and Nutrition Examination Survey, 1999–2000. *American Journal of Epidemiology*, 160(4), 339–349.

Raz, N., Lindenberger, U., Rodrigue, K. M., Kennedy, K. M., Head, D., Williamson, A., Dahle, C., Gerstorf, D., and Acker, J. D. (2005). Regional brain changes in aging healthy adults: General trends, individual differences and modifiers. *Cerebral Cortex*, 15(11), 1676–1689.

Reisberg, B. (2007). Global measures: Utility in defining and measuring treatment response in dementia. *International Psychogeriatrics*, 19(3), 421–456.

Riggs, K. M., Spiro, A. I., Tucker, K., and Rush, D. (1996). Relations of vitamin B-12, vitamin B-6, folate, and homocysteine to cognitive performance in the Normative Aging Study. *American Journal of Clinical Nutrition*, 63(3), 306–314.

Rinaldi, P., Polidori, M. C., Metastasio, A., Mariani, E., Mattioli, P., Cherubini, A., Catani, M., Cecchetti, R., Senin, U., and Mecocci, P. (2003). Plasma antioxidants are similarly depleted in mild cognitive impairment and in Alzheimer's disease. *Neurobiology of Aging*, 24(7), 915–919.

Rusinek, H., De Leon, M. J., George, A. E., Stylopoulos, L. A., Chandra, R., Smith, G., Rand, T., Mourino, M., and Kowalski, H. (1991). Alzheimer disease: Measuring loss of cerebral gray matter with MR imaging. *Radiology*, 178(1), 109–114.

Ryan, J., Croft, K., Mori, T., Wesnes, K., Spong, J., Downey, L., Kure, C., Lloyd, J., and Stough, C. (2008). An examination of the effects of the antioxidant Pycnogenol® on cognitive performance, serum lipid profile, endocrinological and oxidative stress biomarkers in an elderly population. *Journal of Psychopharmacology*, 22, 553–562.

Sachdev, P. S. (2005). Homocysteine and brain atrophy. *Progress in Neuro-Psychopharmacology and Biological Psychiatry*, 29(7), 1152–1161.

Salthouse, T. A. (1990). Working memory as a processing resource in cognitive aging. *Developmental Review*, 10(1), 101–124.

Salthouse, T. A. (1996). The processing-speed theory of adult age differences in cognition. *Psychological Review*, 103(3), 403–428.

Sandstead, H. H. (2000). Causes of iron and zinc deficiencies and their effects on brain. *Journal of Nutrition*, 130(2S Suppl.), 347S–349S.

Selhub, J. (2006). The many facets of hyperhomocysteinemia: Studies from the Framingham cohorts. *Journal of Nutrition*, 136(6), 1726S–1730S.

Selhub, J., Bagley, L. C., Miller, J., and Rosenberg, I. H. (2000). B vitamins, homocysteine, and neurocognitive function in the elderly. *American Journal of Clinical Nutrition*, 71(2), 614S–620S.

Seshadri, S. (2006). Elevated plasma homocysteine levels: Risk factor or risk marker for the development of dementia and Alzheimer's disease? *Journal of Alzheimer's Disease*, 9(4), 393–398.

Shah, J. and Goyal, R. (2010). Comparative clinical evaluation of herbal formulation with multivitamin formulation for learning and memory enhancement. *Asian Journal of Pharmaceutical and Clinical Research*, 3(1), 69–75.

Smith, A. D., Smith, S. M., de Jager, C. A., Whitbread, P., Johnston, C., Agacinski, G., Oulhaj, A. Bradley, K. M., Jacoby, R., and Refsum, H. (2010). Homocysteine-lowering by B vitamins slows the rate of accelerated brain atrophy in mild cognitive impairment: A randomized controlled trial. *PLoS ONE*, 5(9), 1–10.

Snowdon, D. A., Tully, C. L., Smith, C. D., Riley, K. P., and Markesbery, W. R. (2000). Serum folate and the severity of atrophy of the neocortex in Alzheimer disease: Findings from the Nun Study. *American Journal of Clinical Nutrition*, 71(4), 993–998.

Sobów, T. and Kłoszewska, I. (2007). Cholinesterase inhibitors in mild cognitive impairment: A meta-analysis of randomized controlled trials. *Neurologia i Neurochirurgia Polska*, 41(1), 13–21.

Solfrizzi, V., D'Introno, A., Colacicco, A. M., Capurso, C., Todarello, O., Pellicani, V., Capurso, S. A. et al. (2006). Circulating biomarkers of cognitive decline and dementia. *Clinica Chimica Acta*, 364(1–2), 91–112.

Stott, D. J., MacIntosh, G., Lowe, G. D. O., Rumley, A., McMahon, A. D., Langhorne, P., Tait, R. C. et al. (2005). Randomized controlled trial of homocysteine-lowering vitamin treatment in elderly patients with vascular disease. *American Journal of Clinical Nutrition*, 82(6), 1320–1326.

Summers, W. K., DeBoynton, V., Marsh, G. M., and Majovski, L. V. (1990). Comparison of seven psychometric instruments used for evaluation of treatment effect in Alzheimer's dementia. *Neuroepidemiology*, 9(4), 193–207.

Summers, W. K., Martin, R. L., Cunningham, M., Deboynton, V. L., and Marsh, G. M. (2010). Complex antioxidant blend improves memory in community-dwelling seniors. *Journal of Alzheimer's Disease*, 19(2), 429–439.

Tchantchou, F., Graves, M., Falcone, D., and Shea, T. B. (2008). S-Adenosylmethionine mediates glutathione efficacy by increasing glutathione S-transferase activity: Implications for S-adenosyl methionine as a neuroprotective dietary supplement. *Journal of Alzheimer's Disease*, 14(3), 323–328.

Thies, W. and Bleiler, L. (2011). 2011 Alzheimer's disease facts and figures. *Alzheimer's and Dementia*, 7(2), 208–244.

Tucker, K. L., Qiao, N., Scott, T., Rosenberg, I., and Spiro Iii, A. (2005). High homocysteine and low B vitamins predict cognitive decline in aging men: The Veterans Affairs Normative Aging Study. *American Journal of Clinical Nutrition*, 82(3), 627–635.

Valenzuela, M. and Sachdev, P. (2009). Can cognitive exercise prevent the onset of dementia? Systematic review of randomized clinical trials with longitudinal follow-up. *American Journal of Geriatric Psychiatry*, 17(3), 179–187.

Valko, M., Leibfritz, D., Moncol, J., Cronin, M. T. D., Mazur, M., and Telser, J. (2007). Free radicals and antioxidants in normal physiological functions and human disease. *International Journal of Biochemistry and Cell Biology*, 39(1), 44–84.

Vogiatzoglou, A., Refsum, H., Johnston, C., Smith, S. M., Bradley, K. M., de Jager, C., Budge, M. M., and Smith, A. D. (2008). Vitamin B12 status and rate of brain volume loss in community-dwelling elderly. *Neurology*, 71(11), 826–832.

Warsama Jama, J., Launer, L. J., Witteman, J. C. M., Den Breeijen, J. H., Breteler, M. M. B., Grobbee, D. E., and Hofman, A. (1996). Dietary antioxidants and cognitive function in a population-based sample of older persons: The Rotterdam study. *American Journal of Epidemiology*, 144(3), 275–280.

Wengreen, H. J., Munger, R. G., Corcoran, C. D., Zandi, P., Hayden, K. M., Fotuhi, M., Skoog, I. et al. (2007). Antioxidant intake and cognitive function of elderly men and women: The Cache County study. *Journal of Nutrition, Health and Aging*, 11(3), 230–237.

Williams, J. H., Pereira, E. A., Budge, M. M., and Bradley, K. M. (2002). Minimal hippocampal width relates to plasma homocysteine in community-dwelling older people. *Age and Ageing*, 31(6), 440–444.

Wilson, R. S., Bienias, J. L., Evans, D. A., and Bennett, D. A. (2004). Religious orders study: Overview and change in cognitive and motor speed, Psychology Press. *Aging, Neuropsychology, and Cognition*, 11, 280–303.

Wolters, M., Hermann, S., and Hahn, A. (2003). B vitamin status and concentrations of homocysteine and methylmalonic acid in elderly German women. *American Journal of Clinical Nutrition*, 78(4), 765–772.

Wolters, M., Hickstein, M., Flintermann, A., Tewes, U., and Hahn, A. (2005). Cognitive performance in relation to vitamin status in healthy elderly German women—The effect of 6-month multivitamin supplementation. *Preventive Medicine*, 41(1), 253–259.

Yaffe, K., Clemons, T. E., McBee, W. L., and Lindblad, A. S. (2004). Impact of antioxidants, zinc, and copper on cognition in the elderly: A randomized, controlled trial. *Neurology*, 63(9), 1705–1707.

Yurko-Mauro, K., McCarthy, D., Rom, D., Nelson, E. B., Ryan, A. S., Blackwell, A., Salem, N. Jr., and Stedman, M. (2010). Beneficial effects of docosahexaenoic acid on cognition in age-related cognitive decline. *Alzheimer's and Dementia*, 6(6), 456–464.

5 Lipoic Acid as an Anti-Inflammatory and Neuroprotective Treatment for Alzheimer's Disease

*Annette E. Maczurek, Lezanne Ooi,
Mili Patel, and Gerald Münch*

CONTENTS

ALZHEIMER'S DISEASE (AD)

Alzheimer's disease (AD) is a progressive neurodegenerative brain disorder that gradually destroys a patient's memory and ability to learn, make judgments, communicate effectively, and perform day-to-day tasks. The short-term memory is affected first, caused by neuronal dysfunction and degeneration in the hippocampus and amygdala. As the disease progresses further, neurons also degenerate and die in other cortical regions of the brain (Stuchbury and Münch, 2005). Sufferers then often experience dramatic changes in personality and behavior, such as anxiety, paranoia, or agitation, as well as delusions or hallucinations (Cummings, 2004). The prevalence of AD in the age bracket of 65–69 years is 1%; 70–74 years, 3%; 75–79 years, 6%; 80–84 years, 12%, and for people aged 85 and over the prevalence is 25%.

AD is further characterized by two major neuropathological hallmarks. The deposition of neuritic, β-amyloid (Aβ) peptide-containing senile plaques in hippocampal and cerebral cortical regions of AD patients is accompanied by the presence of intracellular neurofibrillary tangles that occupy most of the cytoplasm of pyramidal neurons. Inflammation, as evidenced by the activation of microglia and astroglia, is another hallmark of AD. Inflammation, including superoxide production ("oxidative burst"), is a significant source of oxidative stress in AD patients (Münch et al., 1998; Retz et al., 1998). The inflammatory process occurs mainly around the amyloid plaques and is characterized by pro-inflammatory substances that are released from activated microglia and astroglia (Wong et al., 2001b). Cytokines, including interleukin(IL)-1β, IL-6, macrophage colony-stimulating factor (M-CSF), and tumor necrosis factor (TNF-α), are the prominent signaling molecules in the inflammatory process, being responsible for a vicious cycle of microglial and astroglial activation, resulting in the secretion of neurotoxins including superoxide and nitric oxide (Griffin et al., 1995). In addition to those morphological and physiological alterations, AD is also associated with a markedly impaired cerebral glucose metabolism as detected by reduced cortical [^{18}F]-deoxyglucose utilization in positron emission tomography (Ishii and Minoshima, 2005).

CHOLINERGIC DEFICIT IN AD

AD patients show a progressive neuronal cell loss that is associated with region-specific brain atrophy. In particular, the earliest and most severely affected pathway is the cholinergic projection from the nucleus basalis of Meynert to areas of the cerebral cortex (Nordberg et al., 1987). Loss of basal forebrain cholinergic neurons is demonstrated by a reduction in the expression of choline acetyltransferase (ChAT), lower numbers of muscarinic and nicotinic acetylcholine receptors, and lower levels of acetylcholine (ACh) itself (Nordberg and Winblad, 1986). These changes are highly correlated with the degree of dementia in AD. ACh is derived from acetyl-CoA, the final product of the glycolytic pathway. Pyruvate, derived from glycolytic metabolism serves as an important energy source in neurons. Therefore, the inhibition of pyruvate production, for example, by glucose depletion, is considered a crucial factor that leads to acetyl-CoA deficiency in AD brains. ACh is hydrolyzed by acetylcholine esterase (AChE) and since AD patients have reduced levels of ChAT and ACh, compared to healthy elderly people, acetylcholine esterase inhibitors were introduced for the treatment of AD, but they do not delay the progression of the disease.

CURRENT TREATMENT STRATEGIES FOR AD ARE PURELY SYMPTOMATIC

Currently, only symptomatic treatments with AChE inhibitors are approved for mild to moderate forms of AD. The need for an all-encompassing therapy that not only improves cholinergic transmission but also targets other pathological processes in AD is urgent, and we propose Lipoic acid (LA) to be a promising candidate for such multi-target treatment (Holmquist et al., 2007), as will be outlined in the following chapters.

LA: A MULTIMODAL DRUG FOR THE TREATMENT OF AD

POSSIBLE MODES OF ACTION OF LA INTERFERING WITH AD-SPECIFIC DEGENERATION

In vitro and in vivo studies suggest that LA also acts as a powerful micronutri-ent with diverse pharmacological and antioxidant properties (Packer et al., 1995). LA naturally occurs only as the R-form (RLA), but pharmacological formulations in the past have also extensively used a racemic mixture of RLA and S-lipoic acid (SLA) as stereoselective synthesis methods have not been available.

LA has been suggested to have the following properties relevant to AD:

1. Increase of ACh production by activation of ChAT ("LA: An activator of ChAT" section)
2. Chelation of redox-active transition metals, thereby inhibiting the formation of hydrogen peroxide and hydroxyl radicals ("LA: A potent metal chelator" section)
3. Scavenging of Reactive oxygen species (ROS) (thus sparing glutathione) and downregulation of redox-sensitive inflammatory signals ("LA: An anti-inflammatory antioxidant and modulator of redox-sensitive signaling" section)
4. Scavenging of reactive carbonyl compounds including lipid peroxidation products ("LA: A carbonyl scavenger" section)
5. Increase of glucose uptake and utilization ("LA: A stimulator of glucose uptake and utilization ("insulinomimetic")" section)
6. Induction of enzymes for GSH synthesis and other antioxidant protective enzymes ("Upregulation of glutathione synthesis via activation of NRF-2" section)

LA: AN ACTIVATOR OF ChAT

DHLA, the reduced form of LA, is formed by reduction of LA by the pyruvate dehy-drogenase (PDH) complex. Haugaard et al. have demonstrated that DHLA strongly increases the activity of a purified preparation of ChAT (Haugaard and Levin, 2000). In a further publication, the authors showed that removal of DHLA by dialysis from purified ChAT (from rabbit bladder, rat brain, and heart extracts) causes complete disappearance of enzyme activity. The addition of DHLA restored activity toward normal levels while the addition of reduced ascorbic acid or reduced nicotinamide adenine dinucleotide was not effective (Haugaard and Levin, 2002). The authors concluded that DHLA serves an essential function in the action of this enzyme and that the ratio of reduced to oxidized LA plays an important role in ACh synthesis. From these data the authors further conclude that DHLA (1) may act as a coenzyme in the ChAT reaction or (2) is able to reduce an essential functional cysteine residue in ChAT, which cannot be reduced by any other physiological antioxidant, including reduced GSH (Haugaard and Levin, 2002).

LA: A POTENT METAL CHELATOR

There is now compelling evidence that Aβ, the main component of amyloid plaques in the AD-affected brain, does not spontaneously aggregate as was originally thought. Evidence suggests that there is an age-dependent reaction with excess metal ions in the brain (copper, iron, and zinc), which induces the peptide to precipitate and form plaques. Furthermore, the abnormal combination of Aβ with copper or iron ions induces the production of hydrogen peroxide from molecular oxygen (Huang et al., 1999), which subsequently produces the neurotoxic hydroxyl radical by Fenton or Haber–Weiss reactions. Because LA is a potent chelator of divalent metal ions in vitro, the effect of an RLA inclusive diet on cortical iron levels and antioxidant status was investigated in aged rats (Suh et al., 2005). Results show that cerebral iron levels in old LA-fed animals were lower when compared to controls and were similar to levels seen in young rats. These results thus show that chronic LA supplementation may be a means to modulate the age-related accumulation of cortical iron content, thereby lowering oxidative stress associated with aging (Suh et al., 2005). Since amyloid aggregates have been shown to be stabilized by transition metals such as iron and copper, it was also speculated that LA could inhibit aggregate formation or potentially dissolve existing amyloid deposits. Fonte et al. successfully resolubilized Aβ with transition metal ion chelators and showed that LA enhanced the extraction of Aβ from the frontal cortex in a mouse model of AD, suggesting that like other metal chelators, it could reduce amyloid burden in AD patients (Fonte et al., 2001). A potential side effect of a long-term therapy with high doses of a metal chelator such as LA could be the inhibition of metal containing enzymes such as insulin degrading enzyme or superoxide dismutase. Suh et al. investigated whether LA and DHLA remove copper or iron from the active sites of Cu, Zn superoxide dismutase, and aconitase. They found that even at millimolar concentrations neither LA nor DHLA altered the activity of these enzymes (Suh et al., 2004b), providing promising results for the long-term use of LA in AD.

LA: AN ANTI-INFLAMMATORY ANTIOXIDANT AND MODULATOR OF REDOX-SENSITIVE SIGNALING

AD is accompanied by a chronic inflammatory process around amyloid plaques, characterized by the activation of microglia and astrocytes and increased levels of radicals and pro-inflammatory molecules such as iNOS, IL-1 β, IL-6, and TNF-α

(Griffin et al., 1995). AD patients also show increased cytokine levels (e.g., IL-1 β and TNF-α) in the cerebral spinal fluid (CSF), with TNF-α being a good predictor for the progression from mild cognitive impairment to AD. Recently, much attention has been paid to ROS as mediators in signaling processes, termed "redox-sensitive signal transduction." ROS modulate the activity of cytoplasmic signal transducing enzymes by at least two different mechanisms: oxidation of cysteine residues or reaction with iron–sulfur clusters. One widely investigated sensor protein is the p21Ras protein (Lander et al., 1997). Activation of Ras by oxidants is caused by oxidative modification of a specific cysteine residue (Cys118). Ras interacts with PI3-kinase, protein kinase C, diacylglycerol kinase, and MAP-kinase-kinase-kinase, regulating expression of IL-1 β, IL-6, and iNOS. LA can scavenge intracellular free radicals (acting as second messengers), downregulate pro-inflammatory redox-sensitive signal transduction processes including NF-κB translocation, and thus attenuate the release of more free radicals and cytotoxic cytokines (Bierhaus et al., 1997; Wong et al., 2001a).

LA: A CARBONYL SCAVENGER

Cellular and mitochondrial membranes contain a significant amount of arachidonic acid and linoleic acid, precursors of lipid peroxidation products 4-hydroxynonenal (HNE) and acrolein that are extremely reactive. Acrolein decreases pyruvate dehydrogenase (PDH) and α-ketoglutarate dehydrogenase (KGDH) activities by covalently binding to LA, a component in both the PDH and KGDH complexes. Acrolein, which is increased in AD brains, may be partially responsible for the dysfunction of mitochondria and loss of energy found in the AD-affected brain through its inhibition of PDH and KGDH activities, potentially contributing to neurodegeneration (Pocernich and Butterfield, 2003). In a further study, levels of lipid peroxidation, oxidized glutathione (GSSG), and nonenzymatic antioxidants and the activities of mitochondrial enzymes were measured in liver and kidney mitochondria of young and aged rats before and after LA supplementation. In both the liver and kidney, a decrease in the activities of mitochondrial enzymes was observed in aged rats. LA supplemented aged rats showed a decrease in the levels of lipid peroxidation and inhibition of the activities of mitochondrial enzymes like isocitrate dehydrogenase, KGDH, succinate dehydrogenase, NADH dehydrogenase, and cytochrome C oxidase. The authors conclude that LA reverses the age-associated decline in mitochondrial enzymes and therefore may lower the increased risk of oxidative damage that occurs during aging (Arivazhagan et al., 2001).

LA: A STIMULATOR OF GLUCOSE UPTAKE AND UTILIZATION ("INSULINOMIMETIC")

Increased prevalence of insulin abnormalities and insulin resistance in AD may contribute to the disease pathophysiology and clinical symptoms. Insulin and insulin receptors are densely but selectively expressed in the brain, including the medial temporal regions that support the formation of memory. It has recently been demonstrated that insulin-sensitive glucose transporters are localized to the same regions and that insulin plays a role in memory functions. Collectively, these findings suggest that insulin contributes to normal cognitive functioning and that insulin abnormalities may exacerbate cognitive impairments, such as those associated with AD (Watson and Craft, 2003). This view is further supported by the finding that higher fasting plasma insulin levels and reduced CSF: plasma insulin ratios (suggestive of insulin resistance) have also been observed in patients with AD. When AD patients were treated with insulin in a glucose clamp approach, a marked enhancement in memory was observed, whereas normal adults' memory was unchanged (Craft et al., 2003). As previously mentioned, AD is associated with a markedly impaired cerebral glucose metabolism in affected regions. Impaired glucose uptake (partially mediated by insulin resistance) in vulnerable neuronal populations not only compromises production of ACh but also renders neurons vulnerable to excitotoxicity and apoptosis. There is abundant evidence that LA can ameliorate insulin resistance and impaired glucose metabolism in the periphery in type II diabetes mellitus. One study examined the beneficial effects of LA on glucose uptake using soleus muscles derived from nonobese, insulin-resistant type II diabetic Goto-Kakizaki rats, a genetic rat model for human type II diabetes. In this model, chronic administration of LA moderately improved the diabetes-related deficit in glucose metabolism and protein oxidation, as well as the activation of Akt/PKB and PI3K by insulin (Bitar et al., 2004). In a further study, the incorporation of ^{14}C-2-deoxyglucose (2DG) into areas of basal ganglia was investigated in rats treated acutely or for 5 days with RLA or SLA. Following acute administration, RLA was more effective than SLA in increasing ^{14}C-2DG incorporation. For example, acute administration of RLA caused an approximate 40% increase in ^{14}C-2DG incorporation in the substantia nigra while SLA was without effect. However, the effects observed were dependent on basal ^{14}C-DG incorporation in different rat strains. Following subacute administration, the pattern of change in ^{14}C-2DG incorporation was altered and both isomers were equally effective. The effects of RLA were largely maintained with increasing animal age, but the ability of the S-isomer to alter ^{14}C-2DG incorporation was lost by 30 months of age. The authors conclude that RLA has the ability to increase glucose utilization in vivo, which may be relevant to the treatment of neurodegenerative disorders (Seaton et al., 1996). Based on this and similar studies it is quite conceivable that LA might increase glucose uptake in insulin-resistant neurons and thus provide more glycolytic metabolites including acetyl-CoA for these neurons. Since ACh synthesis depends on the availability of acetyl-CoA provided from glucose metabolism, LA might additionally be able to directly increase the concentration of the substrate acetyl-CoA for ACh synthesis (Hoyer, 2003).

UPREGULATION OF GLUTATHIONE SYNTHESIS VIA ACTIVATION OF NRF-2

The tri-peptide γ-L-glutamyl-L-cysteinyl-glycine or glutathione (GSH) is the most abundant nonprotein thiol in animal cells. GSH is required for the maintenance of the thiol redox status of the cell, protection against oxidative damage, detoxification of endogenous and exogenous reactive metal ions and electrophiles, storage and transport of cysteine, as well as protein and DNA synthesis, cell cycle regulation, and cell differentiation (Butterfield et al., 2002). GSH and GSH enzymes play a key role in protecting the cell against the effects of ROS. The key functional element of GSH is the cysteinyl moiety, which provides the reactive thiol group. ROS are reduced by GSH through the enzymatic activity of glutathione peroxidase (GSH-Px) (Butterfield et al., 2002). As a result, GSH is oxidized to GSSG, which is rapidly reduced back to GSH by glutathione reductase (GR) at the expense of NADPH. This is a redox-cycling mechanism to prevent GSH loss (Pocernich et al., 2000). The key role of GSH is that it is a cofactor of glutathione peroxidase and glyoxalase I, important for the detoxification of ROS and methylglyoxal, respectively. In the de novo synthesis, GSH is synthesized from its constituent amino acids by the sequential action of two enzymes, γ-glutamylcysteine synthetase (γ-GCS), which is the rate-limiting enzyme, and glutathione synthetase. γ-GCS catalyzes the formation of the dipeptide γ-glutamylcysteine, which is the rate-limiting substrate in this reaction.

Nuclear factor E2-related factor 2 (Nrf2) is a transcription factor known to induce expression of a variety of cytoprotective and detoxification genes. In recent years, Nrf2 has become a promising novel drug target. Activators of Nrf-2-mediated transcription increase the expression of enzymes involved in GSH synthesis, to maintain sustainable high GSH production and provide protection to neurons against oxidative stress. Nrf-2 activators (a class of potential GSH "boosters") include tert-butylhydroquinone (TBH), sulforaphane (from broccoli), resveratrol, a variety of polyphenols, and α-lipoic acid (Karelson et al., 2001).

In 2002, it was suggested that lipoic acid, like the structurally related dithiolethiones, such as ancthole dithiolethione (ADT), induces phase II detoxification enzymes (which are involved in conjugation reactions) in cultured astroglial cells. LA, like ADT, induced a highly significant, time and concentration dependent, increase in the activity of NAD(P)H dehydrogenase (NQO1) and glutathione-S-transferase (GST) in C6 astroglial cells. The LA- or ADT-mediated induction of NQO1 was further confirmed by quantitative PCR and Western blot analysis. This work for the first time unequivocally demonstrates LA-mediated upregulation of phase II detoxification enzymes, which may highly contribute to the neuroprotective potential of LA. Moreover, the data support the notion of a common mechanism of action of LA and ADT (Flier et al., 2002).

In 2004, Hagen's group at the Linus Pauling Institute at Oregon State University discovered that R-lipoic acid is an in vivo inducer of Nrf2 and increases the enzymatic activity of gamma-glutamylcysteine ligase on a transcriptional level. They observed that the rate-controlling enzyme in GSH, gamma-glutamylcysteine ligase (GCL) loses enzymatic activity with age. With age, the expression of the catalytic (GCLC) and modulatory (GCLM) subunits of GCL decrease by about 50%. In addition, approximately 50% age-related loss in total and nuclear Nrf2 levels was observed, suggesting attenuation of Nrf2-dependent gene transcription. To determine whether the constitutive loss of Nrf2 transcriptional activity also affects the inducible nature of Nrf2 nuclear translocation, old rats were treated with RLA. LA administration increased nuclear Nrf2 levels in old rats and induced Nrf2 binding to the ARE and consequently, higher GCLC levels and GCL activity were observed after LA injection (Suh et al., 2004a).

NEUROPROTECTIVE EFFECTS OF LA IN VITRO AND IN VIVO

PROTECTION OF CULTURED NEURONS AGAINST TOXICITY OF Aβ, IRON, AND OTHER NEUROTOXINS BY LA

Aß, the major component of senile plaques, contributes to neuronal degeneration in AD by stimulating the formation of free radicals. Zhang et al. have investigated the potential efficacy of LA against cytotoxicity induced by Aß (30 μM) and hydrogen peroxide (100 μM) in primary neurons of rat cerebral cortex and found that treatment with LA protected cortical neurons against cytotoxicity induced by both toxins (Zhang et al., 2001). In a similar study, Lovell et al. investigated the effects of LA and DHLA on neuronal hippocampal cultures treated with Aß (25–35) and iron/hydrogen peroxide (Fe/H_2O_2) (Lovell et al., 2003).

In a further study, Müller and Krieglstein tested whether pretreatment with LA can protect cultured neurons against injury caused by cyanide, glutamate, or iron ions. Neuroprotective effects were only significant when the pretreatment with LA occurred for >24 h. The authors conclude that neuroprotection occurs only after prolonged pretreatment with LA and is probably due to the radical scavenger properties of endogenously formed DHLA (Muller and Krieglstein, 1995).

In summary, data from these studies suggest that pretreatment of neurons with LA (or application of DHLA) before exposure to Aß or Fe/H_2O_2 significantly reduces oxidative stress and increases cell survival. However, concomitant application of Aß or Fe/H_2O_2 with LA can temporarily increase oxidative stress as the reduction of LA by the PDH complex consumes reducing equivalents and inhibits energy production.

PROTECTIVE EFFECTS OF LA AGAINST AGE-RELATED COGNITIVE DEFICITS IN AGING RODENTS

Protective effects of LA against cognitive deficits have been shown in several studies in aged rats and mice. In one study, a diet supplemented with RLA was fed to aged rats to determine its efficacy in reversing the decline in metabolism seen with age. Young (3–5 months) and aged (24–26 months) rats were fed for 2 weeks. Ambulatory activity, a measure of general metabolic activity, was almost threefold lower in untreated old rats vs. controls, but this decline was reversed in old rats fed with RLA (Hagen et al., 1999). In a combination treatment study, the effects on cognitive function, brain mitochondrial structure, and biomarkers of oxidative damage were studied after feeding old rats a combination of acetyl-L-carnitine (ALCAR) and/or RLA. Dietary supplementation with ALCAR and/or RLA improved memory, the combination being the most effective for tests of spatial memory and temporal memory. The authors suggest that feeding ALCAR and RLA to old rats improves performance on memory tasks by lowering oxidative damage and improving mitochondrial function. Feeding the substrate ALCAR with RLA restores the velocity of the reaction ($K_{(m)}$) for ALCAR transferase and mitochondrial function. The principle appears to be that, with age, increased oxidative damage to protein causes a deformation of structure of key enzymes with a consequent lessening of affinity ($K_{(m)}$) for the enzyme substrate (Liu et al., 2002).

Similar experiments were performed in the senescence accelerated prone mouse strain 8 (SAMP8), which exhibits age-related deterioration of memory and learning along with increased oxidative markers, and provides a good model for disorders with age-related cognitive impairment. In one study, the ability of LA (and also N-acetyl-L-cysteine) to reverse the cognitive deficits found in the SAMP8 mouse was investigated. Chronic administration of LA improved cognition of 12 month old SAMP8 mice in the T-maze foot-shock avoidance paradigm and the lever press appetitive task. Furthermore, treatment of 12 month old SAMP8 mice with LA reversed all three indexes of oxidative stress. These results provide further support for a therapeutic role for LA in age- and oxidative stress–mediated cognitive impairment, including that associated with AD (Farr et al., 2003).

PROTECTIVE EFFECTS OF LA IN RODENT MODELS OF AD

The effects of LA were also investigated in various animal models of familial AD. For example, 10 month old Tg2576 and wild-type mice were fed an LA-containing diet for 6 months and then assessed for the diet's influence on memory and neuropathology. LA-treated Tg2576 mice exhibited significantly improved learning and memory retention in the Morris water maze task compared to untreated Tg2576 mice. Twenty-four hours after contextual fear conditioning, untreated Tg2576 mice exhibited significantly impaired context-dependent freezing. Assessment of brain soluble and insoluble Aβ levels revealed no differences between LA-treated and untreated Tg2576 mice. The authors conclude that chronic dietary LA can reduce hippocampal-dependent memory deficits of Tg2576 mice without affecting Aβ levels or plaque deposition (Quinn et al., 2007).

CLINICAL TRIALS WITH LA IN AD PATIENTS

Although LA has been used for the treatment of diabetic polyneuropathy in Germany for more than 30 years, no epidemiological study has taken advantage of this large patient population and investigated whether the incidence of AD in the LA-treated patients is lower than in the untreated diabetic and/or untreated nondiabetic population. Therefore, the first indication for a beneficial effect of LA in AD and related dementias came from a rather serendipitous case study. In 1997, a 74 year old patient presented herself at the Department of Medical Rehabilitation and Geriatrics at the Henriettenstiftung Hospital in Hannover with signs of cognitive impairment. Diabetes mellitus and a mild form of polyneuropathy were her main concomitant diseases. With clinical criteria of Diagnostic and Statistical Manual of Mental Disorders (DSM)-III-R, deficits in the neuropsychological tests, an MRI without signs of ischemia, and a typical single photon emission computed tomography showing a decreased bi-temporal and bi-parietal perfusion, early stage AD was diagnosed. Treatment with AChE inhibitors was initiated and the patient received 600 mg LA each day for treatment of her diabetic polyneuropathy. Since 1997, several retests were performed, which showed no substantial decline in the patient's cognitive functions. Therefore, the diagnosis of mild AD was reevaluated several times, but the diagnosis did not change and the neuropsychological tests showed an unusually slow progression of her cognitive impairment. This observation inspired an open pilot trial at the Henriettenstiftung Hospital in Hannover. LA was administered at a dose of 600 mg once daily (in the morning 30 min before breakfast) to nine patients with probable AD (age: 67 ± 9 years, mini-mental state examination (MMSE) score at first visit/start of AChE inhibitor therapy: 23 ± 2 points) receiving a standard treatment with AChE inhibitors over an observation period of 337 ± 80 days. The cognitive performance of the patients before and after addition of LA to their standard medication was compared. A steady decrease in cognitive performance (a 2 points/year decrease in scores in the MMSE and a 4 points/year increase in the AD assessment scale, cognitive subscale (ADAScog) was observed before initiation of the LA regimen. Treatment with LA led to a stabilization of cognitive function, demonstrated by constant scores in two neuropsychological tests for nearly a year. This study was continued and in the end included 43 patients who were followed for an observation period of up to 48 months. In patients with mild dementia (ADAScog < 15), the disease progressed extremely slowly (ADAScog: +1.2 points = year, MMSE: −0.6 points/year). In patients with moderate dementia the disease progressed at approximately twice the rate (Hager et al., 2007). However, this study was small and not randomized. In addition, patients were diagnosed with "probable" AD, and the diagnosis was (in nearly all cases) not confirmed by a neuropathological postmortem analysis. Therefore, a double-blind, placebo-controlled phase II trial is urgently needed before LA could be recommended as a therapy for AD and related dementias. The first randomized, double-blind, placebo-controlled 12 month pilot study involving LA was conducted with 39 subjects diagnosed with mild to moderate AD in Portland, Oregon, USA. Inclusion criteria consisted of the following: aged 55 years or older, diagnosis of probable AD, MMSE score 15–26, Clinical Dementia Rating Scale 0.5–1.0, not depressed, general health status that would not interfere with patients ability to participate and complete the study. Subjects were allowed to continue stable doses of medications for cognitive impairment (e.g., AChE inhibitors, memantine) and stable doses of dietary supplements (e.g., ginkgo biloba, vitamin E). Subjects were excluded if they were eating fish more than one serving per week, taking omega-3 fatty acid supplementation, or taking LA supplementation. Subjects that met inclusion criteria were randomized to one of three groups: (1) placebo, (2) omega-3 fatty acids (ω-3), and (3) omega-3 fatty acids plus lipoic acid (ω-3+LA).

Omega-3 fatty acids were administered in the form of fish oil concentrate at 3 g/day, containing a daily dose of 675 mg DHA and 975 mg EPA (Lynne Shinto, OHSU, personal communication). LA was given as the racemic form with a daily dose of 600 mg. The primary outcome measure was peripheral F2-isoprostane levels to measure lipid peroxidation. Of the 39 randomized subjects, 32 completed their 12 month outcomes visit. There was no baseline difference between groups on medication for the treatment of cognitive impairment (p = 0.50, mean use range 77%–92%) and no difference between groups on 12 month change in peripheral F2-isoprostane levels (p = 0.10). For secondary clinical measures, a significantly delayed decline in MMSE score between groups favoring ω-3 + LA (p = 0.04) over 12 months but no difference between the groups in the ADAS-cog (p = 0.54) was found. In summary, the promising results found for the ω-3+LA group in the double-blind placebo control trial and for LA alone in the open trial for delayed cognitive decline warrant further investigation of LA in combination with other anti-inflammatory compounds for the treatment of AD, but should include longer studies and patients in different stages of the disease (Münch et al., 2010).

COMBINATION TREATMENT OF LA WITH NUTRACEUTICALS

Since AD is a multifactorial disease, it has been suggested that a combination, rather than a single drug, might be most beneficial for AD patients. Among many suggested add-on treatments to LA, nutraceuticals with antioxidant and anti-inflammatory properties could potentially be promising candidates (Steele et al., 2007). Nutraceuticals may be broadly defined as any food substance that offers health or medical benefits (Ferrari and Torres, 2003; Ferrari, 2004). Given that plant foods are derived from biological systems, they contain many compounds in addition to traditional nutrients that can elicit biological responses and are also termed phytonutrients. One of the largest groups of phytonutrients that may confer beneficial health effects are polyphenols (Shanmugam et al., 2008). Over the past decade, polyphenols, which are abundant in fruits and vegetables, have gained recognition for their antioxidant properties and their roles in protecting against chronic diseases such as cancer and cardiovascular diseases (Hertog et al., 1997; Liu, 2004). Consequently, diet is now considered to be an important environmental factor in the development of late-onset AD (Solfrizzi et al., 2003). Polyphenols are therefore beginning to attract increasing interest. Numerous epidemiological studies have suggested a positive association between the consumption of polyphenol-rich foods and the prevention of diseases. A recent epidemiological study reported that consumption of fruit and vegetable juices (high in polyphenols) more frequently than three times a week resulted in a 76% reduction in the risk of developing probable AD over a 9 year period (Dai et al., 2006). Another epidemiological study of 1010 subjects aged 60–93 reported that individuals who consumed curry (containing curcumin) "often" and "very often" had significantly better cognitive test scores as measured via MMSE (Ng et al., 2006). As antioxidants, polyphenols may protect cells against oxidative damage, thereby limiting the risk of AD, which is associated with oxidative stress. Converging epidemiological data also suggests that a low dietary intake of omega-3 (ω-3) essential fatty acids is a candidate risk factor for AD (Morris et al., 2003; Maclean et al., 2005). Docosahexaenoic acid (DHA) is one of the major ω-3 fatty acids in the brain where it is enriched in neurons and synapses. DHA is associated with learning and memory and is also required for the structure and function of brain cell membranes. In the AD-affected brain DHA levels are known to be decreased (Soderberg et al., 1991; Prasad et al., 1998), while people who ingest higher levels of DHA are less likely to develop AD (Morris et al., 2003; Tully et al., 2003). As many Western diets have been reported to be deficient in DHA and also low in polyphenolic content, supplementation with DHA and polyphenols may offer potential preventative treatments for AD. Several of these nutraceuticals/phytonutrients (e.g., (-)-epigallocatechin gallate (EGCG) from green tea, curcumin from the curry spice turmeric and the omega-3 DHA from fish oils) have shown promising results, when used as single therapies in animal studies.

EGCG has previously been shown to prevent neuronal cell death caused by Aβ neurotoxicity in cell cultures (Choi et al., 2001; Levites et al., 2003). A study by Rezai-Zadeh et al. (Rezai-Zadeh et al., 2005) reported that EGCG reduced Aβ generation in vitro in neuronal-like cells and primary neuronal cultures from Tg2576 mice, along with promotion of the nonamyloidogenic α-secretase proteolytic pathway. Furthermore when 12 month old Tg2576 mice were treated with 20 mg/kg EGCG via intra-peritoneal injections for 60 days, it was found that Aβ levels and plaque load in the brain were decreased. Curcumin has been reported to be several times more potent than vitamin E as a free radical scavenger (Zhao et al., 1989) and there is also increasing evidence showing that curcumin can inhibit Aβ aggregation (Yang et al., 2005). In a study by Lim et al. curcumin was tested for its ability to inhibit the combined inflammatory and oxidative damage in Tg2576 transgenic mice. In this study Tg2576 mice aged 10 months old were fed a curcumin diet (160 ppm) for 6 months. Their results showed that the curcumin diet significantly lowered the levels of oxidized proteins, IL-1β, the astrocyte marker glial fibrillary acidic protein (GFAP), soluble and insoluble Aβ, and also plaque burden (Lim et al., 2001). Following on from this work, Yang et al. evaluated the effect of feeding a curcumin diet (500 ppm) to 17 month old Tg2576 mice for 6 months. When fed to the aged Tg2576 mice with advanced amyloid accumulation, curcumin resulted in reduced soluble amyloid levels and plaque burden. These data raise the possibility that dietary supplementation with curcumin may provide a potential preventative treatment for AD by decreasing Aβ levels and plaque load via inhibition of Aβ oligomer formation and fibrillization, along with decreasing oxidative stress and inflammation.

The interest in dietary DHA supplementation has arisen from the approach to protect neurons from neuronal degradation and therefore prevent neurological diseases like AD. Converging epidemiological data suggests that a low dietary intake of ω-3 polyunsaturated fatty acids is a candidate risk factor for AD (Calon et al., 2005). In the AD brain, DHA is known to be decreased (Soderberg et al., 1991; Prasad et al., 1998), while people who ingest higher levels of DHA are less likely to develop AD (Conquer et al., 2000; Barberger-Gateau et al., 2002; Morris et al., 2003). A recent in vitro study by Florent et al. demonstrated that DHA provided cortical neurones with a higher level of resistance to the cytotoxic effects induced by soluble Aβ oligomers (Florent et al., 2006). Lukiw et al. also demonstrated that DHA decreased $A\beta_{40}$ and $A\beta_{42}$ secretion from aging human neuronal cells (Lukiw et al., 2005). A study by Calon et al. showed that a reduction of dietary ω-3 PUFA in Tg2576 transgenic mice resulted in a loss of post-synaptic proteins and behavioral deficits, while a DHA-enriched diet prevented these effects (Calon et al., 2004). Other studies have shown that DHA protects neurons from Aβ accumulation and toxicity and ameliorates cognitive impairment in rodent models of AD (Hashimoto et al., 2005; Lim et al., 2005). A recent study by Cole and Frautschy showed that DHA supplementation in Tg2576 transgenic mice aged 17 months markedly reduced Aβ accumulation, oxidative damage, and also improved cognitive function (Cole and Frautschy, 2006). Thus, dietary supplementation with DHA may also provide a potential preventative treatment for AD via prevention of cognitive deficits and reduction of Aβ accumulation and oxidative stress.

These nutraceuticals have all been demonstrated to have varying mechanisms of action, relating to decreasing cognitive deficits, oxidative stress, inflammation, and Aβ levels. Therefore, combination therapies of these nutraceuticals containing polyphenols (EGCG and curcumin), ω-3 essential fatty acids (DHA), and LA have the potential to provide nutritional supplement therapies for the prevention of AD pathology and cognitive impairments as LA has been previously demonstrated in Tg2576 mice to prevent cognitive deficits(Quinn et al., 2007). The use of LA in addition to other nutraceuticals would provide the means to prevent cognitive deficits in combination with the benefits of curcumin, EGCG, and DHA to decrease oxidative stress, inflammation, Aβ levels, and Aβ plaque load.

ACKNOWLEDGMENTS

This work was supported by the J.O. and J.R Wicking Foundation, Alzheimer's Australia, and the NHMRC.

REFERENCES

Arivazhagan, P., Ramanathan, K., Panneerselvam, C. (2001) Effect of dl-alpha-lipoic acid on mitochondrial enzymes in aged rats. *Chem Biol Interact*, **138**, 189–198.

Barberger-Gateau, P., Letenneur, L., Deschamps, V., Peres, K., Dartigues, J.F., Renaud, S. (2002) Fish, meat, and risk of dementia: Cohort study. *BMJ*, **325**, 932–933.

Bierhaus, A., Chevion, S., Chevion, M., Hofmann, M., Quehenberger, P., Illmer, T., Luther, T., Berentshtein, E., Tritschler, H., Muller, M., Wahl, P., Ziegler, R., Nawroth, P.P. (1997) Advanced glycation end product-induced activation of nf-kappab is suppressed by alpha-lipoic acid in cultured endothelial cells. *Diabetes*, **46**, 1481–1490.

Bitar, M.S., Wahid, S., Pilcher, C.W., Al-Saleh, E., Al-Mulla, F. (2004) Alpha-lipoic acid mitigates insulin resistance in goto-kakizaki rats. *Horm Metab Res*, **36**, 542–549.

Butterfield, D.A., Pocernich, C.B., Drake, J. (2002) Elevated glutathione as a therapeutic strategy in Alzheimer's disease. *Drug Dev Res*, **56**, 428–437.

Calon, F., Lim, G.P., Morihara, T., Yang, F., Ubeda, O., Salem, N., Jr., Frautschy, S.A., Cole, G.M. (2005) Dietary n-3 polyunsaturated fatty acid depletion activates caspases and decreases nmda receptors in the brain of a transgenic mouse model of Alzheimer's disease. *Eur J Neurosci*, **22**, 617–626.

Calon, F., Lim, G.P., Yang, F., Morihara, T., Teter, B., Ubeda, O., Rostaing, P., Triller, A., Salem, N., Jr., Ashe, K.H., Frautschy, S.A., Cole, G.M. (2004) Docosahexaenoic acid protects from dendritic pathology in an Alzheimer's disease mouse model. *Neuron*, **43**, 633–645.

Choi, Y.T., Jung, C.H., Lee, S.R., Bae, J.H., Baek, W.K., Suh, M.H., Park, J., Park, C.W., Suh, S.I. (2001) The green tea polyphenol (-)-epigallocatechin gallate attenuates beta-amyloid-induced neurotoxicity in cultured hippocampal neurons. *Life Sci*, **70**, 603–614.

Cole, G.M., Frautschy, S.A. (2006) Docosahexaenoic acid protects from amyloid and dendritic pathology in an Alzheimer's disease mouse model. *Nutr Health*, **18**, 249–259.

Conquer, J.A., Tierney, M.C., Zecevic, J., Bettger, W.J., Fisher, R.H. (2000) Fatty acid analysis of blood plasma of patients with Alzheimer's disease, other types of dementia, and cognitive impairment. *Lipids*, **35**, 1305–1312.

Craft, S., Asthana, S., Cook, D.G., Baker, L.D., Cherrier, M., Purganan, K., Wait, C., Petrova, A., Latendresse, S., Watson, G.S., Newcomer, J.W., Schellenberg, G.D., Krohn, A.J. (2003) Insulin dose-response effects on memory and plasma amyloid precursor protein in Alzheimer's disease: Interactions with apolipoprotein e genotype. *Psychoneuroendocrinology*, **28**, 809–822.

Cummings, J.L. (2004) Alzheimer's disease. *N Engl J Med*, **351**, 56–67.

Dai, Q., Borenstein, A.R., Wu, Y., Jackson, J.C., Larson, E.B. (2006) Fruit and vegetable juices and Alzheimer's disease: The kame project. *Am J Med*, **119**, 751–759.

Farr, S.A., Poon, H.F., Dogrukol-Ak, D., Drake, J., Banks, W.A., Eyerman, E., Butterfield, D.A., Morley, J.E. (2003) The antioxidants alpha-lipoic acid and n-acetylcysteine reverse memory impairment and brain oxidative stress in aged samp8 mice. *J Neurochem*, **84**, 1173–1183.

Ferrari, C.K. (2004) Functional foods, herbs and nutraceuticals: Towards biochemical mechanisms of healthy aging. *Biogerontology*, **5**, 275–289.

Ferrari, C.K.B., Torres, E.A.F.S. (2003) Biochemical pharmacology of functional foods and prevention of chronic diseases of aging. *Biomed Pharm*, **57**, 251–260.

Flier, J., Van Muiswinkel, F.L., Jongenelen, C.A., Drukarch, B. (2002) The neuroprotective antioxidant alpha-lipoic acid induces detoxication enzymes in cultured astroglial cells. *Free Radic Res*, **36**, 695–699.

Florent, S., Malaplate-Armand, C., Youssef, I., Kriem, B., Koziel, V., Escanye, M.C., Fifre, A., Sponne, I., Leininger-Muller, B., Olivier, J.L., Pillot, T., Oster, T. (2006) Docosahexaenoic acid prevents neuronal apoptosis induced by soluble amyloid-beta oligomers. *J Neurochem*, **96**, 385–395.

Fonte, J., Miklossy, J., Atwood, C., Martins, R. (2001) The severity of cortical Alzheimer's type changes is positively correlated with increased amyloid-beta levels: Resolubilization of amyloid-beta with transition metal ion chelators. *J Alzheimer's Dis*, **3**, 209–219.

Griffin, W.S., Sheng, J.G., Roberts, G.W., Mrak, R.E. (1995) Interleukin-1 expression in different plaque types in Alzheimer's disease: Significance in plaque evolution. *J Neuropathol Exp Neurol*, **54**, 276–281.

Hagen, T.M., Ingersoll, R.T., Lykkesfeldt, J., Liu, J., Wehr, C.M., Vinarsky, V., Bartholomew, J.C., Ames, A.B. (1999) (r)-alpha-lipoic acid-supplemented old rats have improved mitochondrial function, decreased oxidative damage, and increased metabolic rate. *FASEB J*, **13**, 411–418.

Hager, K., Kenklies, M., McAfoose, J., Engel, J., Münch, G. (2007) Alpha-lipoic acid as a new treatment option for Alzheimer's disease— a 48 months follow-up analysis. *J Neural Transm Suppl*, 189–193.

Hashimoto, M., Tanabe, Y., Fujii, Y., Kikuta, T., Shibata, H., Shido, O. (2005) Chronic administration of docosahexaenoic acid ameliorates the impairment of spatial cognition learning ability in amyloid beta-infused rats. *J Nutr*, **135**, 549–555.

Haugaard, N., Levin, R.M. (2000) Regulation of the activity of choline acetyl transferase by lipoic acid. *Mol Cell Biochem*, **213**, 61–63.

Haugaard, N., Levin, R.M. (2002) Activation of choline acetyl transferase by dihydrolipoic acid. *Mol Cell Biochem*, **229**, 103–106.

Hertog, M.G., Feskens, E.J., Kromhout, D. (1997) Antioxidant flavonols and coronary heart disease risk. *Lancet*, **349**, 699.

Holmquist, L., Stuchbury, G., Berbaum, K., Muscat, S., Young, S., Hager, K., Engel, J., Münch, G. (2007) Lipoic acid as a novel treatment for Alzheimer's disease and related dementias. *Pharmacol Ther*, **113**, 154–164.

Hoyer, S. (2003) Memory function and brain glucose metabolism. *Pharmacopsychiatry*, **36(Suppl 1)**, S62–S67.

Huang, X., Atwood, C.S., Hartshorn, M.A., Multhaup, G., Goldstein, L.E., Scarpa, R.C., Cuajungco, M.P., Gray, D.N., Lim, J., Moir, R.D., Tanzi, R.E., Bush, A.I. (1999) The a beta peptide of Alzheimer's disease directly produces hydrogen peroxide through metal ion reduction. *Biochemistry*, **38**, 7609–7616.

Ishii, K., Minoshima, S. (2005) Pet is better than perfusion spect for early diagnosis of Alzheimer's disease—for. *Eur J Nucl Med Mol Imaging*, **32**, 1463–1465.

Karelson, E., Bogdanovic, N., Garlind, A., Winblad, B., Zilmer, K., Kullisaar, T., Vihalemm, T., Kairane, C., Zilmer, M. (2001) The cerebrocortical areas in normal brain aging and in Alzheimer's disease: Noticeable differences in the lipid peroxidation level and in anti-oxidant defense. *Neurochem Res*, **26**, 353–361.

Lander, H.M., Tauras, J.M., Ogiste, J.S., Hori, O., Moss, R.A., Schmidt, A.M. (1997) Activation of the receptor for advanced glycation end products triggers a p21(ras)-dependent mitogen-activated protein kinase pathway regulated by oxidant stress. *J Biol Chem*, **272**, 17810–17814.

Levites, Y., Amit, T., Mandel, S., Youdim, M.B. (2003) Neuroprotection and neurorescue against abeta toxicity and pkc-dependent release of nonamyloidogenic soluble precursor protein by green tea polyphenol (-)-epigallocatechin-3-gallate. *FASEB J*, **17**, 952–954.

Lim, G.P., Calon, F., Morihara, T., Yang, F., Teter, B., Ubeda, O., Salem, N., Jr., Frautschy, S.A., Cole, G.M. (2005) A diet enriched with the omega-3 fatty acid docosahexaenoic acid reduces amyloid burden in an aged Alzheimer mouse model. *J Neurosci*, **25**, 3032–3040.

Lim, G.P., Chu, T., Yang, F., Beech, W., Frautschy, S.A., Cole, G.M. (2001) The curry spice curcumin reduces oxidative damage and amyloid pathology in an Alzheimer transgenic mouse. *J Neurosci*, **21**, 8370–8377.

Liu, R.H. (2004) Potential synergy of phytochemicals in cancer prevention: Mechanism of action. *J Nutr*, **134**, 3479S–3485S.

Liu, J., Killilea, D.W., Ames, B.N. (2002) Age-associated mitochondrial oxidative decay: Improvement of carnitine acetyltransferase substrate-binding affinity and activity in brain by feeding old rats acetyl-l-carnitine and/or r-alpha -lipoic acid. *Proc Natl Acad Sci USA*, **99**, 1876–1881.

Lovell, M.A., Xie, C., Xiong, S., Markesbery, W.R. (2003) Protection against amyloid beta peptide and iron/hydrogen peroxide toxicity by alpha lipoic acid. *J Alzheimer's Dis*, **5**, 229–239.

Lukiw, W.J., Cui, J.G., Marcheselli, V.L., Bodker, M., Botkjaer, A., Gotlinger, K., Serhan, C.N., Bazan, N.G. (2005) A role for docosahexaenoic acid-derived neuroprotectin d1 in neural cell survival and Alzheimer disease. *J Clin Invest*, **115**, 2774–2783.

Maclean, C.H., Issa, A.M., Newberry, S.J., Mojica, W.A., Morton, S.C., Garland, R.H., Hilton, L.G., Traina, S.B., Shekelle, P.G. (2005) Effects of omega-3 fatty acids on cognitive function with aging, dementia, and neurological diseases. *Evid Rep Technol Assess (Summ)*, 1–3.

Morris, M.C., Evans, D.A., Bienias, J.L., Tangney, C.C., Bennett, D.A., Wilson, R.S., Aggarwal, N., Schneider, J. (2003) Consumption of fish and n-3 fatty acids and risk of incident Alzheimer disease. *Arch Neurol*, **60**, 940–946.

Muller, U., Krieglstein, J. (1995) Prolonged pretreatment with alpha-lipoic acid protects cultured neurons against hypoxic, glutamate-, or iron-induced injury. *J Cereb Blood Flow Metab*, **15**, 624–630.

Münch, G., Gerlach, M., Sian, J., Wong, A., Riederer, P. (1998) Advanced glycation end products in neurodegeneration: More than early markers of oxidative stress? *Ann Neurol*, **44**, S85–S88.

Münch, G., Shinto, L., Maczurek, A. (2010) Lipoic acid as a treatment for Alzheimer's disease. *Medicine Today*, **11**, 62–64.

Ng, T.P., Chiam, P.C., Lee, T., Chua, H.C., Lim, L., Kua, E.H. (2006) Curry consumption and cognitive function in the elderly. *Am J Epidemiol*, **164**, 898–906.

Nordberg, A., Nyberg, P., Adolfsson, R., Winblad, B. (1987) Cholinergic topography in Alzheimer brains: A comparison with changes in the monoaminergic profile. *J Neural Transm*, **69**, 19–32.

Nordberg, A., Winblad, B. (1986) Reduced number of [3h]nicotine and [3h]acetylcholine binding sites in the frontal cortex of Alzheimer brains. *Neurosci Lett*, **72**, 115–119.

Packer, L., Witt, E.H., Tritschler, H.J. (1995) Alpha-lipoic acid as a biological antioxidant. *Free Radic Biol Med*, **19**, 227–250.

Pocernich, C.B., Butterfield, D.A. (2003) Acrolein inhibits nadh-linked mitochondrial enzyme activity: Implications for Alzheimer's disease. *Neurotox Res*, **5**, 515–520.

Pocernich, C., La Fontaine, M., Butterfield, D. (2000) In-vivo glutathione elevation protects against hydroxyl free radical-induced protein oxidation in rat brain. *Neurochem Int*, **36**, 185–191.

Prasad, M.R., Lovell, M.A., Yatin, M., Dhillon, H., Markesbery, W.R. (1998) Regional membrane phospholipid alterations in Alzheimer's disease. *Neurochem Res*, **23**, 81–88.

Quinn, J.F., Bussiere, J.R., Hammond, R.S., Montine, T.J., Henson, E., Jones, R.E., Stackman, R.W., Jr. (2007) Chronic dietary alpha-lipoic acid reduces deficits in hippocampal memory of aged tg2576 mice. *Neurobiol Aging*, **28**, 213–225.

Retz, W., Gsell, W., Münch, G., Rosler, M., Riederer, P. (1998) Free radicals in Alzheimer's disease. *J Neural Transm Suppl*, **54**, 221–236.

Rezai-Zadeh, K., Shytle, D., Sun, N., Mori, T., Hou, H., Jeanniton, D., Ehrhart, J., Townsend, K., Zeng, J., Morgan, D., Hardy, J., Town, T., Tan, J. (2005) Green tea epigallocatechin-3-gallate (egcg) modulates amyloid precursor protein cleavage and reduces cerebral amyloidosis in Alzheimer transgenic mice. *J Neurosci*, **25**, 8807–8814.

Seaton, T.A., Jenner, P., Marsden, C.D. (1996) The isomers of thioctic acid alter c-deoxyglucose incorporation in rat basal ganglia. *Biochem Pharmacol*, **51**, 983–986.

Shanmugam, K., Holmquist, L., Steele, M., Stuchbury, G., Berbaum, K., Schulz, O., García, O.B., Castillo, J., Burnell, J., Rivas, V.G., Dobson, G., Münch, G. (2008) Plant-derived polyphenols attenuate lipopolysaccharide-induced nitric oxide and tumour necrosis factor production in murine microglia and macrophages. *Mol Nutr Food Res*, **52**, 427–438.

Soderberg, M., Edlund, C., Kristensson, K., Dallner, G. (1991) Fatty acid composition of brain phospholipids in aging and in Alzheimer's disease. *Lipids*, **26**, 421–425.

Solfrizzi, V., Panza, F., Capurso, A. (2003) The role of diet in cognitive decline. *J Neural Transm*, **110**, 95–110.

Steele, M., Stuchbury, G., Münch, G. (2007) The molecular basis of the prevention of Alzheimer's disease through healthy nutrition. *Exp Gerontol*, **42**, 28–36.

Stuchbury, G., Münch, G. (2005) Alzheimer's associated inflammation, potential drug targets and future therapies. *J Neural Transm*, **112**, 429–453.

Suh, J.H., Moreau, R., Heath, S.H., Hagen, T.M. (2005) Dietary supplementation with (r)-alpha-lipoic acid reverses the age-related accumulation of iron and depletion of antioxidants in the rat cerebral cortex. *Redox Rep*, **10**, 52–60.

Suh, J.H., Shenvi, S.V., Dixon, B.M., Liu, H., Jaiswal, A.K., Liu, R.M., Hagen, T.M. (2004a) Decline in transcriptional activity of nrf2 causes age-related loss of glutathione synthesis, which is reversible with lipoic acid. *Proc Natl Acad Sci USA*, **101**, 3381–3386.

Suh, J.H., Zhu, B.Z., deSzoeke, E., Frei, B., Hagen, T.M. (2004b) Dihydrolipoic acid lowers the redox activity of transition metal ions but does not remove them from the active site of enzymes. *Redox Rep*, **9**, 57–61.

Tully, A.M., Roche, H.M., Doyle, R., Fallon, C., Bruce, I., Lawlor, B., Coakley, D., Gibney, M.J. (2003) Low serum cholesteryl ester-docosahexaenoic acid levels in Alzheimer's disease: A case-control study. *Br J Nutr*, **89**, 483–489.

Watson, G.S., Craft, S. (2003) The role of insulin resistance in the pathogenesis of Alzheimer's disease: Implications for treatment. *CNS Drugs*, **17**, 27–45.

Wong, A., Dukic-Stefanovic, S., Gasic-Milenkovic, J., Schinzel, R., Wiesinger, H., Riederer, P., Münch, G. (2001a) Anti-inflammatory antioxidants attenuate the expression of inducible nitric oxide synthase mediated by advanced glycation endproducts in murine microglia. *Eur J Neurosci*, **14**, 1961–1967.

Wong, A., Lüth, H.J., Deuther-Conrad, W., Dukic-Stefanovic, S., Gasic-Milenkovic, J., Arendt, T., Münch, G. (2001b) Advanced glycation endproducts co-localize with inducible nitric oxide synthase in Alzheimer's disease. *Brain Res*, **920**, 32–40.

Yang, F., Lim, G.P., Begum, A.N., Ubeda, O.J., Simmons, M.R., Ambegaokar, S.S., Chen, P.P., Kayed, R., Glabe, C.G., Frautschy, S.A., Cole, G.M. (2005) Curcumin inhibits formation of amyloid beta oligomers and fibrils, binds plaques, and reduces amyloid in vivo. *J Biol Chem*, **280**, 5892–5901.

Zhang, L., Xing, G.Q., Barker, J.L., Chang, Y., Maric, D., Ma, W., Li, B.S., Rubinow, D.R. (2001) Alpha-lipoic acid protects rat cortical neurons against cell death induced by amyloid and hydrogen peroxide through the akt signalling pathway. *Neurosci Lett*, **312**, 125–128.

Zhao, B.I., Li, X.J., He, R.G., Cheng, S.J., Xin, W.J. (1989) Scavenging effect of extracts of green tea and natural antioxidants on active oxygen radicals. *Cell Biophys*, **14**, 175–185.

6 Metabolic Agents and Cognitive Function

Lauren Owen

CONTENTS

CHAPTER OVERVIEW

There are a number of agents which are believed to impact on metabolic functions which may ultimately impact on neuronal cell survival and cognitive function. Aging is characterized by a progressive deterioration in physiological functions and metabolic processes. Regimes that buffer intracellular energy levels may impede the progression of the neurodegenerative process. This chapter focuses on some of the metabolic agents that may prove to be effective in combating neurodegeneration and lead to better cognitive aging through the life span. The metabolic agents specifically focused on in this chapter are glucose and oxygen, pyruvate, creatine, and L-carnitine. Each of these agents is directly responsible for generating adenosine triphosphate (ATP), the molecular unit of currency of intracellular energy transfer. Their roles as cognitive agents are explored.

METABOLIC AGENTS FOR COGNITIVE FUNCTION

During normal aging neuronal cell injury and death are accelerated and lead to region-specific brain shrinkage. In brain regions particularly important for the formation of memories and decision making, for example, the hippocampus and the prefrontal white matter, shrinkage increases with age (Raz et al. 2005). Reduction in the total number of viable cells may lead to an accelerated decline in brain functioning. A likely cause of reduced neuronal cell number is impaired energy metabolism. Impeded energy metabolism may trigger pro-apoptotic signaling (programmed cell death), oxidative damage, and excitotoxicity and impede mitochondrial DNA repair (Klein and Ferrante 2007). These processes can interact and potentiate one another, which in turn results in a continuation of energy depletion. Reduced energy levels threaten cellular homeostasis and integrity. The brain is the most metabolically active organ in the body and as such is particularly vulnerable to disruption of energy resources. In addition, because of the high levels of oxygen metabolism in brain tissue, mitochondria are highly susceptible to oxidative stress (Chinnery et al. 2006). Therefore interventions that improve mitochondrial function by sustaining ATP levels may have direct and indirect importance for improving neuronal dysfunction and loss. Regimes that buffer intracellular energy levels may significantly impede the progression of neurodegenerative diseases and disorders. Figure 6.1 lists some of the nutritional agents which may impact on mechanisms involved in improving metabolic function and thus may have the potential to improve cognitive function. Metabolic function may be improved in a variety of ways, either by improving the availability of substrates necessary for energy production or by improving the transport and effectiveness of cells involved in the metabolic process, for example, by improving mitochondrial transport and respiration. A summary of the various agents which may improve energy availability is summarized in Figure 6.1. Many of these agents have multifaceted mechanisms of action and may lead to numerous cascades of biological events. It is beyond the scope of this chapter to review all of the possible nutritional contributors to optimal metabolic function; therefore, this chapter will focus on the agents which have a central action of improving availability of energy. These agents are listed next as "energy enhancers."

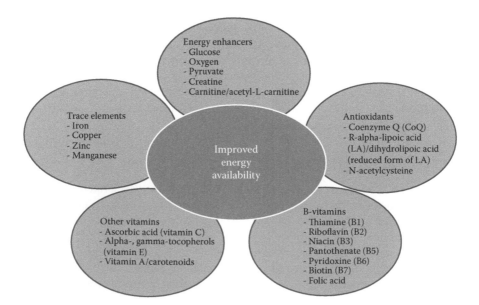

FIGURE 6.1   Metabolic agents with the potential to improve cognitive function.

GLUCOSE AND OXYGEN

The principal source of energy for brain function is derived from the oxidative breakdown of glucose. The human brain is an extremely metabolically active organ accounting for approximately 30% of the total basal energy expenditure. The brain remains metabolically active at all times, including sleep, and is thus entirely dependent on continuous and uninterrupted supply of energy in the form of the substrates glucose and oxygen. Compared to other organs in the body, the brain is particularly vulnerable to small and transient changes in its energy supply. Interrupted delivery leads within seconds to unconsciousness and within minutes may cause irreparable brain damage. Thus, the concentration of glucose in the blood plasma is tightly regulated to stay within the normal range of 60–90 mg/100 mL for humans. When blood glucose drops below 40 mg/100 mL (hypoglycemic condition) in humans, it can cause discomfort, confusion, coma, convulsions, or even death (Lehninger et al. 2005). Beyond infancy, and under normal conditions, the brain's energy requirements are met almost exclusively by the oxidative breakdown of glucose. During times of hypoglycemia other tissues will cease to utilize glucose all together in order to increase glucose availability to the brain (Thomson 1967). Compared with other organs the brain possesses paradoxically limited stores of glycogen, which without replenishment are exhausted in up to 10 min. There is, however, no storage capacity for oxygen; thus, disruption leads to instantaneous effects. Associated measurements of oxygen and glucose levels in blood sampled upon entering and leaving the brain in humans show that almost all the oxygen utilized by the brain can be accounted for by the oxidative metabolism of glucose (McIlwain 1959). Since the brain is clearly susceptible to small changes in energy supply, metabolic activity is limited by glucose and oxygen resources.

A few early studies demonstrated the effects of glucose on cognition around the 1950s. For example, administration of 10 g of glucose to school children every 45 min throughout a morning demonstrated improved mathematical ability and generally improved concentration (Hafermann 1955). However, a more widespread interest in glucose did not occur until the 1980s when the glucose effect was reevaluated by psychopharmacologists examining possible mechanisms of action for neuroendocrine facilitation of memory. Since then there have been increasing reports that cognitive functioning is influenced by the increased availability of glucose provision. Many reports have illustrated the robust association between changes in blood glucose levels and cognition in animals (Gold 1986; Wenk 1989; White 1991), the elderly (Gonder-Frederick et al. 1987; Craft et al. 1992, 1994), and the young (Benton and Sargent 1992; Benton and Owens 1993; Sünram-Lea et al. 2001, 2002a,b, 2004; Riby et al. 2008; Scholey et al. 2009). Thus, the cognition-enhancing action of glucose is well established. In terms of dosing the most optimal glucose dose for cognitive enhancement generally appears to follow the classic Yerkes–Dodson inverted-U dose–response profile (Sunram-Lea et al. 2011). For young adults 25 g seems to most reliably facilitate cognitive performance; however, there is some contention regarding whether the dose–response profile may be dependent upon the cognitive domain being assessed. In rats bimodal response variability was observed when different tasks were used which represented the action of glucose on two different brain substrates: the caudate nucleus and the hippocampus (Packard and White 1990). In humans the inverted-U dose–response profile has been specifically observed for tasks of verbal declarative memory, where other tasks (specifically spatial and numeric working memory) demonstrated slightly different response profiles (cubic and quartic respectively) (Sunram-Lea et al. 2011). The clearest enhancement effects of increased glucose supply have been observed for declarative memory tasks in the form of word and paragraph recall; for a review see Hoyland et al. (2008). These findings have led to the notion that glucose facilitation may be particularly pronounced in tasks which pertain to the hippocampal formation (Sünram-Lea et al. 2001). Furthermore several studies have shown that an important mediating factor for cognitive enhancement by increased energy resources is level of task demands. That is, tasks which are more demanding appear to be more sensitive to the effect of glucose (Kennedy and Scholey 2000; Scholey et al. 2001; Sünram-Lea et al. 2002a). It has also been demonstrated that tasks which are more demanding lead to a significantly accelerated reduction in blood glucose levels compared with a semantically matched task (Scholey et al. 2001). However recent research has shown that at high dosages (60 g) implicit memory which is not regarded as either demanding nor hippocampally mediated may also be enhanced by glucose (Owen et al. 2010), adding further support to the notion that different domains of memory may follow different glucose dose–response profiles.

It is widely acknowledged that oxygen restriction and ischemic deprivation exert marked effects on cognitive function (Volpe and Hirst 1983). Furthermore restriction of oxygen supply due to altitude results in cognitive impairment on a number of cognitive parameters with these effects being instantaneously reversed by the administration of oxygen (Crowley et al. 1992). Evidence suggests that even small fluctuations in cerebral oxygen delivery within normal physiological limits may impact on cognitive performance (Walker and Sandman 1979). While cognitive deficits from oxygen restriction due to altitude (Crowley et al. 1992), carbon monoxide poisoning (Weaver et al. 2002), and isovolemic anemia (Weiskopf et al. 2002) can all be reversed by oxygen administration, impairment effects may be permanent if treatment is not administered in time. Similarly cognitive degeneration due to age is not reversed by oxygen treatment when administered either normobaric or hypobaric oxygen treatment (Raskin et al. 1978). There is very limited research of oxygen administration on cognition in normal healthy individuals. Early research examining the effects of hyperbaric oxygen supplementation demonstrated improved cognitive function (short-term memory and visual organization) in elderly outpatients compared to baseline performance. However this study failed to compare with a control group (Edwards and Hart 1974).

In normal healthy humans research has demonstrated that oxygen administration can improve cognitive functioning compared to air-breathing control conditions. Research has shown that oxygen administration leads to improved long-term memory and reaction times compared to a control group of normal air-breathing (Moss and Scholey 1996; Moss et al. 1998; Scholey et al. 1998). Furthermore, similar to glucose facilitation, oxygen administration appears to facilitate cognition most effectively for tasks with a higher cognitive load (Moss et al. 1998; Scholey et al. 1998). In addition to this finding a further study also examined heart rate during cognitive testing with oxygen versus air-breathing controls. Compared to baseline, heart rate was significantly elevated during cognitive testing tasks in both the air and oxygen groups. In the oxygen group, significant correlations were found between changes in oxygen saturation and cognitive performance. In the air group, greater changes in heart rate were associated with improved cognitive performance (Scholey et al. 1999). These findings suggest that during times of cognitive demand availability of metabolic resources impact on cognitive functioning. A more recent study has further demonstrated the importance of metabolic resources during cognitive demand by manipulating level of cognitive demand during oxygen administration. In this study oxygen administration of 40% versus 21% was examined during completion of an addition task with three levels of difficulty. It was observed that 40% oxygen improved accuracy scores across the task compared to the 21% oxygen dose, with the difference in accuracy rate increasing between the two dosages as the task difficulty level increased (Chung et al. 2008). While cognitive demand is clearly a moderating factor for cognitive enhancement by oxygen, enhancement has been observed on several cognitive domains; for example, oxygen supplementation has been shown to improve everyday memory tasks such as memory for shopping lists and putting names to faces when participants received 100% oxygen compared with air-breathing controls (Winder and Borrill 1998). The dose–response for oxygen administration on performance appears to follow the Yerkes–Dodson inverted-U shape in a similar fashion to glucose facilitation with shorter doses of 30 s to 3 min appearing to be most beneficial while continuous oxygen breathing for longer than 10 min leading to decline in performance (Moss et al. 1998). The window for cognitive improvement through oxygen administration therefore appears to be quite brief, with research demonstrating that administration of oxygen increases blood oxygen levels for only 4–5 min (Moss et al. 1998).

PYRUVATE

Neuronal cell death resulting from hypoglycemia and hypoxia is the result of a series of events triggered by reduced energy availability, and the normalization of blood glucose and oxygen levels does not necessarily block or reverse this cell death process once it has begun. During times of low availability of glucose and oxygen the brain utilizes other, less efficient energy sources that can be produced aerobically. Pyruvate is the end product of glycolysis, which is converted into acetyl coenzyme A that enters the Krebs cycle when there is sufficient oxygen available. When the oxygen is insufficient, pyruvate is broken down anaerobically, creating lactate in humans and animals. Lactate has recently been considered as a central neuroprotective agent (Gladden 2004). The blood-brain barrier normally transports pyruvate at a rate much slower than glucose, but prior work suggests that significant pyruvate entry to the brain can be achieved by elevating plasma pyruvate concentrations (Lee et al. 2001).

During pathological insult or general aging, the main upstream event most responsible for neuronal cell death is excitotoxicity from glutamate receptor activity (Wieloch 1985). Recent research has shown that cells that would otherwise go on to die after the cascade of excitotoxic activity could be rescued by providing pyruvate (Ying et al. 2002).

However, there is remarkably little research evaluating the effects of pyruvate on cognitive function. One recent study assessed the effect of pyruvate administration in rats with hypoglycemia-induced brain injury. Insulin was used to induce hypoglycemia then hypoglycemia was terminated with either glucose alone or with glucose plus pyruvate. They found that in the four brain regions studied (CA1, subiculum, dentate gyrus of the hippocampus, and piriform cortex) the addition of pyruvate reduced neuron death by 70%–90%. Neuron survival was also observed when pyruvate delivery was delayed for up to 3 h. The improved neuron survival was accompanied by a sustained improvement in cognitive function as assessed by the Morris water maze (Suh et al. 2005).

Furthermore recent animal research has demonstrated the potential usefulness of ethyl pyruvate as a stroke therapy. Yu et al. (2005) found that ethyl pyruvate affords the strong protection of delayed cerebral ischemic injury with significant reduction in infarct volume accompanied by the suppression of the clinical manifestations associated with cerebral ischemia, including motor impairment and neurological deficits.

There are, as yet, no studies evaluating the effects of pyruvate administration on cognitive function in humans; however, pyruvate may be a good candidate for further research in those with energetic depletion and neurodegenerative diseases. Impaired energy metabolism is an early, predominant feature in Alzheimer's disease and it is believed that impaired cerebral oxidative glucose metabolism is responsible, at least in part, for cognitive impairment in AD. Research has demonstrated that in both animals and humans increased cerebrospinal pyruvate is a biomarker for AD (Parnetti et al. 1995; Pugliese et al. 2005). Since pyruvate appears to be quite safe, aside from mild side effects, such as occasional stomach upset and diarrhea, pyruvate therapy might represent an excellent candidate for therapy in disease states accompanied by energy depletion.

CREATINE

Creatine (Cr) is a naturally occurring substance found in vertebrates and is essential for maintaining energy homeostasis. Cr participates in metabolic reactions within cells and eventually is catabolized in the muscles creating creatinine, which is then excreted by the kidney in urine. In the average-sized adult (70 kg) Cr store is approximately 120 g, with the daily turnover of Cr to creatinine being estimated to be about 1.6% of the body's total Cr (Balsom et al. 1995). The daily requirement of Cr either through diet or endogenous synthesis is suggested to be approximately 2 g/day (Walker 1979).

Since Cr is concentrated in muscle tissue dietary sources of Cr are fish and red meat, with a much lower concentration found in some plants (Balsom et al. 1995). Unsurprisingly Cr levels of vegetarian or vegan individuals are much lower than omnivores. In a typical omnivorous diet between 0.25 and 1 g of Cr per day is obtained. It appears that Cr derived from the diet, after passing through the intestinal lumen, enters the bloodstream intact (Conway and Clark 1996).

Cr is stored in the high-energy form of phosphocreatine (PCr). PCr acts as a high-energy reserve in a coupled reaction in which energy derived from donating a phosphate group is used to regenerate the compound ATP. PCr plays a particularly important role in tissues that have high, fluctuating energy demands such as muscle and brain. During times of brain activity, brain phospocreatine decrease rapidly in order to maintain constant ATP levels (Sappey-Marinier et al. 1992; Rango et al. 1997). Cr supplementation has pronounced effects on the body including increased muscle mass and improvements in physical performance on exercise tasks (Kreider 2003). Furthermore Cr supplementation can increase brain Cr. Studies using nuclear magnetic resonance spectroscopy have demonstrated that Cr and PCr can be increased in the brains of healthy adults by Cr supplementation (Dechent et al. 1999; Lyoo et al. 2003).

The majority of previous research examining the effects of Cr has focused on muscle mass, body mass index, and physical performance; however, more recently attention has been directed toward Cr's effects on the brain and the metabolic changes therein.

Animal research has shown that Cr is particularly important for normal brain development and function. In its absence deleterious effects on cognition and brain development are observed, in abundance evidence for neuroprotection has been observed. For example, deletion of cytosolic brain-type creatine kinase in mice has been shown to result in slower learning of a spatial task and diminished open-field habituation as well as increased intra- and infra-pyramidal hippocampal mossy fiber area suggesting that the creatine–creatine kinase network is involved in brain plasticity in addition to metabolism (Jost et al. 2002).

Animal research has demonstrated that Cr affords significant neuroprotection against ischemic and oxidative insults (Holtzman et al. 1998; Wilken et al. 1998; Balestrino et al. 1999). One experiment investigated the possible effect of Cr dietary supplementation on brain tissue damage after experimental traumatic brain injury. Results demonstrated that chronic administration of Cr ameliorated the extent of cortical damage by as much as 36% in mice and 50% in rats. The authors suggested that protection is mediated by Cr-induced maintenance of mitochondrial bioenergetics as they observed that mitochondrial membrane potential was significantly increased, intra-mitochondrial levels of reactive oxygen species and calcium were significantly decreased, and ATP levels were maintained. Induction of mitochondrial permeability transition was significantly inhibited in animals fed Cr. The authors further suggested that Cr may be a good candidate as a neuroprotective agent against acute and delayed neurodegenerative processes (Sullivan et al. 2000).

In rodents where neurodegenerative symptoms are induced, Cr attenuated these deficits, for example, rats administered 3-nitropropionic acid (3NP) displayed neuropathological and behavioral abnormalities that are analogous to those observed in Huntington's disease (HD). Rats fed diets containing 1% Cr over an 8 week period showed attenuation of 3NP-induced striatal lesions, striatal atrophy, ventricular enlargement, cognitive deficits, and motor abnormalities on a balance beam task compared to non-Cr supplemented rats. These findings indicate that Cr provides significant protection against neuropathological insult specifically associated with 3NP-induced behavioral and neuropathological abnormalities (Shear et al. 2000).

Clearly Cr plays a fundamental role in brain protection and development in the animal model. Specifically deleterious effects were observed on cognition and brain development when Cr is absent (and/or PCr and creatine kinases), and the neuroprotective attributes of Cr in supplemented animals. These data provide a strong rationale for examination of Cr supplementation on the brain and cognition in the human model.

Despite the obvious impact Cr has on brain development and metabolic actions in the brain, there are relatively few studies assessing the effects of Cr on cognitive performance in humans. One study assessed the effect of 20 g Cr supplementation over 7 days in sleep-deprived individuals, following 24 h sleep deprivation. Individuals who received Cr supplementation demonstrated significantly reduced decrement in performance on a number of mood, cognitive, and physical performance parameters including random movement generation, choice reaction time, balance, and mood state (McMorris et al. 2006). In a further study following 36 h sleep deprivation, Cr-supplemented individuals also demonstrated improved performance on a random number generation task (McMorris et al. 2007b). These studies appear to demonstrate benefits of Cr supplementation in young individuals who are temporarily cognitively impaired through sleep deprivation. However, these studies were considerably underpowered having no higher than 10 participants per group. Cr supplementation has also been demonstrated to improve cognition in individuals who are not cognitively impaired. One study assessed the effects of 8 g Cr per day for 5 days in healthy individuals and demonstrated reduced mental fatigue when subjects repeatedly perform a simple mathematical calculation. After Cr supplementation, task-evoked increase of cerebral oxygenated hemoglobin in the brains of subjects and reduced cerebral oxygenated hemoglobin (measured by near-infrared spectroscopy) was significantly reduced, which is compatible with increased oxygen utilization in the brain (Watanabe et al. 2002). Again, however, this study appeared to be rather underpowered with only 12 participants per group. Nonetheless, it appears that Cr supplementation may impact on cognitive function even over a relatively short period of time as these studies assessed the effects of acute supplementation over periods of 5–7 days. A more recent study assessed the impact of a new form of creatine, creatine ethyl ester, over a 2 week period (5 g/day dose compared to dextrose control group) in healthy 18–24 year old participants. The overall findings demonstrated consistent improvements for reaction time across a range of measures as well as improved accuracy on some and also improved IQ scores. The most modest improvements appeared to be on tasks that were less demanding, indicating that creatine supplementation may be particularly useful when performing particularly demanding or complex cognitive tasks (Ling et al. 2009).

In chronic administration conditions, one study examining Cr supplementation in young healthy adults failed to observe any effect of Cr on cognitive performance (Rawson et al. 2008). In this study 0.03 g/kg was administered daily for 6 weeks and a battery of neurocognitive tests was administered to asses cognitive processing and psychomotor performance including simple reaction time, code substitution, code substitution delayed, logical reasoning symbolic, mathematical processing, running memory, and Sternberg memory recall. No effect of Cr was observed on any of these outcome measures.

However, research examining young adults who only produce Cr endogenously (vegetarian sample), Cr supplementation was shown to improve cognitive performance following chronic administration (6 week period) (Rae et al. 2003). In this work, 5 g Cr supplementation (Cr monohydrate) was administered per day for 6 weeks to 45 young vegetarian adults in a counterbalanced cross-over design. They observed that Cr supplementation had a significant positive effect on both working memory (backward digit span) and intelligence (Raven's Advanced Progressive Matrices).

The pattern emerging from the present literature examining Cr and cognitive function appears to demonstrate that cognition is ameliorated specifically during times of metabolic impairment or depletion, either through low creatine availability (vegan and vegetarian samples) or by inducement (sleep deprivation or high cognitive demand). Furthermore, since there is some evidence that creatine supplementation improves cognitive function in young, non-vegetarian, healthy individuals over shorter periods of administration (5 days to 2 weeks) but not longer periods (6 weeks) it may be the case that creatine supplementation might merely have been redressing nutritional imbalances.

Since elderly populations are generally metabolically impaired and often nutritionally deficient, it seems likely that elderly and degenerative populations would most benefit from creatine interventions over time. To our knowledge only one study has assessed the impact of Cr supplementation in an elderly human population. McMorris et al. (2007a) administered 20 g of Cr per day for 7 days which resulted in improved performance of random number generation, forward and backward number and spatial recall, and long-term memory tasks but no effect on backward recall performance (McMorris et al. 2007b). In terms of neurodegeneration, there has been no research examining the effects of creatine supplementation in dementia sufferers; however, there appears to be some differences in creatine levels in those with genetic risk of developing dementia (apolipoprotein ε4 carriers). Laakso et al. (2003) demonstrated that compared with the noncarriers, the levels of creatine were significantly lower in the ε4 carriers. This finding may suggest increased metabolic demands in the brain of the ε4 carriers. They also observed that the levels of creatine also correlated significantly with age and performance on the Mini-Mental State Examination test in the ε4 carriers, but not in the noncarriers (Laakso et al. 2003). Creatine supplementation in this sample seems like a logical next step for creatine and cognitive function research.

Despite the obvious potential benefits of Cr supplementation, there is considerable lack of research examining the cognitively enhancing capabilities of Cr and a number of questions remain to be answered. Firstly there has been no research examining whether an acute administration of one single dose of Cr can affect cognitive performance. Secondly the only study to examine the effects of Cr on cognition in the elderly was only over a period of 7 days. Further to this there has been no examination of the usefulness of creatine in dementia research where there appears to be some evidence that creatine may be of particular therapeutic value. Since the evidence seems to suggest that Cr acts to buffer intracellular energy levels and potentially impede the progression of neurodegenerative processes a more systematic evaluation of Cr mapping cognitive performance over a more substantial timeframe is required.

CARNITINE/ACETYL-ʟ-CARNITINE

In animals and humans, carnitine is biosynthesized primarily in the liver and kidneys from the amino acids lysine or methionine (Steiber et al. 2004) with Vitamin C (ascorbic acid) being essential to the synthesis of carnitine. In food, the highest concentrations of carnitine are found in red meat and dairy products. Other natural sources of carnitine include nuts and seeds, legumes or pulses, vegetables, and cereals. Carnitine is a quaternary ammonium compound that, in living cells, is required for the transport of fatty acids from the cytosol into the mitochondria during the breakdown of lipids (or fats) for the generation of metabolic energy. Carnitine exists in two stereoisomers: its biologically active form is L-carnitine, while its enantiomer, D-*carnitine*, is biologically inactive (Liedtke et al. 1982). Carnitine transports long-chain acyl groups from fatty acids into the mitochondrial matrix, so that they can be broken down through β-oxidation to acetate to obtain usable energy via the citric acid cycle. Under normal nutritional conditions and in healthy persons, L-carnitine availability is not a limiting step in β-oxidation; however, L-carnitine is required for mitochondrial long-chain fatty acid oxidation (Simon 2005), which is a main source of energy during exercise (Wasserman and Whipp 1975). Furthermore increase in L-carnitine content might increase the rate of fatty acid oxidation, permitting a reduction of glucose utilization, preserving muscle glycogen content, and ensuring maximal rates of oxidative ATP production. In one study L-carnitine improved glucose disposal among 15 patients with type II diabetes and 20 healthy volunteers (Mingrone et al. 1999). Glucose storage increased between both groups and glucose oxidation increased in the diabetic group. Furthermore glucose uptake increased by approximately 8% for both diabetic and non-diabetic groups.

In neuronal cells, the L-carnitine shuttle mediates translocation of the acetyl moiety from mitochondria into the cytosol and contributes to the synthesis of acetylcholine and of acetylcarnitine (Imperato et al. 1989; Nalecz and Nalecz). The neurobiological effects of acetyl carnitine include modulation of brain energy and phospholipids metabolism, cellular macromolecules (such as neurotrophic factors and neurohormones), synaptic morphology, and synaptic transmission of multiple neurotransmitters (see review [Furlong 1996]).

The majority of research assessing the effects of L-carnitine or acetyl-L-carnitine (acetylated derivative of L-carnitine with improved bioavailability) has focused on its benefits to elderly and demented populations. It has been established that acetyl L-carnitine transverses the blood brain-barrier efficiently. With CSF concentrations increasing sufficiently via both intravenous and oral rout in patients with severe dementia (Parnetti et al. 1992). In terms of efficacy, a meta-analysis examining the effects of acetyl-L-carnitine in mild cognitive impairment and mild (early) Alzheimer's disease was conducted (Montgomery et al. 2003). Studies included in the analysis were at least 3 months in duration, with a dosage of 1.5–3.0 g/day. The results showed beneficial effects on both clinical scales and psychometric tests with improvements being observed at the first assessment (3 months) and increasing over time.

In a more recent study, the effects of 2 g of L-carnitine per day for 6 weeks were assessed in centenarians aged between 100 and 106 (Malaguarnera et al. 2007). Those treated with L-carnitine demonstrated significant physiological improvements in fat mass, muscle mass, plasma total carnitine, and plasma long- and short-chain acetylcarnitine. They also showed significantly improved mental fatigue and cognitive function assessed by the Mini-Mental State Examination (MMSE).

There are, as yet, no studies examining the effect of L-carnitine or acetyl-L-carnitine on cognitive function in young human populations. Since the action of L-carnitine availability is not a limiting step in β-oxidation, any beneficial effects are most likely to be observed in populations with depleted energy resources or under physically fatigued conditions. Therefore, the utility of L-carnitine/acetyl-L-carnitine may be more pronounced in age and degenerative disease.

SUMMARY

There are a number of agents with the potential to improve metabolic activity. Research is now beginning to identify these various agents and delineate their potential usefulness for improving cognition in health and disease. Glucose and oxygen are the primary sources of fuel for the brain and body. In metabolic terms these substrates have by far received the most extensive research attention in terms of their effects on cognition. As a result, the optimal dosage, timeframe, and conditions of administration to reveal facilitation of cognition are becoming better delineated. Other substrates reviewed in this chapter include pyruvate, creatine, and L-carnitine. For all of these metabolic agents we are clearly in the early stages of intervention-cognition research. However the evidence thus far provides an excellent basis for future research of the potential benefits and treatment options. A common theme for all of the metabolic agents reviewed here is that cognitive mediation by metabolic agents is most robustly observed in those with some metabolic impairment or low energy availability. This finding indicates two general points: (1) If it is possible to enhance cognitive performance in young healthy adults whose metabolic activity is presumed to be optimal, then all individuals must be subject to energy fluctuations that can be enhanced with supplementation. (2) Since general aging leads to metabolic impairment and age-associated cognitive decline, it is quite plausible that these interventions represent a real and simple strategy for improving cognitive function as we age and possibly slowing or preventing decline into dementia.

REFERENCES

Balestrino, M, Rebaudo, R, and Lunardi, G (1999), Exogenous creatine delays anoxic depolarization and protects from hypoxic damage: Dose–effect relationship, *Brain Research,* 816 (1), 124–130.

Balsom, PD et al. (1995), Skeletal muscle metabolism during short duration high-intensity exercise: Influence of creatine supplementation, *Acta Physiologica Scandinavica,* 154 (3), 303–310.

Benton, D and Owens, DS (1993), Blood glucose and human memory, *Psychopharmacology,* 113 (1), 83–88.

Benton, D and Sargent, J (1992), Breakfast, blood glucose and memory, *Biological Psychology,* 33 (2–3), 207–210.

Chinnery, P et al. (2006), Treatment for mitochondrial disorders, *Cochrane Database of Systematic Reviews,* (1), CD0044261.

Chung, SC et al. (2008), A study on the effects of 40% oxygen on addition task performance in three levels of difficulty and physiological signals, *International Journal of Neuroscience,* 118 (7), 905–916.

Conway, MA and Clark, JF (1996), *Creatine and Creatine Phosphate: Scientific and Clinical Perspectives*, Academic Press, San Diego, CA.

Craft, S, Murphy, CG, and Wemstrom, J (1994), Glucose effects on complex memory and non-memory tasks: The influence of age, sex, and glucoregulatory response, *Psychobilology*, 22, 95–105.

Craft, S, Zallen, G, and Baker, LD (1992), Glucose and memory in mild senile dementia of the Alzheimer type, *Journal of Clinical and Experimental Neuropsychology*, 14 (2), 253–267.

Crowley, JS et al. (1992), Effect of high terrestrial altitude and supplemental oxygen on human performance and mood, *Aviation, Space, and Environmental Medicine*, 63 (8), 696.

Dechent, P et al. (1999), Increase of total creatine in human brain after oral supplementation of creatine-monohydrate, *American Journal of Physiology- Regulatory, Integrative and Comparative Physiology*, 277 (3), 698–704.

Edwards, AE and Hart, GM (1974), Hyperbaric oxygenation and the cognitive functioning of the aged, *Journal of the American Geriatrics Society*, 22 (8), 376.

Furlong, JH (1996), Acetyl-L-carnitine: Metabolism and applications in clinical practice, *Alternative Medicine Review*, 1 (2), 85.

Gladden, LB (2004), Lactate metabolism: A new paradigm for the third millennium, *Journal of physiology*, 558 (1), 5.

Gold, PE (1986), Glucose modulation of memory storage processing, *Behavioral and Neural Biology*, 45 (3), 342–349.

Gonder-Frederick, L et al. (1987), Memory enhancement in elderly humans: Effects of glucose ingestion, *Physiology & Behavior*, 41 (5), 503–504.

Hafermann, G (1955), Schulmudigkeit und Blutzuckerverhalten, Ofentlicher Gesundheitsdienst, *Öffentlicher Gesundheitsdienst* 17, 1.

Holtzman, D et al. (1998), Creatine increases survival and suppresses seizures in the hypoxic immature rat, *Pediatric Research*, 44 (3), 410.

Hoyland, A, Lawton, CL, and Dye, L (2008), Acute effects of macronutrient manipulations on cognitive test performance in healthy young adults: A systematic research review, *Neuroscience & Biobehavioral Reviews*, 32 (1), 72–85.

Imperato, A, Ramacci, MT, and Angelucci, L (1989), Acetyl-L-carnitine enhances acetylcholine release in the striatum and hippocampus of awake freely moving rats, *Neuroscience Letters*, 107 (1–3), 251–255.

Jost, CR et al. (2002), Creatine kinase B-driven energy transfer in the brain is important for habituation and spatial learning behaviour, mossy fibre field size and determination of seizure susceptibility, *European Journal of Neuroscience*, 15 (10), 1692.

Kennedy, DO and Scholey, AB (2000), Glucose administration, heart rate and cognitive performance: Effects of increasing mental effort, *Psychopharmacology*, 149 (1), 63–71.

Klein, AM and Ferrante, RJ (2007), The neuroprotective role of creatine, *Creatine and Creatine Kinase in Health and Disease*, 46, 205–243.

Kreider, RB (2003), Effects of creatine supplementation on performance and training adaptations, *Molecular and Cellular Biochemistry*, 244 (1), 89–94.

Laakso, MP et al. (2003), Decreased brain creatine levels in elderly apolipoprotein E 4 carriers, *Journal of Neural Transmission*, 110 (3), 267–275.

Lee, JY, Kim, YH, and Koh, JY (2001), Protection by pyruvate against transient forebrain ischemia in rats, *Journal of Neuroscience*, 21 (20), 171.

Lehninger, AL, Nelson, DL, and Cox, MM (2005), *Lehninger Principles of Biochemistry*, WH Freeman, New York.

Liedtke, AJ et al. (1982), Metabolic and mechanical effects using L-and D-carnitine in working swine hearts, *American Journal of Physiology: Heart and Circulatory Physiology*, 243 (5), H691.

Ling, J, Kritikos, M, and Tiplady, B (2009), Cognitive effects of creatine ethyl ester supplementation, *Behavioural Pharmacology,* 20 (8), 673.

Lyoo, IK et al. (2003), Multinuclear magnetic resonance spectroscopy of high-energy phosphate metabolites in human brain following oral supplementation of creatinemonohydrate, *Psychiatry Research: Neuroimaging,* 123 (2), 87–100.

Malaguarnera, M et al. (2007), L-carnitine treatment reduces severity of physical and mental fatigue and increases cognitive functions in centenarians: A randomized and controlled clinical trial, *American Journal of Clinical Nutrition,* 86 (6), 1738.

McIlwain, H (1959), Thiols and the control of carbohydrate metabolism in cerebral tissues, *Biochemical Journal,* 71 (2), 281.

McMorris, T et al. (2006), Effect of creatine supplementation and sleep deprivation, with mild exercise, on cognitive and psychomotor performance, mood state, and plasma concentrations of catecholamines and cortisol, *Psychopharmacology,* 185 (1), 93–103.

McMorris, T et al. (2007a), Creatine supplementation and cognitive performance in elderly individuals, *Aging, Neuropsychology, and Cognition,* 14, 517–528.

McMorris, T et al. (2007b), Creatine supplementation, sleep deprivation, cortisol, melatonin and behavior, *Physiology & Behavior,* 90 (1), 21–28.

Mingrone, G et al. (1999), L-carnitine improves glucose disposal in type 2 diabetic patients, *Journal of the American College of Nutrition,* 18 (1), 77.

Montgomery, SA, Thal, LJ, and Amrein, R (2003), Meta-analysis of double blind randomized controlled clinical trials of acetyl-L-carnitine versus placebo in the treatment of mild cognitive impairment and mild Alzheimer's disease, *International Clinical Psychopharmacology,* 18 (2), 61.

Moss, MC and Scholey, AB (1996), Oxygen administration enhances memory formation in healthy young adults, *Psychopharmacology,* 124 (3), 255–260.

Moss, MC, Scholey, AB, and Wesnes, K (1998), Oxygen administration selectively enhances cognitive performance in healthy young adults: A placebo-controlled double-blind crossover study, *Psychopharmacology,* 138 (1), 27–33.

Nalecz, K and Nalecz, M (1996), Carnitine: A known compound, a novel function in neural cells, *Acta Neurobiol. Exp (Wars.),* 56, 597–609.

Owen, L et al. (2010), Glucose effects on long-term memory performance: Duration and domain specificity, *Psychopharmacology,* 211 (2), 131–140.

Packard, MG and White, NM (1990), Effect of posttraining injections of glucose on acquisition of two appetitive learning tasks, *Psychobiology,* 18, 282–286.

Parnetti, L et al. (1992), Pharmacokinetics of IV and oral acetyl-L-carnitine in a multiple dose regimen in patients with senile dementia of Alzheimer type, *European Journal of Clinical Pharmacology,* 42 (1), 89–93.

Parnetti, L et al. (1995), Increased cerebrospinal fluid pyruvate levels in Alzheimer's disease, *Neuroscience Letters,* 199 (3), 231–233.

Pugliese, M et al. (2005), Severe cognitive impairment correlates with higher cerebrospinal fluid levels of lactate and pyruvate in a canine model of senile dementia, *Progress in Neuro-Psychopharmacology and Biological Psychiatry,* 29 (4), 603–610.

Rae, C et al. (2003), Oral creatine monohydrate supplementation improves brain performance: A double-blind, placebo-controlled, cross-over trial, *Proceedings of the Royal Society B: Biological Sciences,* 270 (1529), 2147.

Rango, M, Castelli, A, and Scarlato, G (1997), Energetics of 3.5 s neural activation in humans: A 31P MR spectroscopy study, *Magnetic Resonance in Medicine,* 38 (6), 878–883.

Raskin, A et al. (1978), The effects of hyperbaric and normobaric oxygen on cognitive impairment in the elderly, *Archives of General Psychiatry,* 35 (1), 50–56.

Rawson, ES et al. (2008), Creatine supplementation does not improve cognitive function in young adults, *Physiology & Behavior,* 95 (1–2), 130–134.

Raz, N et al. (2005), Regional brain changes in aging healthy adults: General trends, individual differences and modifiers, *Cerebral Cortex,* 15 (11), 1676.

Riby, LM et al. (2008), P3b versus P3a: An event-related potential investigation of the glucose facilitation effect, *Journal of Psychopharmacology,* 22 (5), 486.

Sappey-Marinier, D et al. (1992), Effect of photic stimulation on human visual cortex lactate and phosphates using 1H and 31P magnetic resonance spectroscopy, *Journal of Cerebral Blood Flow and Metabolism: Official Journal of the International Society of Cerebral Blood Flow and Metabolism,* 12 (4), 584.

Scholey, AB et al. (1999), Cognitive performance, hyperoxia, and heart rate following oxygen administration in healthy young adults, *Physiology & Behavior,* 67 (5), 783.

Scholey, AB et al. (2009), Glucose administration prior to a divided attention task improves tracking performance but not word recognition: Evidence against differential memory enhancement?, *Psychopharmacology,* 202 (1), 549–558.

Scholey, AB, Harper, S, and Kennedy, DO (2001), Cognitive demand and blood glucose, *Physiology and Behavior,* 73 (4), 585–592.

Scholey, AB, Moss, MC, and Wesnes, K (1998), Oxygen and cognitive performance: The temporal relationship between hyperoxia and enhanced memory, *Psychopharmacology,* 140 (1), 123–126.

Shear, DA, Haik, KL, and Dunbar, GL (2000), Creatine reduces 3-nitropropionic-acid-induced cognitive and motor abnormalities in rats, *Neuroreport,* 11 (9), 1833.

Simon, EO (2005), Fatty acid oxidation defects as a cause of neuromyopathic disease in infants and adults, *Clinical Laboratory,* 51 (5–6), 289–306.

Steiber, A, Kerner, J, and Hoppel, CL (2004), Carnitine: A nutritional, biosynthetic, and functional perspective, *Molecular Aspects of Medicine,* 25 (5–6), 455–473.

Suh, SW et al. (2005), Pyruvate administered after severe hypoglycemia reduces neuronal death and cognitive impairment, *Diabetes,* 54 (5), 1452.

Sullivan, PG et al. (2000), Dietary supplement creatine protects against traumatic brain injury, *Annals of Neurology,* 48 (5), 723–729.

Sünram-Lea, SI et al. (2001), Glucose facilitation of cognitive performance in healthy young adults: Examination of the influence of fast-duration, time of day and pre-consumption plasma glucose levels, *Psychopharmacology,* 157 (1), 46.

Sünram-Lea, SI et al. (2002a), Investigation into the significance of task difficulty and divided allocation of resources on the glucose memory facilitation effect, *Psychopharmacology,* 160 (4), 387–397.

Sünram-Lea, SI et al. (2002b), The effect of retrograde and anterograde glucose administration on memory performance in healthy young adults, *Behavioural Brain Research,* 134 (1–2), 505–516.

Sünram-Lea, SI et al. (2004), The influence of fat co-administration on the glucose memory facilitation effect, *Nutritional Neuroscience,* 7 (1), 21–32.

Sunram-Lea, S et al. (2011), Dose-response investigation into glucose facilitation of memory performance and mood in healthy young adults, *Journal of Psychopharmacology,* 25: 1076–1087.

Thomson, JL (1967), Effect of inhibitors of carbohydrate metabolism on the development of preimplantation mouse embryos, *Experimental Cell Research,* 46 (2), 252.

Volpe, BT and Hirst, W (1983), Amnesia following the rupture and repair of an anterior communicating artery aneurysm, *British Medical Journal,* 46 (8), 704.

Walker, JB (1979), Creatine: Biosynthesis, regulation, and function [Chick embryo experiments, dietary aspects], *Advances in Enzymology and Related Areas of Molecular Biology,* 50, 177–242.

Walker, B and Sandman, CA (1979), Human visual evoked responses are related to heart rate. *J. Comp. Physiol. Psychol.* 1979, 93, 717–729.

Wasserman, K and Whipp, BJ (1975), Excercise physiology in health and disease, *American Review of Respiratory Disease,* 112 (2), 219.

Watanabe, A, Kato, N, and Kato, T (2002), Effects of creatine on mental fatigue and cerebral hemoglobin oxygenation, *Neuroscience Research,* 42 (4), 279–285.

Weaver, LK et al. (2002), Hyperbaric oxygen for acute carbon monoxide poisoning, *New England Journal of Medicine,* 347 (14), 1057.

Weiskopf, RB et al. (2002), Oxygen reverses deficits of cognitive function and memory and increased heart rate induced by acute severe isovolemic anemia, *Anesthesiology,* 96 (4), 871.

Wenk, GL (1989), An hypothesis on the role of glucose in the mechanism of action of cognitive enhancers, *Psychopharmacology,* 99 (4), 431–438.

White, NM (1991), Peripheral and central memory enhancing actions of glucose, *Peripheral Signalling of the Brain: Role in Neural-Immune Interactions, Learning and Memory.* (eds) DL Felten, JL MacGaugh, and Frederickson RCA, Hogrefe and Huber, Toronto, Ontario, Canada, pp. 421–443.

Wieloch, T (1985), Hypoglycemia-induced neuronal damage prevented by an N-methyl-D-aspartate antagonist, *Science,* 230 (4726), 681.

Wilken, B et al. (1998), Creatine protects the central respiratory network of mammals under anoxic conditions, *Pediatric Research,* 43 (1), 8.

Winder, R and Borrill, J (1998), Fuels for memory: The role of oxygen and glucose in memory enhancement, *Psychopharmacology,* 136 (4), 349–356.

Ying, W et al. (2002), Tricarboxylic acid cycle substrates prevent PARP-mediated death of neurons and astrocytes, *Journal of Cerebral Blood Flow & Metabolism,* 22 (7), 774–779.

Yu, YM, Kim JB, Lee KW, Kim SY, Han PL and Lee, JK (2005), Inhibitions of the cerebral ischemic injury by ethyl pyruvate with a wide therapeutic window. *Stoke,* 36, 2238–2243.

7 Vitamins, Cognition, and Mood in Healthy Nonelderly Adults

Crystal F. Haskell and David O. Kennedy

CONTENTS

INTRODUCTION

Vitamins are essential micronutrients obtained, at least in part, from dietary sources. These compounds are necessary for normal cell function, physiological processes, growth, and development. By definition they are not synthesized in the amounts required by an organism and; as a consequence of the differing needs of different organisms and differences in abilities to synthesize specific compounds, vitamins differ between different species. In the case of humans, four fat-soluble vitamins (A, D, E, K) and nine water-soluble vitamins (B1, B2, B3, B5, B6, B7, B9, B12, C) have been identified. Micronutrients are the most widely taken food supplements throughout the developed world with a recent study reporting usage in approximately 30% of a European cohort, with higher and more consistent consumption by women (Li et al. 2010). These findings are supported by studies in the United States demonstrating that approximately a third of the adult population had recently consumed vitamin/mineral supplements for their purported health benefits (Radimer et al. 2004; Rock 2007). Among this broad class of supplements, multi-vitamin products are consumed most frequently (Timbo et al. 2006).

Humans lost the ability to synthesize these vitamins in sufficient quantities several hundred million years ago. It has been suggested that this was because the supply of these compounds within the food available made it disadvantageous to continue to synthesize them (Pauling 1970). For example, despite being synthesized by all but a few species of animals, the loss of the ability to synthesize vitamin C (ascorbate) by humans has been proposed to have beneficial effects in terms of the energetic and cellular cost, as well as the oxidative cost associated with synthesizing vitamin C, which in turn leads to a cost in terms of endogenous antioxidant requirements (Banhegyi et al. 1997). It has also been suggested that not only has vitamin synthesis been the subject of evolutionary pressure but that vitamins have also shaped aspects of human evolution (Milton 2000). For instance, the loss of the ability to synthesize vitamin C is purported to have led to an increase in free radical–induced mutations and a subsequent acceleration of primate evolution (Challem, 1997). Many of the "diseases of civilization," such as diabetes, obesity, and cardiovascular disease, may be predicated on the shift away from our evolutionarily determined, largely herbivorous diet, to the high-energy, highly digestible, micronutrient-depleted diet of modern humans in westernized societies (Benzie 2003; Milton 2000). As a single example of this, reference to the rate of endogenous synthesis of vitamin C in other mammals, the diet of gorillas (Pauling 1970), and the vitamin constituents of what might be assumed to be a typical pre-agriculture diet (Benzie 2003; Eaton and Konner 1997; Pauling 1970), all suggest that our consumption of vitamin C should be at least ten times the recommended daily allowance espoused by most authorities (e.g., 60 mg in the European Union).

Several indirect strands of evidence that exist to support the efficacy of supplementation with vitamins in improving cognition and/or mood will be outlined in the following chapter, as well as evidence from randomized controlled trials. A large proportion of research in this area has been carried out with elderly cohorts suffering defined cognitive decrements and dementia, which will be covered in Chapter 4 of this book. However, it is plausible that interventions of this nature may result in very different effects when studied in a non-elderly population. This review will therefore focus on studies that have utilized healthy, non-elderly samples.

MECHANISMS OF ACTION OF VITAMINS RELATED TO BRAIN FUNCTION

While vitamins are intrinsic to all physiological processes within the body, both fat-soluble and water-soluble vitamins also contribute directly to optimal brain function via a plethora of mechanisms. The two groups of vitamins differ in terms of storage.

Fat-soluble vitamins are stored largely in the liver and adipose tissue, and therefore they persevere for potentially long periods in the body. When taken in excessive quantities, they can also build up to toxic levels. The water-soluble vitamins, on the other hand, are readily excreted from the body in urine, and therefore require more regular and consistent replacement.

FAT-SOLUBLE VITAMINS

Vitamin A

Vitamin A (and its bioactive derivative retinoic acid [RA]) plays an important role in embryonic neural development (reviewed by Niederreither and Dolle 2008). Its contribution includes, for instance, neuronal differentiation and neurite outgrowth (for details, see Lane and Bailey 2005; McCaffery et al. 2006). Moreover, recent evidence suggests that vitamin A and its metabolites (RA and all-trans retinoic acid [ATRA]) also have important functions in the adult nervous system (Lane and Bailey 2005; Luo et al. 2009; McCaffery et al. 2006; Olson and Mello 2010). For instance, they have roles in the dopaminergic system (Krezel et al. 1998; Valdenaire et al. 1994, 1998) and hippocampal synaptic plasticity (e.g., long-term potentiation [LTP] and long-term depression [LTD]; Etchamendy et al. 2003; McCaffery et al. 2006; Mey and Mccaffery 2004). Vitamin A has direct effects on cognition, e.g., retinoid signaling influences vocal memory in song birds (Olson and Mello 2010) and in retinoid receptor–deficient mice, learning and memory deficits have been observed alongside changes in LTP and LTD (Chiang et al. 1998; Wietrzych et al. 2005). Animal studies have demonstrated that vitamin A deficiency (VAD) is associated with impaired hippocampal LTP and LTD in mice (Misner et al. 2001), which is reversed by vitamin A administration. In addition, administration of vitamin A or retinoid derivatives ameliorates VAD-mediated learning impairments (e.g., Bonnet et al. 2008; Cocco et al. 2002). High doses, however, can be detrimental (Crandall et al. 2004). Evidence also suggests a role in a number of neurodegenerative and psychiatric disorders, e.g., Parkinson's disease, Huntington's disease, schizophrenia, and depression.

Vitamin D

Until comparatively recently, the steroid hormone, vitamin D, was largely associated with its established roles in the regulation of bone mineralization and levels of calcium and phosphorus. More recent evidence has shown that the active form of vitamin D, $1,25(OH)_2D_3$, plays a plethora of roles related to health (Holick 2008), which include a number that are specific to brain function. In vitro/vivo evidence suggests these include a host of neuroprotective and neurotransmission functions, including homeostatic regulation of neuronal calcium, modulation of inducible nitric oxide synthase (iNOS) and upregulation of the endogenous antioxidant glutathione. It also upregulates neurotrophin factors, including neurotrophin-3 (NT-3) and glial cell line–derived neurotrophic factor (GDNF), which in turn play a role in synapticity and nerve transmission in the neocortex and hippocampus (Buell and Dawson-Hughes 2008; McCann and Ames 2008). Recent evidence has also confirmed the presence of vitamin D receptors and the catalytic enzymes involved in the synthesis of $1,25(OH)_2D_3$ throughout the human brain, including cognition-relevant areas (Buell and Dawson-Hughes 2008; McCann and Ames 2008). To date, a wealth of evidence has suggested behavioral decrements in rodents either bred with vitamin D receptor dysfunction or deprived of vitamin D during brain development and beyond (McCann and Ames 2008).

Vitamin E

Vitamin E is composed of the tocotrienols and tocopherols, which are closely related groups of compounds, each of which possesses four analogs, α, β, γ, and δ. It is generally accepted that α-tocopherol has the highest biological activity among the compounds (Yang and Wang 2008). On the basis of copious in vitro evidence, it is generally accepted that vitamin E is the brain's most prevalent lipophilic antioxidant, and in support of this it is notable that it is transported across the blood–brain barrier (BBB) by lipoproteins (Goti et al. 2002; Mardones and Rigotti 2004). However, direct in vivo evidence to support this role is scarce, and the molecular mechanisms underlying vitamin E's physiological roles are still largely undelineated (Brigelius-Flohe 2009). Animal studies indicate a beneficial effect of vitamin E on cognitive function (Fukui et al. 2002; Jhoo et al. 2004; Joseph et al. 1998); and it has been suggested that, other than its potential antioxidant properties, vitamin E might owe its beneficial effects to an indirect role in the inhibition and activation of a raft of essential enzymatic processes, and gene expression (Brigelius-Flohe 2009). Vitamin E deficiency also results in a host of neurological deficits, and supplementation has been suggested as a potential treatment for a variety of neurodegenerative disorders (Ricciarelli et al. 2007).

WATER-SOLUBLE VITAMINS

B Vitamins

The B vitamins play key roles in brain function as co-enzymes and precursors of co-factors in enzymatic processes. In this respect they contribute at some level to all physiological processes within the brain. However, they also have a number of specific roles that might be expected to directly affect aspects of brain function, which in turn may modify behavior, either in the short or long term. For instance, folate (vitamin B9) supplies the methyl group for the conversion of methionine to S-adenosylmethionine (SAMe), and therefore plays a role in the synthesis and integrity of DNA and the methylation of proteins, phospholipids, and monoamine and catecholamine neurotransmitters (Mattson and Shea 2003). Similarly, adequate levels of folate and vitamin B12 are required for the remethylation of homocysteine (Hcy), which is a potentially toxic amino acid by-product of one carbon metabolism. Vitamin B6 also plays a key role in this process as a coenzyme of cystathionine synthase and cystathioninelyase, which are required for the metabolism of homocysteine to cysteine (Mattson and Shea 2003). Blocking of the conversion process (e.g., by deficiencies of folate or vitamins B12 and B6) leads to elevated levels of homocysteine, which in turn may contribute to a range of deleterious effects on cellular, hemodynamic, oxidative, and vascular parameters, and ultimately may contribute to a range of neurodegenerative and psychiatric disorders (Reynolds 2006). Vitamin B6 also plays a raft of roles in metabolic processes and is integral to the synthesis of a range of neurotransmitters, including dopamine and serotonin, in its role as a cofactor for aromatic l-amino acid decarboxylase (AADC); an enzyme that catalyzes the decarboxylation of a variety of aromatic l-amino acids. It has also been shown to regulate levels of serotonin (Boadlebiber 1993; Calderon-Guzman et al. 2004).

As well as playing a role in the structure and function of central nervous system, cellular membranes vitamin B1 (thiamine) plays a key role as a co-factor in several enzymatic processes essential for the cerebral metabolism of glucose. At its most extreme, thiamine deficiency, including as a consequence of chronic alcoholism, leads to Wernicke's encephalopathy; a condition involving selective neuropathological lesions related to dysfunction in neuronal metabolic pathways (Ba 2008).

Vitamin C

Vitamin C (ascorbate) is transported into the brain against a steep concentration gradient and accumulates in neuron-rich areas such as the hippocampus, cortex, and cerebellum at high concentrations (Mefford et al. 1981; Mun et al. 2006). This suggests that it plays a pivotal role in brain function. Within the central nervous system vitamin C plays a plethora of roles, in all of which it functions as a single electron donor. In vitro evidence suggests that vitamin C is a powerful antioxidant. In this respect its roles include reducing oxygen, sulfur, and nitrogen–oxygen radicals generated during normal cellular metabolism, and the recycling of other radicals to their previous forms. An important example of the latter is the reduction of tocopheroxyl radical back to alpha tocopherol (Padayatty et al. 2003). Vitamin C also acts as an essential electron donor for a number of separate enzymes, and, among many other effects, contributes to the synthesis of tyrosine, carnitine, catecholamine neurotransmitters, and peptide hormones (Padayatty et al. 2003). Harrison and May (2009) also elaborate roles for vitamin C in neural maturation, and the neuromodulation of the activity of acetylcholine and the catecholamine neurotransmitters, with resultant direct impacts on behavior in animal models.

VITAMIN DEFICIENCY IN DEVELOPED SOCIETIES

Governmentally dictated recommended dietary allowances (RDAs), or similar, of vitamins exist in most developed nations. These RDAs are estimated from the average requirement of individuals within a group/population and the variability in the need for the nutrient among individuals to prevent specific vitamin deficiency diseases such as pellagra, rickets, beri-beri or scurvy, and more general chronic diseases such as osteoporosis and heart disease, in the vast majority (97%–98%) of the population. More recently, most authorities have moved toward adopting dietary reference values (or dietary reference intakes), which incorporate RDAs and simply expand on them to include an "estimated average requirement" and tolerable upper limit.

The U.K. "National Diet and Nutrition Survey" (NDNS) (Ruston et al. 2004) aimed to assess the incidence of biochemical vitamin deficiencies in cross sections of the population aged 19–64 years. The survey reported the results of blood analyte samples for B vitamins and vitamins A, C, D, and E taken from a representative cross section of 1347 respondents. The percent incidence within the population that had abnormally low levels of each vitamin indicative of biochemical depletion/deficiency as defined by a number of established criteria, and which may predispose the individual to specific diseases related to deficiency of the vitamin in question, were presented. This survey showed that 5% of men and 3% of women were biochemically depleted in terms of vitamin C; 2% of men and 4% of women were deficient in vitamin B12; 5% of males and females were marginally deficient in red blood cell folate (B9); 10% of males and 11% of females were deficient in vitamin B6, and 66% of both males and females showed marginal or deficient status in vitamin B2 (potentially due to methodological issues). Similarly, averaged across the year, 14% of men and 15% of women had deficient status in vitamin D, with this peaking in the winter months and attenuating during summer. Very similar percentages of prevalence with regard to vitamin C (Schleicher et al. 2009) and vitamin B12 (Evatt et al. 2010) deficiencies have also recently been reported from the United States using National Health and Nutrition Examination Survey (NHANES) data.

However, inconsistencies across studies have arisen as a consequence of differing definitions of deficiency. For instance, the NDNS data show low levels (<1%) of deficiency in terms of serum folate levels when a definition of deficiency of <6.3 nmol/L is applied. This contrasts with serum folate deficiencies of 6% in the French population (sample size = 2102) when a cutoff of 7 nmol/L is applied (Castetbon et al. 2009), and a previous figure of 16% in the U.S. adult population (sample size = 7300) when a cutoff of 6.8 nmol/L was applied. In the latter case, this percentage decreased to 0.5% following the start of mandatory fortification of cereal-grain products with folic acid in 1998 (Pfeiffer et al. 2005). Similar inconsistencies exist in the data for vitamin E status. Ford et al. (2006) note that a wide range of lower cutoff points (from 7 to 28 µmol/L) in serum alpha-tocopherol levels have been employed to indicate vitamin E deficiency. Their own data from 4087 participants would give rates of deficiency ranging from 0.5% of the population at a cutoff point of 11.6 µmol/L to more than 20% of adults in the United States when this is raised to 20 µmol/L. This latter figure would seem to be in better agreement with data showing that the habitual intake of vitamin E is below the "estimated average requirement" in 90% of the adult population of the United States (Ahuja et al. 2004). Similarly, while the NDNS (Ruston et al. 2004) used a figure of 25 nmol/L of 25-hydroxyvitamin D to indicate biochemical deficiency in vitamin D, the current consensus is that a much higher cutoff of <50 nmol/L is indicative of deficiency (Bischoff-Ferrari et al. 2006; Holick 2007), suggesting a much higher prevalence of deficiency in the U.K. population than previously reported.

These figures highlight that for most nutrients information regarding RDAs is either unknown or incomplete, and the recommendations are made on the basis of a number of assumptions and considerations that can lead to large variations in the eventual RDA (Levine et al. 1996; Young 1996). Not only is there a certain amount of inconsistency in definitions of deficiency, but the data also suggest that a sizeable minority of the population may be suffering levels of deficiency that could, at the least, dispose them to a variety of chronic diseases. Given that those in the deficient category represent the tail end of a distribution, and that optimum nutrition must lie some way above the cutoff for insufficiency, it would appear from this that there may be room for improvement in micronutrient status throughout a substantial segment of the population.

EVIDENCE FROM EPIDEMIOLOGICAL STUDIES

The intrinsic importance of vitamins to many aspects of brain function would suggest that a relationship might exist between elements of psychological functioning and vitamin status. This relationship could be seen both in terms of the accrual of physiological damage due to the long-term effects of vitamin status, for instance as a consequence of systemic damage related to oxidative stress or homocysteine levels, or alternatively in terms of effects directly related to current circulating levels. The latter situation may be more amenable to vitamin-related improvements, with the timescale of effects potentially ranging from almost immediately up to the maximum length of time it might take to fully replete physical stores.

The majority of studies in this area have concentrated on cognitive decline and dementia in elderly, at risk or diagnosed cohorts, that may well have suffered an accumulation of systemic damage over many years or decades. However, given the possibility that many healthy adults would be classified as deficient in one or more vitamins, and the fact that the, as yet unidentified, optimum level of vitamins must lie some distance above current definitions of deficiency, it would also seem likely that a relationship between vitamin consumption/status and psychological functioning should also be evident in cohorts of non-elderly adults assumed to be cognitively intact or representative of the general population. A relatively small number of epidemiological studies have investigated this question in such cohorts. The studies published after 1994 that employed sample sizes >100 which included, but were not limited to, healthy volunteers under the age of 55 years within their cohort are described in the following and shown in Tables 7.1 (B vitamins and/or homocysteine) and 7.2 (vitamins A, C, D, and E).

B VITAMINS AND HOMOCYSTEINE LEVELS

Biochemical Status

The most research attention in this general area has been focused on the relationship between circulating levels of B vitamins and aspects of cognitive function, including cognitive decline, and mood, including depression and anxiety. In terms of healthy, non-elderly cohorts this relationship has been examined with regard to circulating levels of folate, with or without vitamin B12.

Despite a relatively large number of studies exploring this relationship in elderly samples, only three cross-sectional studies (Bjelland et al. 2003; Krieg Jr and Butler 2009; Morris et al. 2003) and two longitudinal studies (Teunissen et al. 2003; Tucker et al. 2005), with follow ups at 6 and 3 years, respectively, met our inclusion criteria. In terms of cognitive performance, Krieg and Butler (2009) found no relationship between folate or B12 and cognition, whereas Teunissen et al. (2003) found a positive relationship between circulating levels of folate and delayed recall but only at baseline. Tucker et al. (2005) found positive relationships between baseline plasma

TABLE 7.1

Epidemiological Research Examining the Relationship between Biochemical Levels and/or Dietary Intake of B Vitamins and/or Homocysteine and Cognitive Function and Mood

References	Design	Sample	Methods Nutritional Status Measures	Methods Cognition and Mood Outcome Measures	Results	Comments
Bjelland et al. (2003)	Cross sectional	N = 5948, age = 46–49 and 70–74 years	Plasma tHcy, serum folate, serum B12, MTHFR C677T genotype	HADS-A and HADS-D (anxiety and depression defined as corresponding HADS-score > 8)	High Hcy related to depression. Folate inversely related to depression only in middle-aged women	—
Bryan and Calvaresi (2004)	Cross sectional, population-based	N = 1183, age = 39–65 years	Folate, B12, and B6 intake estimated using FFQ	Self-report cognitive and memory function (CFQ, MFQ). Psychological well-being (CESD, PSS, STAI-Y, RSE-B)	In men B12 and B6 intake were positively related to memory. Folate and B6 negatively related to memory and perceived stress in women	No objective cognitive measures
Bryan et al. (2002)	Cross sectional	N = 221, age = 20–30, 45–55, and 65–92 years	Dietary intake of folate, B6, and B12 assessed by FFQ	FFQ, digit–symbol coding, symbol search, digit span-backward, letter–number sequencing, vocabulary, recall of symbols from digit-symbol coding (WAIS-III), RAVLT, Boxes test, activity recall, Stroop, self-ordered pointing test, uses for common objects, trail making test, initial letter fluency, excluded letter fluency, spot-the-word, CESD, POMS	Folate was negatively correlated with speed of processing (Boxes). Positive effects seen in the younger group only: association between folate and excluded letter fluency; folate, B6, and B12 associated with recall; B6 associated with short-delay recall. In the older group B6 was positively correlated with long-delay recall	Education controlled for

Study	Design	N, age	Variables	Tests	Findings	Notes
Elias et al. (2005)	Prospective (follow-up 7.6 years)	N = 2096 age = 40–49, 50–59, 60–82 years	Plasma Hcy under fasting conditions	Subtests from WAIS-R: Similarities; paired associates; logical memory: immediate recall, delayed recall, delayed recognition; visual reproductions: immediate recall, delayed recall, delayed recognition. Also, TMT-A and B, Hooper visual organization test, Boston naming test	Hcy levels associated with cognitive impairments on all tasks (except for: logical memory-delayed, visual reproductions—immediate and delayed) only in those aged 60 or over	—
Elias et al. (2008)	Cross sectional	N = 911, age = 26–98 years	ApoE4 (carriers and non-carriers), plasma Hcy	MMSE and Maine-Syracuse Neuropsychological test battery (global composite score calculated based on averaged z-scores of subtests)	Inverse relationship between Hcy and all cognitive measures which was greater in ApoE4 group. With adjustment for demographic characteristics alone or in combination with CVD and B-vitamins, inverse relationship between Hcy and MMSE, global cognitive function, similarities and working memory observed only in apoE4 group	ApoE and non-ApoE groups unbalanced (n = 667 and n = 224, respectively)
Krieg and Butler (2009)	Cross sectional	N = 2911, age = 20–59 years	Blood level of lead, serum: folate, B12, and Hcy concentrations and RBC folate	SRT, DSST, and serial digit learning	No association between cognitive performance and lead, serum folate, or B12. In 20–39 year olds digit learning enhanced with increased Hcy	—

(continued)

TABLE 7.1 (continued)

Epidemiological Research Examining the Relationship between Biochemical Levels and/or Dietary Intake of B Vitamins and/or Homocysteine and Cognitive Function and Mood

			Methods			
			Nutritional Status Measures	**Cognition and Mood Outcome Measures**		
References	**Design**	**Sample**			**Results**	**Comments**
Mishra et al. (2009)	Prospective (follow-up 53 years)	N = 636 women, age = 53 years	Dietary intake of thiamine, B2, B3, folate, B6, and B12 assessed based on parental 24 h food recall (age 4) and 5 day food record (age 36, 43 and 53)	GHQ-28	Low B12 at age 53 years associated with increased psychological distress	—
Morris et al. (2003)	Cross sectional	N = 2948, age = 15–39 years	Serum folate, RBC folate, and serum Hcy concentrations after variable fasting states	DIS examining lifetime assessment of major depression and dysthymia	Folate status higher in subjects who had never been depressed	—

Study	Design	Sample	Measures	Tests	Findings
Schafer et al. (2005)	Cross sectional, population-based	N = 1140, age = 50–70 years	Serum Hcy, ApoE4 genotype	BNT, category fluency, letter fluency, RPM, finger tapping, SRT, purdue pegboard, purdue pegboard assembly, Stroop, TMT-A and B, RAVLT recall and recognition, ROCF, symbol digit paired associate learning	Hcy level associated with worse neurobehavioral test performance in all tests. Following adjustment effect maintained in tests relating to: simple motor and psychomotor speed, eye-hand coordination/manual dexterity, and verbal memory and learning. Effects of Hcy worse for ApoE ε4/ε4 haplotype
Teunissen et al. (2003)	Prospective (follow up 6 years)	N = 144 age = 30–80 years	Serum concentrations of Hcy, folate, and B12	Letter–digit coding, Stroop, word learning test, and delayed recall	Hcy negatively correlated with Stroop and word learning throughout. Folate correlated to delayed recall only at baseline
Tucker et al. (2005)	Prospective (follow up 3 years)	N = 321 men, age = 50–85 years (mean age = 67 years)	Plasma total Hcy, folate, B6, and B12 measured at baseline under fasting conditions. Plus dietary intake folate, B6, and B12	MMSE, BDS, word list recall, verbal fluency, constructional praxis	Change in constructional praxis positively associated with plasma and dietary vitamin levels. Hcy negatively correlated with constructional praxis and word recall at 3 years. Dietary folate positively associated with verbal fluency at 3 years. Plasma and dietary folate effects independent of other vitamins and Hcy

(continued)

TABLE 7.1 (continued)

Epidemiological Research Examining the Relationship between Biochemical Levels and/or Dietary Intake of B Vitamins and/or Homocysteine and Cognitive Function and Mood

References	Design	Sample	Methods		Results	Comments
			Nutritional Status Measures	Cognition and Mood Outcome Measures		
Wright et al. (2004)	Cross sectional	N = 2871 age ≥ 40 years	Fasting Hcy levels	MMSE	Only in older (≥65 years) increased Hcy associated with lower MMSE score	Not all healthy: subjects were only excluded if they were not stroke-free, study included because sample contains young adults

Apo-E, apolipoprotein E; AVLT, auditory verbal learning test; BDS, backward digit span; BNT, Boston naming test; CESD, Centre for Epidemiologic Studies Depression Scale; CFQ, cognitive failures questionnaire; CVD, cardiovascular disease; DIS, diagnostic interview schedule; DSST, digit symbol substitution test; FFQ, food frequency questionnaire; GHQ-28, general health questionnaire; Hcy, homocysteine; MFQ, memory functioning questionnaire; MMSE, mini mental state exam; POMS, profile of mood states; PSS, perceived stress scale; RAVLT, Rey auditory verbal learning test; RBC, red blood cell; ROCF, Rey–Osterreith complex figure; RPM, Raven's progressive matrices; RSE-B, Rosenberg self-esteem scale; STAI-Y, state trait anxiety inventory; SRT, simple reaction time; TMT-A and B, trail-making test parts A and B; WAIS-III, Wisconsin adult intelligence scale-III; WAIS-R, Wisconsin adult intelligence scale-revised.

TABLE 7.2
Epidemiological Research Examining Relationship between Biochemical Levels of Vitamins A, C, D, and E and Cognitive Function and Mood

References	Design	Sample	Methods		Results	Comments
			Nutritional Status Measures	Cognition and Mood Outcome Measures		
Lee et al. (2009)	Cross sectional	N = 3133 males, age = 40–79 years	Serum 25(OH)D	ROCF, CTRM, DSST, PASE, PPT, and BDI	Lower levels of 25(OH)D associated with worse DSST performance	—
McGrath et al. (2007)	Cross sectional	N = 9556, age = 20–60 years (n = 4747), 60–90 years (n = 4809)	Serum 25(OH)D (vitamin D)	Adults: (From NES): RT, DSST, serial digit learning. Elderly: Memory of a story	Association between serum 25(OH)D and performance only observed in elderly group (age 60–90 years) with highest serum levels associated with largest learning and memory impairment	
Pan et al. (2009)	Cross sectional, population-based	N = 3262, age = 50–70 years	Plasma 25(OH)D	CESD	No association between CESD and plasma 25(OH)D concentration	

25(OH)D, 25 hydroxyvitamin D; BDI, Beck depression inventory; CESD, Centre for Epidemiological Studies Depression Scale; CTRM, Camden topographical recognition memory; DSST, digit symbol substitution test; NES, neurobehavioral evaluation system; PASE, physical activity scale for the elderly; PPT, physical performance test; ROCF, Rey-osterrieth complex figure; RT, reaction time.

vitamin levels (folate, B6 and B12) and change in constructional praxis as measured by spatial copying at 3 years follow-up. However, only the effects of folate were demonstrated to be independent when effects of the other vitamins were adjusted for.

Morris et al. (2003) explored the role of folate and B12 in depression in those with major depression, those with dysthymia and those with no depression. They found significantly lower levels of red blood cell and serum folate in those with major depression and lower levels of serum folate in those with dysthymia as compared to those with no depression after adjustment for sociodemographic factors. Partial support for these findings comes from Bjelland et al. (2003) who demonstrated an inverse relationship between depression and serum folate, but only in middle-aged women.

Given the intrinsic role that folate and vitamin B12 play in the metabolism of homocysteine, which is a potentially toxic amino acid by-product of one carbon metabolism and is implicated in a number of diseases, all of the studies mentioned earlier also included an assessment of blood levels of homocysteine (Bjelland et al. 2003; Krieg Jr and Butler 2009; Morris et al. 2003; Teunissen et al. 2003; Tucker et al. 2005). Four additional studies also investigated the relationship between homocysteine and cognitive performance in sizeable cohorts independently of circulating vitamin levels (Elias et al. 2005, 2008; Schafer 2005; Wright et al. 2004). Six of the studies assessing this relationship reported that higher homocysteine levels were associated in some way with aspects of poorer cognitive function or cognitive decline (Elias et al. 2005, 2008; Schafer 2005; Teunissen et al. 2003; Tucker et al. 2005; Wright et al. 2004), but two of these were only in a subset of elderly participants (Elias et al. 2005; Wright et al. 2004). These effects were also shown to be worse in apolipoprotein E-ε 4 carriers in two (Elias et al. 2008; Schafer 2005) out of three studies (Wright et al. 2004). The only other study of cognition and homocysteine found a positive association between homocysteine and digit learning in 20–39 year olds, with no effect in 40–59 year olds (Krieg and Butler 2009). In terms of mood, Morris et al. (2003) found no relationship between homocysteine and depression/depressive symptoms, whereas Bjelland et al. (2003) demonstrated a positive relationship between homocysteine and depression. The individual study details and results are presented in Table 7.1.

Dietary Intake

Although there are several cross-sectional studies that include an examination of the associations between dietary intake of B vitamins and cognitive performance or mood in elderly populations, reference to the dietary intake studies presented in Table 7.1 shows that only four studies met our inclusion criteria for non-elderly participants. In a large cross-sectional study, Bryan and Calvaresi (2004) found that intake of vitamins B6, and B12, as assessed by a food frequency questionnaire, were positively associated with subjective perceptions of memory in men; but B12 and folate were negatively associated with subjective memory in women. However, no objective measures of cognitive performance were employed in the study. In an earlier study, using objective measures of cognition, Bryan et al. (2002) found that folate was negatively associated with speed of processing, but positively associated with verbal fluency in a younger group (20–30 years) only. Folate, B12, and B6 were all positively associated with recall in the younger group, and B6 was positively associated with short-delay recall in the younger group and with long-delay recall in an older group (65–92). No age-specific effects were found in the middle group (45–55). Tucker et al. (2005) found that dietary levels of folate, B12, and B6 were positively associated with spatial copying, and that dietary folate was significantly positively correlated with verbal fluency at 3 years. These effects of dietary folate were independent of the effects of homocysteine or B12 or B6.

Turning to mood measures, one prospective study (Mishra et al. 2009) examined associations between long-term B vitamin intake and subjective mental health (general health questionnaire [GHQ]-28) in a group (N = 636) of 53 year old women, who had been followed-up since birth. The authors found that low B12 intake at 53 years of age was associated with increased psychological distress, and that those with low, as opposed to high, B12 intake throughout adulthood had poorer subjective perceptions of their overall mental health as assessed by the GHQ-28. However, Bryan and Calvaresi (2004) found that higher dietary levels of folate and B6 were related to higher perceived stress levels in women aged 39–65 years. No effects were shown in men. Finally, Bryan et al. (2002) found no effect of folate, B6, or B12 on depression or mood.

Vitamins A, C, D, and E

Biochemical Status

Very few studies have assessed the circulating levels of analytes for vitamins A, C, D, and E in non-elderly cohorts (see Table 7.2). Only three cross-sectional studies (Lee et al. 2009; McGrath et al. 2007; Pan et al. 2009) met our criteria, all had sizeable sample sizes and examined the relationship between levels of vitamin D (25(OH)D) and cognitive function and/or mood. Lee et al. (2009) reported a beneficial relationship between vitamin D levels and performance, with low levels of 25(OH)D being associated with reduced digit symbol substitution task (DSST) performance in 40–79 year olds. However, McGrath et al. (2007) reported no such effect, with a negative association between vitamin D and learning and memory in their oldest group of participants (60–90 year olds). Perhaps surprisingly no association of 25(OH)D and mood has been established, with both studies within our sample that examined this showing no relationship (Lee et al. 2009; Pan et al. 2009).

Dietary Intake

In terms of dietary intake of vitamins A, C, D, or E and its effects on cognitive function and mood, no studies met our inclusion criteria.

EVIDENCE FROM INTERVENTION STUDIES

Recent randomized controlled trials examining the effect of vitamin supplementation on cognitive function and mood in healthy non-elderly cohorts are shown in Tables 7.3 (B vitamins) and 7.4 (multi-vitamins). Studies were included that were

TABLE 7.3
Randomized Controlled Trials Assessing the Effects of B Vitamins on Psychological Functioning in Cognitively Intact Samples

References	Sample	Manipulation	Measures	Results	Comments
Benton et al. (1997)	117 females; mean age: 20.3 years	50 mg thiamine (B1); or placebo for 2 months	Blood: B1, B2, B6. SRT, CRT, familiar faces test, word recall, POMS, GHQ-30, assessed at 3, 6/9 and 12 months	Vitamin supplementation improved thiamine status, mood (POMS), and decision times	
Bryan et al. (2002)	Women: 20–30 years (n = 56), 45–55 years (n = 80), 65–92 years (n = 75)	750 µg folate, 15 µg B12, 75 mg B6; or placebo for 5 weeks	FFQ, digit–symbol coding, symbol search, digit span-backward, letter–number sequencing, vocabulary, recall of symbols from digit symbol coding (WAIS-III), RAVLT, Boxes test activity recall, Stroop, self-ordered pointing test, uses for common objects, trail making test, initial letter fluency, excluded letter fluency, spot-the-word, CESD, POMS	Older folate group significantly better on RAVLT recognition list B than placebo. B6 and placebo groups out-performed B12 and folate groups in initial letter fluency	Education controlled for
Durga et al. (2007)	818 men and women, 50–70 years with Hcy levels between 13 and 26 µmol/L	800 µg folic acid; or placebo for 3 years	Serum: folate, erythrocyte folate, B12, creatinine and lipids. tHcy, B6. C67TT polymorphism, apoE genotype, blood pressure, FFQ Word learning, concept shifting, Stroop, verbal fluency, letter–digit substitution	Folic acid significantly increased serum folate and decreased tHcy at years 1, 2, and 3 when compared to placebo. Folic acid significantly improved global cognition, word learning and letter–digit substitution at 3 years vs. placebo	—

apoE, apolipoprotein E; CESD, Centre for Epidemiological Studies Depression Scale; CRT, choice reaction time; FFQ, food frequency questionnaire; GHQ, general health questionnaire; POMS, profile of mood states; RAVLT, Rey auditory verbal learning test; SRT, simple reaction time; tHcy, total homocysteine; WAIS-III, Wechler adult intelligence scale-revised.

TABLE 7.4

Randomized Controlled Trials Assessing the Effects of Multi-Vitamins (Plus Minerals in Some Cases) on Psychological Functioning in Cognitively Intact Samples

References	Sample	Manipulation	Measures	Results	Comments
Benton et al. (1995a)	127 men and women aged 17–27 years	3334 IE vitamin A, 14 mg thiamine (B1), 16 mg riboflavin (B2), 22 mg pyridoxine (B6), 0.03 mg B12, 600 mg vitamin C, 100 mg vitamin E, 4 mg folic acid (B9), 2 mg biotin (B7), and 180 mg nicotinamide; or placebo for 12 months	Blood: B1, B2, B6. Plasma: vitamin C, B12, folic acid, biotin. vitamin A, vitamin E, β-carotene. SRT, CRT, DSS, CAT assessed at 3, 6/9, and 12 months	Vitamin supplementation decreased the intercept of the regression line in females after 12 months administration, indicating improved attentional processing	—
Benton et al. (1995b)	119 men and women aged 17–27 years	3334 IE vitamin A, 14 mg thiamine (B1), 16 mg riboflavin (B2), 22 mg pyridoxine (B6), 0.03 mg B12, 600 mg vitamin C, 100 mg vitamin E, 4 mg folic acid (B9), 2 mg biotin (B7), and 180 mg nicotinamide; or placebo for 12 months	Blood: B1, B2, B6. Plasma: vitamin C, B12, folic acid, biotin, vitamin A, vitamin E, β-carotene. POMS, GHQ-30, assessed at 3, 6/9, and 12 months	Vitamin supplementation increased agreeable scores, levels of B1, B2, B6, and plasma: vitamin C, B12, folic acid, biotin, vitamin E. In females vitamins significantly improved mental health (GHQ) and 'composed' ratings at 12 months vs. placebo	—
Carroll et al. (2000)	80 males aged 18–42 years	15 mg B1, 15 mg B2, 50 mg niacin (B3), 23 mg pantothenic acid (B5), 10 mg B6, 150 µg biotin (B7), 400 µg folic acid, 10 µg B12, 500 mg vitamin C, 100 mg calcium, 100 mg magnesium, 10 mg zinc; or placebo for 28 days	Plasma zinc, GHQ-28. HADS. PSS, rating scales, physical symptom checklist	Vitamin supplementation improved GHQ scores, reduced anxiety (HADS), stress (PSS) and physical symptoms vs. placebo	—

(continued)

TABLE 7.4 (continued)

Randomized Controlled Trials Assessing the Effects of Multi-Vitamins (Plus Minerals in Some Cases) on Psychological Functioning in Cognitively Intact Samples

References	Sample	Manipulation	Measures	Results	Comments
Haskell et al. (2010)	216 females, aged 25–50 years	0.8 mg vitamin A, 4.2 mg B1, 4.8 mg B2, 54 mg niacin (B3), 18 mg pantothenic acid (B5), 6 mg B6, 0.45 mg biotin (B7), 0.6 mg folic acid, 0.003 mg B12, 180 mg vitamin C, 5 μg vitamin D3, 10 mg vitamin E, 30 μg vitamin K1, 120 mg calcium, 126 mg phosphorus, 0.025 mg chromium, 0.9 mg copper, 1.5 mg fluoride, 0.075 mg iodine, 8 mg iron, 45 mg magnesium, 1.8 mg manganese, 0.045 mg molybdenum, 0.045 mg selenium, 8 mg zinc; or placebo for 9 weeks	Serum: Hcy, 8-OHdG, SF-36, Chalder fatigue scale, POMS, STAI, Bond-Lader mood scales, VAS, MTF (maths processing, memory search, Stroop, high number tap)	Hcy was significantly lower in the vitamins group following supplementation. Vitamins attenuated physical tiredness prior to and 2 hours following acute supplementation. Overall MTF performance improved and accuracy of memory search was impaired prior to acute supplementation only. Speed and accuracy of maths processing and accuracy of Stroop improved both prior to and following acute supplementation. Speed of Stroop responses and memory search improved following acute supplementation only.	

Study	Sample	Intervention	Measures	Results	Notes
Kennedy et al. (2010)	210 males, aged 30–55 years	15 mg B1, 15 mg B2, 50 mg niacin (B3), 23 mg pantothenic acid (B5), 10 mg B6, 150 μg biotin (B7), 400 μg folic acid, 10 μg B12, 500 mg vitamin C, 100 mg calcium, 100 mg magnesium, 10 mg zinc; or placebo for 33 days	GHQ-12, PSS, POMS, Bond-Lader mood scales, energy VAS, serial 3 subtractions, serials 7s, RVIP, mental fatigue VAS, Stroop, peg-and-ball, WCS	Vitamins improved GHQ-12, PSS, POMS vigor, correct serial 3s, mental tiredness VAS	Corrected for reported fruit/vegetable intake
Kennedy et al. (2011)	210 males, aged 30–55 years	15 mg B1, 15 mg B2, 50 mg niacin (B3), 23 mg pantothenic acid (B5), 10 mg B6, 150 μg biotin (B7), 400 μg folic acid, 10 μg B12, 500 mg vitamin C, 100 mg calcium, 100 mg magnesium, 10 mg zinc; or placebo for 33 days	Mobile phone assessments before and after a day's work prior to and 7, 14, 21 and 28 days after treatment: Bond-Lader mood scales, VAS (stress, concentration, mental stamina, physical stamina), arrows flankers CRT, 2-back	Vitamins improved 'alert' ratings during evening testing on day 14 and during morning and evening assessments on day 28. Concentration and mental stamina ratings were also improved during evening assessments across all days and physical stamina was improved during both assessments across each study day.	Corrected for reported fruit/vegetable intake
Schlebusch et al. (2000)	300 men and women, aged 18–65 with predetermined high stress scores	15 mg B1, 15 mg B2, 50 mg niacin (B3), 23 mg pantothenic acid (B5), 10 mg B6, 150 μg biotin (B7), 10 μg B12, 1000 mg vitamin C, 100 mg calcium, 100 mg magnesium; or placebo for 30 days	HARS, PGWS, stress VAS, BSI	Vitamin supplementation led to significant improvements in BSI, HARS, PGWS, and VAS vs. placebo	—

8-OHdG, 8-hydroxy-2-deoxyguanosine; BSI, Berocca stress index; CAT, continuous attention test; CRT, choice reaction time; DSS, digit-symbol substitution; GHQ, general health questionnaire; HADS, hospital anxiety depression scale; HARS, Hamilton anxiety rating scale; Hcy, homocysteine; MTF, multi-tasking framework; PGWS, psychological general well-being schedule; POMS, profile of mood states; PSS, perceived stress scale; RVIP, rapid visual information processing; SF-36, medical outcome study quality of life short-form 36; SRT, simple reaction time; STAI, state trait anxiety inventory VAS, visual analog scale; WCS, Wisconsin card sort.

published after 1994, were undertaken in a non-elderly adult population, had a minimum of 30 participants per treatment arm, and employed appropriate placebo and double-blind methodology. Findings are summarized in the following.

B Vitamins

Three trials of B vitamin supplementation were identified that conformed to the inclusion criteria (see Table 7.3). Improvements in psychological functioning were seen in all studies (Benton et al. 1997; Bryan et al. 2002; Durga et al. 2007). Benton et al. (1997) administered vitamin B1 (50 mg thiamine) to young females for 2 months and found improvements in total profile of mood states (POMS) score, and faster decision times on 2, 4, and 8 choice reaction time tasks. Two studies also demonstrated improved function following folic acid alone, with Durga et al. (2007), in the largest (N = 818) and longest (3 years) study reported here, demonstrating improvements across cognitive domains in a cohort of 50–70 year old participants with raised homocysteine levels at the outset. Bryan et al. (2002) also found improvements on a single task (Rey auditory verbal learning test recognition list B) from within a large selection of tasks, but only in an older subsection of their female cohorts that were administered folate for 5 weeks. They also found decrements on a letter fluency task, in comparison to placebo, in the same group and another group administered B12 alone.

Vitamins A, C, D, and E

No studies that involved the administration of vitamins A, C, D, or E which met the inclusion criteria were identified.

Multi-Vitamins

Seven studies assessing the cognitive/mood effects of multi-vitamins (plus minerals in several cases) are presented in Table 7.4. Five studies reported unambiguous improvements in psychological functioning (Carroll et al. 2000; Haskell et al. 2010; Kennedy et al. 2010, 2011; Schlebusch et al. 2000) across their cohorts and two further studies described improvements that were seen in subsamples, as a function of gender (Benton et al. 1995a,b). Of the five studies demonstrating effects across the cohorts, four administered a similar B vitamin complex plus vitamin C, calcium, and magnesium. In the first two of these studies (Carroll et al. 2000; Schlebusch et al. 2000), supplementation for ~30 days led to improved well-being (psychological general well-being schedule and GHQ-28, respectively), reduced stress (perceived stress scale [PSS] and Berocca stress index), and anxiety (HADS and Hamilton anxiety rating scale) on validated psychometric measures in non-elderly participants. Kennedy et al. (2010) included computerized cognitive testing and replicated the findings with regard to stress (PSS) and general psychological functioning (GHQ-12), and additionally reported increased vigor (POMS), reduced mental tiredness, and improved serial subtraction task performance in non-elderly males. In a methodologically distinct study, a mobile phone assessment was used in the same cohort as

Kennedy et al. (2010) to measure effects before and after a day's work prior to and 7, 14, 21, and 28 days after treatment (Kennedy et al. 2011). Although there were no effects on cognition, the study demonstrated increased "alert" ratings following multi-vitamin supplementation during evening testing on day 14 and during morning and evening assessments on day 28. Concentration and mental stamina ratings were also improved during evening assessments across all days following vitamin supplementation. Physical stamina was improved in the active group during both assessments across each study day. Similarly, using a computerized assessment of cognitive function, Haskell et al. (2010) demonstrated improved cognitive functioning and reduced physical tiredness following extended multi-tasking in 220 females aged 25–50 years who took a broad multi-vitamin/mineral for 9 weeks.

In terms of the studies that saw differential effects according to gender, Benton et al. found improved mood (Benton et al. 1995b) and improved attentional processing (Benton et al. 1995a) in the females within their young adult cohort, with these effects only becoming apparent following 12 months administration of their high-dose multi-vitamin.

DISCUSSION

In simple terms, the rationale for the current review was that vitamins are intrinsically involved in every aspect of brain function, and that our modern, micronutrient-poor diets predispose us to consume less than the optimal levels of vitamins; therefore raising the possibility that, firstly, intake/levels of vitamins might be related to cognitive function, and, secondly, that supplementation with vitamins might improve psychological functioning.

The epidemiological evidence relating biochemical levels and dietary intake of vitamins to psychological functioning is currently insufficient to make any conclusion on this in this population at present. By far, the most research in this area has concentrated on the B vitamins, most notably folate, and vitamins B6 and B12. In terms of folate, Tucker et al. (2005) found improvements to spatial copying in 50–85 year olds were related to both baseline biochemical and dietary folate. Improvements to verbal fluency were also related to baseline dietary folate intake and these effects were independent of intake of other vitamins and homocysteine levels. Teunissen et al. (2003) also found a positive association between serum folate levels and delayed recall at baseline in 30–80 year olds, but this effect was no longer apparent at the 6 year follow-up and there were no effects on word learning, Stroop, or letter–digit coding. No relationship between serum folate and simple reaction time, digit–symbol substitution, or serial digit learning was observed by Krieg and Butler (2009) in 20–59 year olds, whereas Bryan and Calvaresi (2004) found an inverse relationship between dietary folate and subjective memory in women aged 39–65. Bryan et al. (2002) also found a negative relationship between dietary folate and speed of information processing in 20–92 year olds. However, a positive relationship was shown for the younger group only (20–30 year olds) in terms of verbal fluency and recall.

Tucker et al. (2005) found improvements to spatial copying in 50–85 year olds were related to both baseline biochemical and dietary B6 and B12 levels. Similarly, Bryan and Calvaresi (2004) found a positive relationship with subjective memory for

B6 and B12 in men. Bryan et al. (2002) found that B6 was positively correlated with short-delay recall, and B6 and B12 were positively associated with recall in younger group (20–30 year olds). In an older group (65–92 year olds) B6 was positively correlated with long-delay recall. No specific effects were discovered in a middle age group. Teunissen et al. (2003) found no effects of B12 on word learning, Stroop or letter–digit coding and no relationship between B12 and simple reaction time, digit–symbol substitution or serial digit learning was observed by Krieg and Butler (2009) in 20–59 year olds.

Seven of the studies identified included a measure of the relationship between homocysteine and cognition, of which six showed some evidence for a negative relationship between the two. Teunissen et al. (2003) demonstrated that higher homocysteine levels were related to impaired Stroop and word learning. Tucker et al. (2005) observed that homocysteine was inversely related to spatial copying and word recall. The results of two cross-sectional studies supported this negative relationship and indicated that this was stronger in ApoE4 carriers (Elias et al. 2008; Schafer 2005). Two other cross-sectional studies only demonstrated these effects in elderly sub-groups (Elias et al. 2005; Wright et al. 2004), and one final study observed a positive relationship between homocysteine levels and digit learning in 20–39 year olds.

In terms of mood there were five studies that met our inclusion criteria, of which three showed some element of a positive association, one showed no effects, and one demonstrated a negative relationship between B vitamins and mood. Morris et al. (2003) observed higher serum folate in 15–39 year olds who had never been depressed than those with major depression or dysthymia. Bjelland et al. (2003) also found that serum folate was inversely related to depression score but only in middle aged women (46–49 years), with no effect in men or women aged 70–74 years. In a prospective study with 53 year follow-up, Mishra et al. (2009) found that low dietary B12 intake at age 53 was associated with increased psychological distress. However, Bryan et al. (2002) failed to find any relationship between dietary folate, B6, or B12 and depression and mood in participants aged 20–92 years. Bryan and Calvaresi (2004) also failed to find any relationship between folate, B6, or B12 and depression, anxiety, or self-esteem. However, perceived stress was inversely correlated with dietary folate and B6, but only in women. Only two studies explored the relationship between homocysteine and mood with one of these finding a positive correlation between homocysteine levels and depression in 46–74 year olds (Bjelland et al. 2003), and the other finding no relationship in 15–39 year olds (Morris et al. 2003).

The evidence with regard to the other vitamins (A, C, D, and E) is very limited in this population. Reference to Table 7.2 shows that only three epidemiological studies met our inclusion criteria, and these all examined the effects of biochemical levels of vitamin D. Two studies explored the relationship between serum 25(OH)D and cognition, with one of these finding a positive relationship between vitamin D and digit symbol substitution performance in 40–79 year old males (Lee et al. 2009) and the other finding an impairment to learning and memory associated with higher levels of vitamin D in an older subset of participants (60–90 years). There were no effects in the younger subset (20–60 years). Surprisingly, no relationship between vitamin D and mood was established. Serum 25(OH)D was not associated with physical activity, mood, or depression in 40–79 year old males (Lee et al. 2009); nor was plasma 25(OH)D related to depression in 50–70 year olds (Pan et al. 2009).

It is difficult to make any firm conclusions on the basis of the findings from epidemiological research. Naturally, any interpretation of epidemiological evidence is also complicated by the inability either to attribute cause and effect or rule out the influence of a plethora of other potential factors that might co-vary with vitamin status, which may not have been identified in the statistical models employed. Examples of the latter may include, for instance, aspects of socio-economic status, education, or healthy living practices. One particular weakness in this area is also the predominant use of elderly cohorts, making any meaningful assessment of the relationship in non-elderly populations very difficult. While the cognitive decline associated with old age may provide a sensitive backdrop for examining the effects of dietary habits in the context of a large part of the lifespan, investigations in these age groups are also complicated by a number of factors. These include age-related changes in the ability to absorb and metabolize vitamins (Wolters et al. 2004), which may render the results of analyte studies less meaningful, and a relationship between age, declining health, and cognitive function (Payette and Shatenstein 2005; Shatenstein et al. 2007), which suggests that the elderly develop atypical diets, and may well predispose the less cognitively able individual to seek and consume a poorer diet.

In comparison to the epidemiological evidence, research investigating the effects of vitamin supplementation in non-elderly cohorts is even sparser. Only three studies of B vitamins and no studies of the other vitamins (A, C, D, and E) were identified that met our inclusion criteria. Of the studies covering B vitamins, all three discovered a positive impact of supplementation on at least one aspect of cognition and/or mood. Benton et al. (1997) observed improvements to decision times and mood following 2 month supplementation with thiamine in healthy young females. Folic acid supplementation for 3 years in 50–70 year olds was also shown to improve cognition and to decrease homocysteine at years 1, 2, and 3 (Durga et al. 2007). However, Bryan et al. (2002) found that 5 week supplementation with folate only improved recognition in an older subset (65–92 years) and folate and B12 impaired verbal fluency in all participants (20–92 years).

Naturally, while attractive from the point of view of attributing any treatment-related effects, the focus of these studies on individual vitamins lowers the likelihood of targeting a cohort with a specific requirement for increased levels of the vitamin being supplemented, or alternatively raises the possibility of failing to increase the reduced levels of other micronutrients that might coexist in the individual as a consequence of poor general diet. The evidence with regard to multi-vitamins offers some support for this suggestion and shows some promise with regard to supplementation, particularly in non-elderly populations. Across the seven studies identified in Table 7.4, all reported straightforward benefits to psychological functioning. Interestingly, the multi-vitamins studies included have tended to be conducted solely in younger cohorts rather than the broad range used in epidemiological studies. The pattern of results may reflect the intactness of the cohorts in terms of the physiological mechanisms that might be modulated, and this would provide an argument for focusing on supplementation/dietary improvements in younger cohorts as a means of preventing cognitive decline rather than attempting to "cure" the effects of sub-optimal nutrition at a stage where it may be too late. The effects seen with multi-vitamins may also relate to the broad nature of the treatments, in that the use of multiple vitamins should be more likely to bolster an individual's requirement for one or more vitamins. It is interesting to note in this respect that there are marked inter-individual differences in the absorption and excretion of vitamins (Shibata et al. 2005, 2009) as a consequence of a number of factors, including genetic makeup, gender, and ethnicity (Caudill 2009; Kauwell et al. 2000). In this respect, it should be noted that RDAs are merely population statistics and of very little use in identifying the required minimum daily intake of a nutrient for any individual.

Interestingly, while the strongest relationship seen in the epidemiological studies reviewed earlier is that between circulating levels of homocysteine and cognitive function, homocysteine levels were only measured in two of the intervention studies. Both studies showed a reduction in homocysteine levels and an improvement in cognition following vitamin supplementation. Although neither study correlated the improvements to cognition with the changes in homocysteine levels, Durga et al. (2007) found the greatest improvement to cognition in those who had homocysteine levels ≥12.9 μmol/L at baseline. However, the inter-relationships between B vitamins and homocysteine levels, and the causality of the relationships with brain function, are far from clear (Elias et al. 2006; Krieg and Butler 2009), and it remains a possibility that homocysteine levels represent an epiphenomenon reflecting other, as yet un-delineated factors. There is also the possibility that high homocysteine levels do not represent a problem until older age and may also be in part related to ApoE-4 status.

Given that the optimum level of vitamin consumption must reside some way above deficiency levels, it is possible that a large subsection of the general, non-elderly population must have less than optimal micronutrient status. More research might therefore be usefully directed toward delineating the optimal levels of vitamins and their relationships with brain function in non-elderly humans. This research should take advantage of the many sensitive computerized measures of cognitive function that are now readily available. The one area where the preponderance of studies has been conducted in younger samples is that assessing the effects of multi-vitamins. The literature here suggests that non-elderly healthy adults might derive benefits from supplementation in terms of psychological functioning. This in itself supports the notion of less than optimal nutritional status in the population, but also suggests that broad multi-vitamin treatments, which might provide different benefits to individuals on the basis of their personal nutritional status, might be more useful than single/several vitamin supplements.

REFERENCES

Ahuja JKC, Goldman JD, and Moshfegh AJ (2004) Current status of vitamin E nutriture. In: Kelly F, Meydani M, Packer L (eds) *Annals of the New York Academy of Sciences* 1031: 387–390.

Ba A (2008) Metabolic and structural role of thiamine in nervous tissues. *Cellular and Molecular Neurobiology* 28: 923–931.

Banhegyi G, Braun L, Csala M, Puskas F, and Mandl J (1997) Ascorbate metabolism and its regulation in animals. *Free Radical Biology and Medicine* 23: 793–803.

Benton D, Fordy J, and Haller J (1995a) The impact of long term vitamin supplementation on cognitive functioning. *Psychopharmacology* 117: 298–305.

Benton D, Griffiths R, and Haller J (1997) Thiamine supplementation mood and cognitive functioning. *Psychopharmacology* 129: 66–71.

Benton D, Haller J, and Fordy J (1995b) Vitamin supplementation for 1 year improves mood. *Neuropsychobiology* 32: 98–105.

Benzie IFF (2003) Evolution of dietary antioxidants. *Comparative Biochemistry and Physiology A: Molecular & Integrative Physiology* 136: 113–126.

Bischoff-Ferrari HA, Giovannucci E, Willett WC, Dietrich T, and Dawson-Hughes B (2006) Estimation of optimal serum concentrations of 25-hydroxyvitamin D for multiple health outcomes. *American Journal of Clinical Nutrition* 84: 18–28.

Bjelland I, Tell GS, Vollset SE, Refsum H, and Ueland PM (2003) Folate, vitamin B-12, homocysteine, and the MTHFR 677C → T polymorphism in anxiety and depression— The Hordaland Homocysteine Study. *Archives of General Psychiatry* 60: 618–626.

Boadlebiber MC (1993) Regulation of serotonin synthesis. *Progress in Biophysics & Molecular Biology* 60: 1–15.

Bonnet E, Touyarot K, Alfos S, Pallet V, Higueret P, and Abrous DN (2008) Retinoic acid restores adult hippocampal neurogenesis and reverses spatial memory deficit in vitamin A deprived rats. *PLoS ONE* 3: e3487.

Brigelius-Flohe R (2009) Vitamin E: The shrew waiting to be tamed. *Free Radical Biology and Medicine* 46: 543–554.

Bryan J and Calvaresi E (2004) Associations between dietary intake of folate and vitamins B-12 and B-6 and self-reported cognitive function and psychological well-being in Australian men and women in midlife. *Journal of Nutrition Health and Aging* 8: 226–232.

Bryan J, Calvaresi E, and Hughes D (2002) Short-term folate, vitamin B-12 or vitamin B-6 supplementation slightly affects memory performance but not mood in women of various ages. *Journal of Nutrition* 132: 1345–1356.

Buell JS and Dawson-Hughes B (2008) Vitamin D and neurocognitive dysfunction: Preventing "D"ecline? *Molecular Aspects of Medicine* 29: 415–422.

Calderon-Guzman D, Hernandez-Islas JL, Espitia-Vazquez I, Barragan-Mejia G, Hernandez-Garcia E, Angel DSD, and Juarez-Olguin H (2004) Pyridoxine, regardless of serotonin levels, increases production of 5-hydroxytryptophan in rat brain. *Archives of Medical Research* 35: 271–274.

Carroll D, Ring C, Suter M, and Willemsen G. (2000) The effects of an oral multivitamin combination with calcium, magnesium, and zinc on psychological well-being in healthy young male volunteers: A double-blind placebo-controlled trial. *Psychopharmacology* 150: 220–225.

Castetbon K, Vernay M, Malon A, Salanave B, Deschamps V, Roudier C, Oleko A, Szego E, and Hercberg S (2009) Dietary intake, physical activity and nutritional status in adults: the French nutrition and health survey (ENNS, 2006–2007). *British Journal of Nutrition* 102: 733–743.

Caudill MA (2009) Folate bioavailability: Implications for establishing dietary recommendations and optimizing status. *American Journal of Clinical Nutrition* 91: 1455S–1460S.

Challem JJ (1997) Did the loss of endogenous ascorbate propel the evolution of Anthropoidea and Homo sapiens? *Medical Hypotheses* 48: 387–392.

Chiang M-Y, Misner D, Kempermann G, Schikorski T, Giguère V, Sucov HM, Gage FH, Stevens CF, and Evans RM (1998) An essential role for retinoid receptors RAR[beta] and RXR[gamma] in long-term potentiation and depression. *Neuron* 21: 1353–1361

Cocco S, Diaz G, Stancampiano R, Diana A, Carta M, Curreli R, Sarais L, and Fadda F (2002) Vitamin A deficiency produces spatial learning and memory impairment in rats. *Neuroscience* 115: 475–482.

Crandall J, Sakai Y, Zhang J, Koul O, Mineur Y, Crusio WE, and McCaffery PJ (2004) 13-cis-Retinoic acid suppresses hippocampal cell division and hippocampal-dependent learning in mice retinoic acid suppresses hippocampal cell division and hippocampal-dependent learning in mice. *Proceedings of the National Academy of Sciences of the United States of America* 101: 5111–5116.

Durga J, van Boxtel MPJ, Schouten EG, Kok FJ, Jolles J, Katan MB, and Verhoef P (2007) Effect of 3-year folic acid supplementation on cognitive function in older adults in the FACIT trial: A randomised, double blind, controlled trial. *Lancet* 369: 208–216.

Eaton SB and Konner MJ (1997) Paleolithic nutrition revisited: A twelve-year retrospective on its nature and implications. *European Journal of Clinical Nutrition* 51: 207–216.

Elias MF, Robbins MA, Budge MM, Elias PK, Brennan SL, Johnston C, Nagy Z, and Bates CJ (2006) Homocysteine, folate, and vitamins B-6 and B-12 blood levels in relation to cognitive performance: The maine-syracuse study. *Psychosomatic Medicine* 68: 547–554.

Elias MF, Robbins MA, Budge MM, Elias PK, Dore GA, Brennan SL, Johnston C, and Nagy Z (2008) Homocysteine and cognitive performance: Modification by the ApoE genotype. *Neuroscience Letters* 430: 64–69.

Elias MF, Sullivan LM, D'Agostino RB, Elias PK, Jacques PF, Selhub J, Seshadri S, Au R, Beiser A, and Wolf PA (2005) Homocysteine and cognitive performance in the Framingham offspring study: Age is important. *American Journal of Epidemiology* 162: 644–653.

Etchamendy N, Enderlin V, Marighetto A, Pallet V, Higueret P, and Jaffard R (2003) Vitamin A deficiency and relational memory deficit in adult mice: Relationships with changes in brain retinoid signalling. *Behavioural Brain Research* 145: 37–49.

Evatt ML, Terry PD, Ziegler TR, Oakley GP (2010) Association between vitamin B-12-containing supplement consumption and prevalence of biochemically defined B-12 deficiency in adults in NHANES III (Third National Health and Nutrition Examination Survey). *Public Health Nutrition* 13: 25–31.

Ford ES, Schleicher RL, Mokdad AH, Ajani UA, and Liu SM (2006) Distribution of serum concentrations of alpha-tocopherol and gamma-tocopherol in the US population. *American Journal of Clinical Nutrition* 84: 375–383.

Fukui K, Omoi NO, Hayasaka T, Shinnkai T, Suzuki S, Abe K, and Urano S (2002) Cognitive impairment of rats caused by oxidative stress and aging, and its prevention by vitamin E. In: Harman D (ed.) *Increasing Healthy Life Span: Conventional Measures and Slowing the Innate Aging Process (Annals of the New York Academy of Sciences).* New York Academic Sciences, New York, pp. 275–284.

Goti D, Balazs Z, Panzenboeck U, Hrzenjak A, Reicher H, Wagner E, Zechner R, Malle E, and Sattler W (2002) Effects of lipoprotein lipase on uptake and transcytosis of low density lipoprotein (LDL) and LDL-associated α-tocopherol in a porcine in vitro blood-brain barrier model. *Journal of Biological Chemistry* 277: 28537–28544.

Harrison FE, and May JM (2009) Vitamin C function in the brain: Vital role of the ascorbate transporter SVCT2. *Free Radical Biology and Medicine* 46: 719–730.

Haskell CF, Robertson B, Jones E, Forster J, Jones R, Wilde A, Maggini S, and Kennedy DO (2010) Effects of a multi-vitamin/mineral supplement on cognitive function and fatigue during extended multi-tasking. *Human Psychopharmacology: Clinical and Experimental* 25: 448–461.

Holick MF (2007) Vitamin D deficiency. *New England Journal of Medicine* 357: 266–281.

Holick MF (2008) Vitamin D: A D-Lightful health perspective. *Nutrition Reviews* 66: S182–S194.

Jhoo JH, Kim HC, Nabeshima T, Yamada K, Shin EJ, Jhoo WK, Kim W, Kang KS, Jo SA, and Woo JI (2004) Beta-amyloid (1–42)-induced learning and memory deficits in mice: Involvement of oxidative burdens in the hippocampus and cerebral cortex. *Behavioural Brain Research* 155: 185–196.

Joseph JA, Shukitt-Hale B, Denisova NA, Prior RL, Cao G, Martin A, Taglialatela G, and Bickford PC (1998) Long-term dietary strawberry, spinach, or vitamin E supplementation retards the onset of age-related neuronal signal-transduction and cognitive behavioral deficits. *Journal of Neuroscience* 18: 8047–8055.

Kauwell GPA, Wilsky CE, Cerda JJ, Herrlinger-Garcia K, Hutson AD, Theriaque DW, Boddie A, Rampersaud GC, and Bailey LB (2000) Methylenetetrahydrofolate reductase mutation (677C → T) negatively influences plasma homocysteine response to marginal folate intake in elderly women. *Metabolism: Clinical and Experimental* 49: 1440–1443.

Kennedy DO, Veasey R, Watson A, Dodd F, Jones E, Maggini S, and Haskell CF (2010) Effects of high-dose B vitamin complex with vitamin C and minerals on subjective mood and performance in healthy males. *Psychopharmacology* 211: 55–68.

Kennedy DO, Veasey R, Watson A, Dodd F, Jones E, Tiplady B, and Haskell CF (2011) Vitamins and psychological functioning: A mobile phone assessment of the effects of a B vitamin complex, vitamin C and minerals on cognitive performance and subjective mood and energy. *Human Psychopharmacology: Clinical and Experimental* 26: 338–347.

Krezel W, Ghyselinck N, Samad TA, Dupe V, Kastner P, Borrelli E, and Chambon P (1998) Impaired locomotion and dopamine signaling in retinoid receptor mutant mice. *Science* 279: 863–867.

Krieg EF and Butler MA (2009) Blood lead, serum homocysteine, and neurobehavioral test performance in the third National Health and Nutrition Examination Survey. *Neurotoxicology* 30: 281–289.

Lane MA and Bailey SJ (2005) Role of retinoid signalling in the adult brain. *Progress in Neurobiology* 75: 275–293.

Lee DM, Tajar A, Ulubaev A, Pendleton N, O'Neill TW, O'Connor DB, Bartfai G et al. (2009) Association between 25-hydroxyvitamin D levels and cognitive performance in middle-aged and older European men. *Journal of Neurology, Neurosurgery and Psychiatry* 80: 722–729.

Levine M, Conry-Cantilena C, Wang Y, Welch RW, Washko PW, Dhariwal KR, Park JB et al. (1996) Vitamin C pharmacokinetics in healthy volunteers: Evidence for a recommended dietary allowance. *Proceedings of the National Academy of Sciences of the United States of America* 93: 3704–3709.

Li KR, Kaaks R, Linseisen J, and Rohrmann S (2010) Consistency of vitamin and/or mineral supplement use and demographic, lifestyle and health-status predictors: Findings from the European Prospective Investigation into Cancer and Nutrition (EPIC)-Heidelberg cohort. *British Journal of Nutrition* 104: 1058–1064.

Luo T, Wagner E, and Dräger UC (2009) Integrating retinoic acid signaling with brain function. *Developmental Psychology* 45: 139–150.

Mardones P and Rigotti A (2004) Cellular mechanisms of vitamin E uptake: Relevance in [alpha]-tocopherol metabolism and potential implications for disease. *Journal of Nutritional Biochemistry* 15: 252–260.

Mattson MP and Shea TB (2003) Folate and homocysteine metabolism in neural plasticity and neurodegenerative disorders. *Trends in Neurosciences* 26: 137–146.

McCaffery P, Zhang J, and Crandall JE (2006) Retinoic acid signaling and function in the adult hippocampus. *Journal of Neurobiology* 66: 780–791.

McCann JC and Ames BN (2008) Is there convincing biological or behavioral evidence linking vitamin D deficiency to brain dysfunction? *FASEB Journal* 22: 982–1001.

McGrath J, Scragg R, Chant D, Eyles D, Burne T, and Obradovic D (2007) No association between serum 25-hydroxyvitamin D 3 level and performance on psychometric tests in NHANES III. *Neuroepidemiology* 29: 49–54.

Mefford IN, Oke AF, and Adams RN (1981) Regional distribution of ascorbate in human-brain. *Brain Research* 212: 223–226.

Mey J and Mccaffery P (2004) Retinoic acid signaling in the nervous system of adult vertebrates. *Neuroscientist* 10: 409–421.

Milton K (2000) Back to basics: Why foods of wild primates have relevance for modern human health. *Nutrition* 16: 480–483.

Mishra GD, McNaughton SA, O'Connell MA, Prynne CJ, and Kuh D (2009) Intake of B vitamins in childhood and adult life in relation to psychological distress among women in a British birth cohort. *Public Health Nutrition* 12: 166–174.

Misner DL, Jacobs S, Shimizu Y, de Urquiza AM, Solomin L, Perlmann T, De Luca LM, Stevens CF, and Evans RM (2001) Vitamin A deprivation results in reversible loss of hippocampal long-term synaptic plasticity. *Proceedings of the National Academy of Sciences of the United States of America* 98: 11714–11719.

Morris MS, Fava M, Jacques PF, Selhub J, and Rosenberg IH (2003) Depression and folate status in the US population. *Psychotherapy and Psychosomatics* 72: 80–87.

Mun GH, Kim MJ, Lee JH, Kim HJ, Chung YH, Chung YB, Kang JS et al. (2006) Immunohistochemical study of the distribution of sodium-dependent vitamin C transporters in adult rat brain. *Journal of Neuroscience Research* 83: 919–928.

Niederreither K and Dolle P (2008) Retinoic acid in development: Towards an integrated view. *Nature Reviews. Genetics* 9: 541–553.

Olson CR and Mello CV (2010) Significance of vitamin A to brain function, behaviour and learning. *Molecular Nutrition & Food Research* 54: 489–495.

Padayatty SJ, Katz A, Wang YH, Eck P, Kwon O, Lee JH, Chen SL, Corpe C, Dutta A, Dutta SK, and Levine M (2003) Vitamin C as an antioxidant: Evaluation of its role in disease prevention. *Journal of the American College of Nutrition* 22: 18–35.

Pan A, Lu L, Franco OH, Yu Z, Li H, and Lin X (2009) Association between depressive symptoms and 25-hydroxyvitamin D in middle-aged and elderly Chinese. *Journal of Affective Disorders*, 118: 240–243.

Pauling L (1970) Evolution and need for ascorbic acid. *Proceedings of the National Academy of Sciences of the United States of America* 67: 1643–1648.

Payette H and Shatenstein B (2005) Determinants of healthy eating in community-dwelling elderly people. *Canadian Journal of Public Health-Revue Canadienne De Sante Publique* 96: S27–S31.

Pfeiffer CM, Caudill SP, Gunter EW, Osterloh J, and Sampson EJ (2005) Biochemical indicators of B vitamin status in the US population after folic acid fortification: Results from the National Health and Nutrition Examination Survey 1999–2000. *American Journal of Clinical Nutrition* 82: 442–450.

Radimer K, Bindewald B, Hughes J, Ervin B, Swanson C, and Picciano MF (2004) Dietary supplement use by US adults: Data from the National Health and Nutrition Examination Survey, 1999–2000. *American Journal of Epidemiology* 160: 339–349.

Reynolds E (2006) Vitamin B12, folic acid, and the nervous system. *Lancet Neurology* 5: 949–960.

Ricciarelli R, Argellati F, Pronzato MA, and Domenicotti C (2007) Vitamin E and neurodegenerative diseases. *Molecular Aspects of Medicine* 28: 591–606.

Rock C (2007) Multivitamin-multimineral supplements: Who uses them? *American Journal of Clinical Nutrition* 85: 277S–279S.

Ruston D, Hoare J, Henderson L, Gregory J, Bates C, Prentice A, Birch M, Swan G, and Farron M (2004) Nutritional Status (anthropometry and blood analytes), blood pressure and physical activity. In: *National Diet and Nutrition Survey: Adults Aged 19–64 Years*, Vol. 4. TSO, London, U.K.

Schafer (2005) Homocysteine and cognitive function in a population-based study of older adults. *Journal of the American Geriatrics Society* 53: 381–388.

Schlebusch L, Bosch B, Polglase G, Kleinschmidt I, Pillay BJ, and Cassimjee MH. (2000) A double-blind, placebo-controlled, double-centre study of the effects of an oral multi-vitamin-mineral combination on stress. *South African Medical Journal* 90: 1216–1223.

Schleicher RL, Carroll MD, Ford ES, and Lacher DA (2009) Serum vitamin C and the preva-lence of vitamin C deficiency in the United States: 2003–2004 National Health and Nutrition Examination Survey (NHANES). *American Journal of Clinical Nutrition* 90: 1252–1263.

Shatenstein B, Kergoat M-J, and Reid I (2007) Poor nutrient intakes during 1-year follow-up with community-dwelling older adults with early-stage alzheimer dementia compared to cognitively intact matched controls. *Journal of the American Dietetic Association* 107: 2091–2099.

Shibata K, Fukuwatari T, Ohta M, Okamoto H, Watanabe T, Fukui T, Nishimuta M et al. (2005) Values of water-soluble vitamins in blood and urine of Japanese young men and women consuming a semi-purified diet based on the Japanese dietary reference intakes. *Journal of Nutritional Science and Vitaminology* 51: 319–328.

Shibata K, Fukuwatari T, Watanabe T, and Nishimuta M (2009) Intra- and inter-individual variations of blood and urinary water-soluble vitamins in Japanese young adults con-suming a semi-purified diet for 7 Days. *Journal of Nutritional Science and Vitaminology* 55: 459–470.

Teunissen CE, Blom AH, Van Boxtel MP, Bosma H, de Bruijn C, Jolles J, Wauters BA, Steinbusch HW, and de Vente J (2003) Homocysteine: A marker for cognitive per-formance? A longitudinal follow-up study. *Journal of Nutrition Health and Aging* 7: 153–159.

Timbo BB, Ross MP, McCarthy PV, and Lin C-TJ (2006) Dietary supplements in a National Survey: Prevalence of use and reports of adverse events. *Journal of the American Dietetic Association* 106: 1966–1974.

Tucker KL, Qiao N, Scott T, Rosenberg I, and Spiro A, III (2005) High homocysteine and low B vitamins predict cognitive decline in aging men: The veterans affairs normative aging study. *American Journal of Clinical Nutrition* 82: 627–635

Valdenaire O, Maus-Moatti M, Vincent JD, Mallet J, and Vernier P (1998) Retinoic acid regu-lates the developmental expression of dopamine D2 receptor in rat striatal primary cul-tures. *Journal of Neurochemistry* 71: 929–936.

Valdenaire O, Vernier P, Maus M, Dumas Milne Edwards JB, Mallet J (1994) Transcription of the rat dopamine-D2-receptor gene from two promoters. *European Journal of Biochemistry* 220: 577–584.

Wietrzych M, Meziane H, Sutter A, Ghyselinck N, Chapman PF, Chambon P, and Krezel W (2005) Working memory deficits in retinoid X receptor Î³-deficient mice. *Learning & Memory* 12: 318–326.

Wolters M, Strohle A, and Hahn A (2004) Age-associated changes in the metabolism of vitamin B-12 and folic acid: Prevalence, etiopathogenesis and pathophysiological con-sequences. *Zeitschrift fur Gerontologie und Geriatrie* 37: 109–135.

Wright CB, Lee HS, Paik MC, Stabler SP, Allen RH, and Sacco RL (2004) Total homocysteine and cognition in a tri-ethnic cohort. The Northern Manhattan Study. *Neurology* 63: 254–260.

Yang T-T, Wang S-J (2008) Facilitatory effect of glutamate exocytosis from rat cerebrocortical nerve terminals by [alpha]-tocopherol, a major vitamin E component. *Neurochemistry International* 52: 979–989.

Young VR (1996) Evidence for a recommended dietary allowance for vitamin C from pharma-cokinetics: A comment and analysis. *Proceedings of the National Academy of Sciences of the United States of America* 93: 14344–14348.

Part III

Essential Fatty Acids and Neurocognition

8 Omega-3 Polyunsaturated Fatty Acids and Behavior

Philippa Jackson

CONTENTS

ESSENTIAL FATTY ACID NOMENCLATURE, STRUCTURE, AND METABOLISM

Humans typically consume about 20 different types of fatty acids in the diet, which can be grouped as either saturated or unsaturated fatty acids. Saturated fatty acids have single bonds between the carbon atoms and are rigid in nature. Unsaturated fatty acids may have one (monounsaturated) or more (polyunsaturated) double bonds and the position of the first double bond in relation to the omega end determines whether a polyunsaturated fatty acid is termed an omega-3 (n-3) or an omega-6 (n-6) fatty acid. Mammals are capable of manufacturing every fatty acid required for biological processes except for two; namely linoleic acid (LA, n-6) and α-linolenic acid (ALA, n-3). These are termed the "essential" fatty acids and must be acquired via the diet

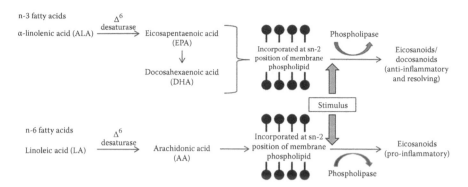

FIGURE 8.1 Metabolic pathway of n-3 and n-6 PUFAs. (Adapted from Sanders, T. and Emery, P., *Molecular Basis of Human Nutrition*, Taylor & Francis Group, London, U.K., 2003.)

(Simopoulos 2000). LA and ALA are sometimes referred to as "parent" fatty acids as it is from these that their respective long-chain biologically active metabolites are derived. Arachidonic acid (AA, n-6) is the major metabolite of LA, whereas eicosapentaenoic acid (EPA, n-3) and docosahexaenoic acid (DHA, n-3) are the major metabolites of ALA (Figure 8.1). AA, EPA, and DHA are synthesized from their respective precursor parent fatty acids by a series of elongations and desaturations that, despite the fact that the conversion pathways for n-6 and n-3 fatty acids are entirely independent, require the same enzymes at each step. There is also some evidence to suggest that DHA can be "retro-converted" into EPA, although rates of only 20% have been observed (Gronn et al. 1991). The metabolism of LA and ALA is predominantly carried out in the endoplasmic reticulum of the liver, in certain structures in the central nervous system such as glial cells (Moore 2001) and the choroid plexus vasculature (Bourre et al. 1997), and has also been observed at low rates in the placenta (Haggarty 2004).

The process of conversion of ALA to DHA is extremely inefficient in humans with most studies reporting rates of less than 0.05% (Burdge et al. 2003). Although a considerable amount of variability exists between individuals, it rarely exceeds 9% in women and 4% in men (Burdge and Calder 2005), and is hindered further by the fact that between 15% and 35% of the ALA provided by dietary sources is immediately converted to carbon dioxide for energy (Burdge et al. 2002). In addition, not only is this process limited in efficiency but as previously mentioned, the n-6 and n-3 biosynthetic pathways compete for enzymes at every step; Δ^6 desaturase, in particular, has a preference to convert ALA to DHA but high dietary intake of LA has been shown to reduce this conversion by 40%–50%, resulting in a preferential shift toward the metabolism of LA to AA (Gerster 1998). So despite the fact that ALA can be used as a source of EPA and DHA, it is more efficient if they are supplied directly via dietary sources.

DIETARY SOURCES, CONSUMPTION, AND CELLULAR INCORPORATION OF n-3 PUFAs

ALA is highly concentrated in selected seed oils such as linseed and canola and also in the chloroplasts of green leafy vegetables. The DHA and EPA that are prevalent in some species of fish are actually produced by the algae they consume, which become incorporated into their flesh, and hence DHA and EPA are sometimes referred to as the "fish oils." Fatty fish that are enriched with DHA and EPA include salmon, mackerel, sardines, trout, and fresh tuna. In lesser quantities, DHA and EPA can also be found in chicken and their eggs and other livestock if they have been fed a diet enriched with n-3 PUFAs.

The consumption of n-3 PUFAs has been falling gradually over the past 100–150 years; the typical "Western" diet of today is characterized by a marked decrease in overall fish consumption and increased intake of n-6 PUFAs that are abundant in cooking oils and processed foods (Simopoulos 2008). There is evidence to suggest that humans evolved on a diet where n-6 and n-3 PUFAs were consumed in approximately equal amounts (1–4:1) (Simopoulos 1991), whereas the consumption ratio of n-6 and n-3 PUFAs in the current Western diet is estimated anywhere between 10: and as much as 25:1 (Simopoulos 2000). There is also mounting evidence to suggest that decreased dietary intake of n-3 PUFAs, DHA and EPA in particular, is a risk factor for a plethora of different diseases including cardiovascular disease (Mori and Woodman 2006), inflammatory disease (De Caterina and Basta 2001), and many neurodevelopmental and psychiatric conditions such as attention-deficit hyperactivity disorder (ADHD), dyslexia, depression, schizophrenia, and dementia (Bourre 2005). It follows that for these two n-3 PUFAs to be implicated in such a range of seemingly unrelated conditions, they are likely to influence fundamental processes common to most cells.

Indeed, once consumed (or metabolized) DHA and EPA are incorporated at the sn-2 position of cellular membrane phospholipids in every type of tissue, where they compete for incorporation at the same position with AA (Calder 2006a). Under certain conditions, DHA and EPA (and AA) are released from the cell membrane by the action of several phospholipases (Farooqui et al. 1997), where they are metabolized further to form potent secondary signaling molecules classed as either eicosanoids (from EPA) or docosanoids (from DHA—Figure 8.1) (Tassoni et al. 2008). The dietary intake of n-3 PUFAs is, therefore, reflected in the composition of all cell membranes, which can impact a number of varied cellular processes, described in the following.

FUNCTIONS OF n-3 PUFAs

Communication between neurons relies on the exchange of ions across the cellular membrane, with maximum efficiency occurring at an "optimal" value where the physical state of the membrane is neither too rigid nor too fluid (Yehuda et al. 1999). The structure of the cell membrane varies greatly, depending on the fatty acids that make up the hydrophobic "tail" of the phospholipids. For example, rigid saturated fatty acids allow phospholipids to pack tightly together, whereas the insertion of double bonds along the hydrocarbon chain alters the properties of the fatty acid. Therefore, as the degree of unsaturation increases, the chain becomes more flexible and starts to "kink." DHA, which has six double bonds and is preferentially incorporated at the sn-2 position of the phospholipids phosphatidylethanolamine and phosphatidylserine, in particular, can adopt countless looped and helical conformations and, thus, tight packing of these DHA-rich phospholipids is prevented, consequently increasing the fluidity of the membrane (Feller et al. 2002). EPA, possessing five double bonds can also adopt multiple conformations, but the extra double bond present in DHA renders this fatty acid unique and highly specialized, as evidenced by its high density in selected tissues (Stillwell and Wassall 2003). More specifically, DHA is heavily concentrated in the cerebral frontal cortex of mammals and comprises anywhere between 10% and 20% of total fatty acids of the brain (McNamara and Carlson 2006) and represents around 30%–40% of the PUFAs found in the retinal rod outer segment (Makrides et al. 1994). Modulation of membrane fluidity in these tissues occurs with dietary manipulation of n-3 PUFAs (Connor et al. 1990; Anderson et al. 2005), and variations in concentrations of n-3 PUFAS in the cell membrane have been shown to impact a number of different cellular processes, all of which have the potential to impact upon brain function and hence behavior. For example, both DHA and EPA have been shown to affect the activities of membrane bound enzymes (e.g., Slater et al. 1995; Turner et al. 2003), ion channels (e.g., Kang and Leaf 1996; Xiao et al. 1997; Seebungkert and Lynch 2002), and gene expression (e.g., Kitajka et al. 2002; Barcelo-Coblijn et al. 2003), which can in turn influence signal transduction and neuronal transmission. In addition, levels of dopamine (Zimmer et al. 2000a), serotonin (de la Presa Owens and Innis 1999), and acetylcholine (Aid et al. 2003) have been observed to either increase or decrease following either an n-3-enriched or n-3-deficient diet. Further to this, DHA in particular has been shown to have a number of neuroprotective properties. These include preventing apoptosis when DHA is metabolized into phosphatidylserine (Kim et al. 2000) and reducing oxidative stress (Mori et al. 2000). In addition, the docosanoid derivatives of DHA, described later, have also been shown to be neuroprotective.

Cell membrane incorporation of DHA and EPA also has an effect on the production of two classes of secondary signaling molecules, namely eicosanoids or docosanoids. These molecules are powerful biological compounds responsible for mediating many aspects of the inflammatory response (Calder 2006a). Eicosanoids—categorized further as either leukotrienes, thromboxanes, or prostaglandins—can also be derived from AA upon its release from the cell membrane (see Figure 8.1), and tend to be more potent and pro-inflammatory than those originating from EPA (Schmitz and Ecker 2008). However, higher intake of dietary EPA leads to increased incorporation of these molecules into membrane phospholipids in a dose response manner and at the expense of membrane incorporation of AA (Calder 2007). Consequently, there is a shift away from production of pro-inflammatory, vaso-constricting, and platelet-aggregating AA-derived eicosanoids, and an increase in the production of anti-inflammatory EPA-derived ones (Gibney and Hunter 1993).

Like eicosanoids, docosanoids are chemical signaling molecules, produced via controlled oxidative degeneration of DHA within or adjacent to the cell membrane (Kidd 2007). Three classes of docosanoids have been identified—docosatrienes, resolvins, and protectins—and have been shown to have neuroprotective qualities. The novel neuroprotectin D1 (NPD1) has been shown to attenuate apoptosis in the presence of oxidative stress and provides protection to neuronal cells in animal models of brain ischemia and neurodegeneration (reviewed in Bazan 2006). More specifically, in Alzheimer's disease (AD) rat models NPD1 repressed the expression of pro-inflammatory β-amyloid-activated genes. Moreover, the recently discovered E-series and D-series resolvins, derived from EPA and DHA, respectively, have also been identified as having anti-inflammatory properties that are not related to altering lipid mediator profiles (i.e., inhibited production of AA-derived eicosanoids), but by inhibiting the expression of pro-inflammatory cytokine genes such as nuclear factor κ B and/or peroxisome proliferator–activated receptor (Calder 2006b). Taken together, the modulation of eicosanoid and docosanoid production is one potential mechanism by which dietary DHA and EPA could prevent the occurrence or ameliorate the symptoms of inflammatory diseases linked to n-3 PUFA intake, including depression (Das 2007), ADHD (Richardson 2006), schizophrenia (Yao and van Kammen 2004), AD (Pratico and Trojanowski 2000), atherosclerosis (von Schacky 2000), rheumatoid arthritis (Kremer 2000), inflammatory bowel disease (De Caterina et al. 2000), and possibly some bronchial diseases such as asthma (Belluzzi et al. 2000).

A final function of n-3 PUFAs relates to their effects on various aspects of cardiovascular function. Given that cerebrovascular events are a risk factor for neurode generation, along with the fact that the cardiovascular system is responsible for the delivery of nutrients to the brain, it follows that any compound that modulates cardiovascular parameters could exert a secondary effect on brain function and behavior. Indeed, a number of different cardiovascular parameters have been shown to be modified by dietary n-3 PUFAs including increased arrhythmic threshold via modulation of sodium and calcium ion channels (Kang and Leaf 1996), decreased platelet aggregation (Mori et al. 1997), lowered triglycerides (Nestel 2000), lowered blood pressure (Morris et al. 1993; Geleijnse et al. 2002), and improved arterial and endothelial function via increased nitric oxide synthesis (Harris et al. 1997; Armah et al. 2008).

In summary, DHA and EPA are involved in a number of varied fundamental func tions at the cellular level. In the brain, DHA is heavily enriched in the cerebral cortex where its incorporation into the phospholipid bilayer of neural cell membranes confers optimal membrane fluidity, resulting in improved membrane function as regards signal transduction and neurotransmission. Furthermore, there is evidence to suggest that the expression of a number of genes and the production of various neurotransmitters is sensitive to dietary intake of n-3 PUFAs, suggesting a role for n-3 PUFAs in these processes. In addition, the DHA and EPA incorporated into cell membranes throughout the body can be subsequently released and metabolized further to produce potent secondary signaling molecules that are essential in the resolution of the immune response and may also be neuroprotective. Finally, dietary n-3 PUFAs modulate a number of cardiovascular parameters, which may contribute to reduced risk of cardiovascular events. Given the fundamental nature of n-3 PUFAs and DHA and EPA in particular, it is plausible that alterations in dietary intake could potentially impact upon brain function and behavior. The following section reviews the current literature on the behavioral effects of n-3 PUFAs in animals and humans.

BEHAVIORAL EFFECTS OF n-3 PUFAs

ANIMAL EVIDENCE

Our knowledge of the impact dietary n-3 PUFAs have upon cognitive function has been greatly extended by the investigation of their effects in animals, the majority of which have been conducted using rodents. Overall, the evidence from these studies indicates that carefully controlled n-3-deficient diets lead to a decrease in levels of brain DHA, which is associated with poorer performance on a selection of learning and memory tasks such as Morris Water Maze (Moriguchi et al. 2000; Fedorova and Salem 2006), avoidance learning (Garcia-Calatayud et al. 2005), and olfactory discrimination tasks (Greiner et al. 2001). In addition, third-generation rats (87% reduction in brain DHA) have been found to perform worse than second-generation rats (83% reduction in brain DHA) (Moriguchi et al. 2000). Interestingly, in both sets of animals, performance was inversely related to levels of docosapentaenoic acid (DPA, n-6) in the frontal cortex, suggesting that the reciprocal replacement of DHA with DPA has significant consequences.

In older rats, impairments in tasks that involve complex motor skills and spatial memory decline throughout the lifespan (Shukitt-Hale et al. 1998), which may be attributable to the observed reductions in brain lipids, have been consistently observed in aged animals (e.g., Ulmann et al. 2001). Long-term potentiation (LTP), commonly thought to be the biological process underlying learning and memory, is reduced in aged rats (Landfield et al. 1978). In addition, both AA and DHA are significantly decreased in these animals (McGahon et al. 1999). Interestingly, the ability of rat hippocampal dentate gyrus cells to sustain LTP is negatively correlated with the concentration of both AA and DHA in these cells, suggesting a link between the prevalence of long-chain PUFAs and learning and memory (McGahon et al. 1999). Eight weeks of n-3 PUFA supplementation (10 mg/day DHA) is sufficient to restore membrane DHA, which is accompanied by a reversal of the deficits in the ability to sustain LTP (McGahon et al. 1999). Other studies have shown that DHA supplementation can restore radial arm maze task performance in both n-3-deficient (Gamoh et al. 2001) and n-3-adequate (Carrie et al. 2000) aged rats. Together these investigations in aged animals suggest a theoretical basis for and observable benefit of n-3 PUFA supplementation in reducing or reversing age-related impairments.

HUMAN EVIDENCE

In humans, n-3 PUFA deficiency to the extent that is observed in animals is extremely rare and only a handful of cases have ever been reported, most commonly as the result of administration of total parenteral nutrition (feeding exclusively via intravenous drip) containing very little or no ALA. Rough, dry skin and hair, excessive thirst and abnormal vision are common features of this type of deficiency; symptoms can be reversed once ALA is reintroduced to the diet (Holman et al. 1982). n-3 PUFA status can be determined in humans by measuring the concentrations of ALA, DHA, and EPA in peripheral tissues such as serum/plasma or erythrocytes. By comparing the n-3 status of healthy normal volunteers to those of various patient groups, it has been revealed that individuals diagnosed with several neurodevelopmental disorders such as ADHD and autism (Bell et al. 2000; Burgess et al. 2000; Schuchardt et al. 2009), along with a number of psychiatric conditions including depression (Edwards et al. 1998), schizophrenia (Assies et al. 2001), and AD and dementia (Conquer et al. 2000), have significantly lower levels of n-3 PUFAs. Collectively, these findings again suggest that adequate intake and incorporation of n-3 PUFAs is a requirement for normal functioning. The results from studies that have used n-3 supplementation as treatment for symptoms of these conditions have been mixed, however, and further investigation is required. In the next section the role of n-3 PUFAs in a number of neuropsychiatric and developmental conditions is outlined, along with an evaluation of the current evidence of their use in the treatment of these conditions. The section will end with a review of the current knowledge of the effects of n-3 PUFA supplementation on behavioral outcomes in healthy individuals.

ROLE OF n-3 PUFAs IN NEUROPSYCHIATRIC CONDITIONS

DEPRESSION

There is growing evidence to support a link between depressed mood and dietary n-3 PUFAs; however, intervention studies that have utilized n-3 PUFAs as either adjunctive treatment or monotherapy have been met with mixed success. It has been shown that fish consumption is inversely related to prevalence of major depression across different countries worldwide (Hibbeln 1998); in countries such as Japan, Norway, and Iceland, where intake of fish and dietary n-3 PUFA are high (0.24%–0.44% energy, as opposed to 0.10% [the United Kingdom, the United States] or 0.08% [Germany]), major, bipolar, and postpartum depression are less prevalent than in countries such as the United States and the United Kingdom, where depression is one of the leading causes of disability (Hibbeln et al. 2006). Two other population-based cross-sectional studies in Finland (Tanskanen et al. 2001) and New Zealand (Silvers and Scott 2002), along with two epidemiological studies of older adults aged >65 years (Mamalakis et al. 2006; Bountziouka et al. 2009) also support these findings. In addition, lower levels of adipose DHA have been reported in mildly depressed healthy adults compared to controls (Tanskanen et al. 2001), and adipose n-3 PUFAs are also negatively related to depression in elderly individuals (Mamalakis et al. 2002) and in adolescents (Mamalakis et al. 2004b). The same is true of erythrocyte membrane levels, where total n-3 PUFAs and DHA are depleted in depressed patients taking medication compared to controls (Edwards et al. 1998; Peet et al. 1998). Further, the DHA content of the orbitofrontal cortex in patients with major depressive disorder was found to be significantly lower (by 22%) than matched controls (McNamara et al. 2007).

In the face of the mounting epidemiological and peripheral tissue evidence implicating n-3 PUFAs in the pathophysiology of depression, intervention trials using n-3 have produced mixed results (e.g., Edwards et al. 1998; Peet et al. 1998; Tanskanen et al. 2001; Mamalakis et al. 2004a, 2006), although the populations studied, length of treatment regimen (4–12 weeks), and dose (1–9.6 g/day DHA + EPA) and formulation (ratio of DHA:EPA) of treatment have varied widely between studies. Similarly, two meta-analyses of the extant literature published around the same time are not in agreement about the efficacy of n-3 PUFAs in the treatment of depressive disorders (Appleton et al. 2006; Freeman et al. 2006b). Among the possible reasons for this discrepancy are the choice of analysis and the inclusion of large (nonsignificant) studies that are diverse in both the type of treatment used and the population studied (Richardson 2008). The largest intervention trial to date, however, examined the effects of 12 weeks' dietary supplementation with 1.5 g DHA + EPA in 190 mild to moderately depressed volunteers, but found no evidence to support the use of n-3 PUFAs in the treatment of depressive symptoms in this population (Rogers et al. 2008), raising the possibility that the benefit of taking n-3 PUFAs for is negligible, at least in mild to moderately depressed individuals. Only further large randomized controlled trials (RCTs) will be able to resolve this issue.

Postpartum Depression

About 10%–20% of postpartum women are diagnosed with postpartum depression (PPD). As maternal stores of fatty acids are depleted during pregnancy to ensure an adequate supply for central nervous system development of the growing neonate, some researchers have explored the hypothesis that without sufficient dietary intake of fatty acids, mothers may increase their risk of suffering from PPD (Holman et al. 1991). In rats, it has been observed that an inadequate supply of dietary DHA is enough to result in a 21% decrease in brain DHA in just one reproductive cycle (Levant et al. 2006), but the extent and possible consequences of depletion in humans has yet to be established. In a cross-national study, Hibbeln (2002) discovered that seafood intake and levels of DHA in breast milk were inversely associated with depressive symptoms as measured by the Edinburgh Postnatal Depression Scale (EPDS) in 22 countries world-wide, but another study of 80 new mothers found no relationship between postnatal n-3 fatty acid status and postnatal depression (Browne et al. 2006). In addition, the results from the few intervention trials that have been conducted in this population generally do not support n-3 PUFAs as a treatment of PPD, although large RCTs are still required. In a small open-label trial, supplementation of 2.96 g/day DHA and EPA starting at between 34 and 36 weeks' gestation did not prevent PPD in four out of seven participants (Marangell et al. 2004). Freeman and colleagues have conducted two intervention trials in women who have been diagnosed with depression following birth. The first of these studies was an open-label pilot trial where participants ($N = 15$) received approximately 1.9 g/day EPA + DHA for 8 weeks (Freeman et al. 2006a). Authors reported a 40.9% decrease in depressive symptoms on the EPDS but in a second randomized dose-ranging study where treatments ranged from 0.5 to 2.8 g/day as adjunctive treatment to supportive psychotherapy, the authors found no difference between groups, with all groups reporting reduced scores on the EPDS and Hamilton Depression Rating Scale (Freeman et al. 2008). It is possible that the association between maternal intake of n-3 PUFAs and PPD has been overestimated; results from the Danish National Birth Cohort, a large prospective study, reveal little evidence to support a link between maternal fish and n-3 PUFA intake and rates of PPD (Strom et al. 2009). A review of the extant evidence in this area concluded that the results are not conclusive overall, but do warrant further investigation (Borja-Hart and Marino 2010).

Other Neuropsychiatric Conditions

Dietary n-3 PUFAs have also been implicated in other neuropsychiatric conditions such as bipolar disorder (BD) and schizophrenia. The similarities between the effects of mood stabilizers such as lithium and valproate—commonly used in the treatment of BD—and DHA and EPA, on the enzyme protein kinase C (PKC) have led researchers to consider n-3 PUFAs as an alternative to standard pharmacological treatment for BD. Further, epidemiological studies have revealed an inverse relationship between seafood consumption and lifetime prevalence rates of BD (Noaghiul and Hibbeln 2003). However, evidence from intervention trials is inconclusive, with some published trials reporting a benefit of n-3 PUFAs (Stoll et al. 1999; Osher et al. 2005; Sagduyu et al. 2005; Frangou et al. 2006), while others do not (Marangell et al. 2003; Keck et al. 2006). A systematic review of the extant literature in this area concluded that although n-3 PUFAs are well tolerated by patients with BD and the evidence seems to show an association between n-3 use and symptom reduction, further studies are required in order to confirm their efficacy in the treatment of BD (Turnbull et al. 2008).

A similar pattern of findings is observed in schizophrenia. The popular "dopamine hypothesis" of schizophrenia proposes that negative symptoms (flat affect) result from reduced activity of the dopamine systems in the prefrontal area, and positive symptoms (delusions and thought disorder) from increased activity of the dopamine systems in the limbic system (Davis et al. 1991). This theory can explain the relationship between dopamine kinetics and the psychiatric symptoms of schizophrenia, but fails to address the cause of the abnormal activities of dopaminergic neurons (Ohara 2007). Zimmer and colleagues discovered that rats who had been fed an n-3-deficient diet suffered a reduction in the number of presynaptic dopamine vesicles and also that basal dopamine metabolism was increased (Zimmer et al. 2000a,b). Dietary n-3 deficiency has also been shown to reduce the number of D2-receptors in the frontal lobe in both rats (Delion et al. 1994) and piglets (de la Presa Owens and Innis 1999). It has also been observed that compared to controls, schizophrenia patients have lower levels of plasma n-3 PUFAs (Assies et al. 2001). Therefore, in an attempt to integrate all of the evidence, Ohara (2007) proposed that the n-3 PUFA abnormalities found in schizophrenia stem from the dysfunction of the enzyme phospholipase A_2 (PLA$_2$). It follows that increased activation of PLA$_2$ observed in patients suffering from schizophrenia may cause the excessive depletion of PUFA from the sn-2 position of cell membrane phospholipids in the body and brain. Dopamine concentration, the number of dopamine vesicles, and the number of D2 receptors are decreased in the prefrontal presynaptic terminals (resulting in the negative symptoms) and these decreases have a knock-on effect for the limbic dopamine system (resulting in the positive symptoms) (Ohara 2007).

Despite the apparent plausibility of this integrated theory, a Cochrane review of PUFA supplementation in schizophrenia concluded that data from the six trials that met the inclusion criteria were inconclusive, and the value of treating schizophrenia with PUFA remains unfounded (Joy et al. 2006). This conclusion was formed largely on the basis that of the six trials, only one enrolled more than 100 participants (Peet and Horrobin 2002) and in only one study did the intervention period exceed 3 months (Fenton et al. 2001). Neither of these studies produced compelling evidence to support the use of n-3 in the treatment of schizophrenia. Only large, longitudinal RCTs will be able to provide sufficient evidence as to whether n-3 PUFAs have a clinically significant and positive impact in the treatment of this illness.

AGE-RELATED COGNITIVE DECLINE AND DEMENTIA

Cognitive function naturally declines with age and has been attributed to a number of factors including reduced synaptic plasticity, decreased membrane fluidity, and increased oxidative damage (Willis et al. 2008). There is growing evidence, however, that various lifestyle factors can either promote or attenuate cognitive aging. These include smoking (Swan and Lessov-Schlaggar 2007), alcohol consumption (Peters et al. 2008), exercise (Colcombe et al. 2003), and diet (Del Parigi et al. 2006; Barberger-Gateau et al. 2007). In particular, one of the dietary factors that have been explored in detail is intake of fatty acids. For example, the Dutch prospective population-based Zutphen Elderly Study identified that LA was positively associated with cognitive decline over a 3 year period (defined as a >2 point drop in Mini Mental State Examination) in 476 men aged 69–89 years (Kalmijn et al. 1997). A recent reanalysis of the same data was able to identify that in this sample of elderly men, those who did not eat fish observed a 1.2 point decline in MMSE score at the 5 years follow-up, as opposed to only a 0.3 point decline in men who reported eating fish (van Gelder et al. 2007). Additionally, a cross-sectional study by the same group identified that oily fish consumption (measured using a FFQ) was significantly associated with a reduced risk of global cognitive function impairment and psychomotor speed in participants of 45–70 years, independent of other confounding factors (e.g., age, sex, education, smoking, alcohol consumption, energy intake) (Kalmijn et al. 2004). Findings from the Chicago Health and Aging Project (CHAP), conducted in 2560 participants aged 65 years and older over a period of 6 years, also discovered that fish intake was associated with a slower rate of cognitive decline at the 6 years follow-up. More specifically, among those who consumed one fish meal per week, decline was 10% slower than those who consumed fish less than weekly and 13% slower for those who consumed two or more fish meals per week, adjusted for age, sex, race, education, cognitive activity, physical activity, alcohol consumption, and total energy intake. What the authors could not conclude is whether it was n-3 PUFAs that were the relevant dietary constituent in fish accountable for this finding (Morris et al. 2005). The prospective population-based Etude du Vieillissement Ateriel (EVA) study evaluated fatty acids in erythrocyte membranes and performance on the MMSE in a sample of 246 63–74 year olds (Heude et al. 2003). These authors found that higher proportions of stearic acid (a saturated fatty acid) and total n-6 PUFAs (LA, AA, γ-linolenic acid (GLA), DPAn-6) were associated with greater risk of cognitive decline and that a higher proportion of total n-3 PUFAs (ALA, DHA, EPA, DPAn-3) was associated with a lower risk of cognitive decline over a 4 year period. Similarly, intake of EPA and DHA (estimated via a food frequency questionnaire) was inversely associated with cognitive impairment (MMSE). Finally, higher plasma n-3 PUFA proportions in a sample of 807 healthy participants aged 50–70 years predicted less decline in sensorimotor speed and complex speed over a 3 year period, although there were no associations between n-3 PUFA proportions and memory, information processing speed or word fluency, and no significant associations were detected at baseline between n-3 status and performance in any of the five assessed cognitive domains (Dullemeijer et al. 2007).

It is only recently that data from large-scale prospective randomized intervention trials evaluating the effects of n-3 PUFAs on cognitive function in older adults have been available; however, results from these trials have been conflicting. The OPAL (Older People And n-3 Long-chain polyunsaturated fatty acids) study assessed the effects of a daily fish oil supplement containing 0.5 g DHA and 0.2 g EPA on cognitive performance on the California Verbal Learning test and other measures of memory and attention in 867 men and women aged 70–79 years (at baseline), but did not find any significant effects of the treatment. Similarly, the 26 weeks intervention trial in 302 healthy older adults reported by van de Rest et al. (2008) also did not find any effects of either a high (1.8 g EPA + DHA) or lower dose (0.4 g EPA + DHA) compared to placebo on a range of cognitive assessments. On the other hand, the memory improvement with docosahexaenoic acid study (MIDAS) intervention trial found a significant effect of 24 weeks supplementation with 0.9 g DHA on 485 healthy adults (≥55 years) who were classified as having age-related cognitive decline (ARCD) on learning and episodic memory tasks, but not working memory or executive function tasks (Yurko-Mauro et al. 2010). This latter study may have potentially highlighted a subgroup of healthy older adults in which administration of n-3 PUFAs has beneficial effect. Further research would need to confirm this hypothesis.

The progression of ARCD to cognitive impairment is rising dramatically the world over and currently around 24.2 million people are affected by dementia, with 4.6 million new cases reported each year; AD accounts for about 60% of cases (Ferri et al. 2005). A number of observational studies in humans have examined the relationship between intakes of n-3 PUFAs, as measured by various food frequency questionnaires (FFQ), and diagnosis of dementia or AD, but overall the results are conflicting. Barberger-Gateau et al. (2002) found in their analysis of the PAQUID epidemiological study (*N* = 1674 aged 68 years or more) that those participants who consumed fish or seafood at least once a week were at a lower risk of developing dementia, including AD at the 7 year follow-up. However, after adjusting for education level, which was positively correlated with fish intake, the strength of the association diminished somewhat. A publication from the CHAP cohort demonstrated, after a mean follow-up of 3.9 years, that a higher intake of DHA and weekly fish consumption reduced the risk of AD, although EPA was not associated with a reduced risk (Morris et al. 2003). Conversely, results from the prospective population-based Rotterdam study (*N* = 5395) found no association between n-3 intake and risk for any type of dementia (Engelhart et al. 2002). Similarly, results from the Canadian Study of Health and Aging also do not suggest that an association between total n-3 PUFAs, DHA, or EPA and incidence of dementia or AD (Kroger et al. 2009). In addition, the results from two other large-scale studies that initially indicated an inverse association between n-3 PUFAs and incidence of AD and dementia were attenuated once sex, age, and education were adjusted for (Huang et al. 2005; Schaefer et al. 2006).

Despite these mixed reports, the biological basis for pursuing research in this area is compelling; n-3 PUFAs possess three properties by which they may protect against the development of dementia, which include increasing cerebral blood flow, attenuating inflammation, and reducing amyloid production (reviewed in Fotuhi et al. 2009). Results from animal studies are indeed encouraging; in their review of the protective effects of n-3 PUFAs in AD, Boudrault et al. (2009) conclude that treatment with

DHA in rodent models of AD consistently protects against the development of AD, with a number of observable effects in the brains of animals fed DHA compared to controls including decreased pro-apoptotic proteins and secretion of amyloid beta (Aβ) and increased activity in the PI-3 kinase cascade, a neuroprotective pathway shown to be reduced in AD. Coupled with these physiological changes are studies showing improvements in cognitive function. One group from Japan have focused particularly on this issue, and have consistently shown protective effects of n-3 PUFA administration on spatial learning ability in Aβ-infused rats (Hashimoto et al. 2002; Hashimoto et al. 2005a,b, 2008). However, it is worth noting that the quantity of n-3 PUFAs given to these animals is two to four times greater than the current intake in humans (Boudrault et al. 2009). Interestingly, in humans, levels of DHA in the brains of AD patients do not significantly differ from those that are normal, although levels of stearic acid (frontal and temporal cortex) and AA (temporal cortex) are reduced, and oleic acid is increased (frontal and temporal cortex), indicating some differences in brain fatty acid composition (Fraser et al. 2009). Compared to animal studies, intervention trials in humans, however, have not been met with the same success. A dose-ranging intervention in 302 participants aged 65 years or older with an MMSE score of >21 found no effect of either dose of fish oil containing either 400 or 1800 mg DHA + EPA on cognitive function (memory, sensorimotor speed, attention, executive function) compared with placebo following 26 weeks of dietary supplementation (van de Rest et al. 2008). Similarly, the OmegaAD clinical trial examined the effects of n-3 PUFA supplementation in 174 patients with mild to moderate AD. In this one-way crossover trial, the active treatment consisted of daily dietary supplementation with 1.6 g of DHA and 0.6 g EPA. At 6 months there was no difference between groups on either the MMSE or the AD Assessment Scale. However, in a subgroup of participants with very mild cognitive dysfunction there was a significant reduction in MMSE decline rate, and this was replicated in the crossover group at 12 months (Freund-Levi et al. 2006). These authors also suggest that in terms of the neuropsychiatric symptoms of AD, carriers of the APOε4 gene might be more susceptible to the effects of treatment with n-3 PUFAs, although this is an avenue of investigation that needs to be pursued further (Freund-Levi et al. 2007). Lim et al. (2006) conclude in their Cochrane review that there is a growing body of evidence from biological, observational, and epidemiological studies suggesting a protective effect of n-3 PUFAs against dementia. The level of this effect remains unclear, however, and, to date, dietary recommendations in relation to fish and n-3 PUFA consumption and risk of dementia cannot be made. It is hoped that the results of the DHA in Slowing the Progression of AD study, a prospective 18 months intervention trial in 400 participants aged 50 or older with mild to moderate cognitive impairment, could be used to inform the efficacy of n-3 PUFA in the prevention of dementia (Quinn 2007).

Role of n-3 PUFAs in Neurodevelopmental Disorders

Richardson and Ross (2000) were among the first researchers to link neurodevelopmental disorders such as ADHD, dyslexia, developmental coordination disorder (DCD), and autism with n-3 PUFA deficiency. These authors noted clinical commonalities between these conditions such as the preponderance of males that were affected, apparent links between allergies and other immune system disorders such as proneness to infections and atopic conditions, abnormalities of mood, arousal and sleep, as well as cognitive impairments in attention and working memory, which suggest disruptions of visual or auditory processing (Richardson 2006). It had also been observed some 25 years previously that individuals with these conditions also shared physical characteristics seen in animals specifically bred on n-3-deficient diets such as excessive thirst, frequent urination, rough, dry hair and skin, and follicular keratosis (Colquhoun and Bunday 1981). Indeed, several studies in children with ADHD have demonstrated that these children have lower blood concentrations of PUFAs, namely AA, DHA, and overall concentrations of n-3 PUFAs (Bekaroglu et al. 1996; Stevens et al. 1996; Burgess et al. 2000; Stevens et al. 2003). Given that there is no evidence to suggest that n-3 PUFA intakes are lower in children with ADHD than in healthy children (Ng et al. 2009), the low levels of n-3 PUFAs found in the blood of children with ADHD have been attributed to either inefficient conversion of ALA to EPA and DHA or enhanced metabolism of these fatty acids (Stevens et al. 1995; Burgess et al. 2000). There have been five widely cited intervention trials investigating the effectiveness of n-3 PUFA treatment on symptoms in children with ADHD and related developmental disorders. These studies have varied in design but interestingly the three experiments that report a positive effect of treatment all used a daily treatment regimen lasting 12 weeks or longer and the treatments themselves originated from fish oil, and, therefore, contained both DHA and EPA (Richardson and Puri 2002; Stevens et al. 2003; Richardson and Montgomery 2005). The study by Voigt et al. (2001) found no effect of 345 mg/day DHA for 16 weeks on a wide range of behavioral and computerized measures of ADHD-related symptoms in 54 children diagnosed with ADHD, and Hamazaki and Hirayama (2004) found no effect of treatment on behavioral symptoms of ADHD with a daily fish oil supplement for 8 weeks, suggesting the possibility that both the composition of the n-3 PUFA treatment and duration of regimen are key factors in ameliorating symptoms of ADHD and related disorders.

Similarly, a relationship between n-3 fatty acid status and autism has also been demonstrated, although intervention trials showing a pronounced benefit of treatment with n-3 PUFAs are lacking. Vancassel et al. (2001) discovered that DHA was decreased by 23% in the plasma phospholipids of autistic children and total fatty acids by 20%. In contrast, a more recent study found that in 16 high-functioning males with autism, DHA and the ratio between total n-3:n-6 PUFAs were increased in plasma phospholipids compared to 22 matched controls, and consequently the authors advised serious caution against treating this condition with n-3 PUFAs (Sliwinski et al. 2006). Despite this, Amminger et al. (2007) published results from a pilot trial wherein they administered seven diagnosed with autistic disorder 7 g/day fish oil for 6 weeks. When compared to matched controls who received a placebo treatment for the same duration, the only significant difference found between groups was on an irritability scale; no differences were found between groups on the social withdrawal, stereotypy, hyperactivity, or inappropriate speech measures. The authors are quick to note the small sample size and the relatively short duration of the trial. No adverse effects on behavior were observed.

n-3 PUFAs have also been linked to dyslexia, and to this end Richardson et al. (2000) examined the associations between the clinical signs of n-3 fatty acid deficiency (excessive thirst, frequent urination, rough, dry hair and skin, etc.) and reading ability, spelling, and auditory working memory in 97 dyslexic children. The authors detected inverse associations between signs of n-3 deficiency and reading and overall ability, and in boys alone, poorer spelling and auditory working memory. This finding was reflected in a study of dyslexic adults who filled out two self-report questionnaires; one on signs of fatty acid deficiency and another concerning signs and severity of dyslexia. The authors reported that the signs of fatty acid deficiency were significantly elevated in dyslexic participants and that this reached higher significance in males (Taylor et al. 2000). Cyhlarova et al. (2007) also examined the link between fatty acid status and literacy skills in 32 dyslexic individuals and 20 matched controls. For both groups, better word reading was associated with higher total n-3 concentrations, although it was only in dyslexic participants that a negative correlation was found between reading performance and the ratio of AA:EPA and with total n-6 concentrations, despite there being no significant differences in membrane fatty acid levels between groups, suggesting that, as in ADHD, the ratio of n-6:n-3 PUFAs or a intrinsic disruption in the metabolism of these fatty acids may be a contributing factor in the etiology of these conditions. A collection of preliminary studies reported by Stordy (2000) seems to indicate that impairments of the visual system can be improved with a high-DHA supplement in dyslexic participants, although larger RCTs have yet to be carried out investigating the full extent of the efficacy of n-3 PUFAs in the treatment of dyslexia.

n-3 PUFA SUPPLEMENTATION AND BEHAVIOR IN HEALTHY INDIVIDUALS

INFANT DEVELOPMENT

The developing fetus requires a supply of both AA and DHA for structural and metabolic functions (Haggarty 2004). The brain and retina require a high concentration of DHA to function optimally and as such, it is thought that the n-3 PUFA composition of the maternal diet can affect visual and intellectual development (Innis 1991). DHA is deposited in fetal fat stores in the last 10 weeks of pregnancy in the quantity of around 10 g. If the diet is devoid of preformed DHA in the first 2 months of life, then this store is mobilized and would be largely used up, supporting critical developmental processes (Farquharson et al. 1993). While the level of AA in breast milk has been found to remain constant at about 0.45% of total fatty acids, the level of DHA, on the other hand, varies with the mother's diet from about 0.1%–3.8% of total fatty acids. Unlike breast milk, until relatively recently both term and preterm infant formulas did not contain any n-6 or n-3 PUFAs and it was observed that formula-fed infants have significantly lower levels of DHA in plasma, erythrocytes, and brain cortex compared to breast-fed infants, and lower levels of AA in plasma and erythrocytes (Menon and Dhopeshwarkar 1983). n-3 PUFA supplemented formulas have indeed been shown to be effective in successfully raising infant's levels of AA and DHA to that of infants who have been fed human milk, within about 10%.

Carlson et al. (1996) were effective in mimicking the levels of AA and DHA in American women's milk, and when the formula contained 0.1% DHA and 0.43% AA, there were no significant differences in plasma levels of AA and DHA between the breast- and formula-fed groups of infants. Both AA and DHA have to be present in the formula, however, as supplementation with DHA alone has been shown to result in lower levels of AA between 15% and 40% (Auestad et al. 1997). By altering the levels of the longer-chain fatty acids in supplemented formulas and using unsupplemented formulas (usually containing only LA and ALA) as a reference group, any developmental effects of these manipulations can be investigated.

Carlson et al. (1996) found only a transient benefit of a supplemented formula (0.1% DHA + 0.43% AA) over an unsupplemented formula (LA:ALA = 22:2.2) on visual acuity, which was only present at 2 months but not at 4, 6, 9, and 12 months. In a study using a very similar design and levels of DHA and AA, no advantage was seen in the supplemented group at any testing point (1, 2, 4, 6, 9, and 12 months), although the disparity in results could possibly be due to a different source of fatty acids, i.e., egg phospholipids versus fish oil, respectively (Auestad et al. 2001). Makrides et al. (1995), on the other hand, found that infants fed for 4 months on a supplemented formula (0.36% DHA, 0.58% EPA, 1.52% ALA, and 0.27% γ-linolenic acid, n-6) had better transient visual evoked potentials (VEP) at 4 and 7.5 months than the standard 1.6% ALA formula, and the same as the infants fed human milk. Birch et al. (1998) also found that infants given higher levels of DHA in two separate supplemented formulas (0.35% DHA and 0.36% DHA + 0.72% AA) had similar steady-state VEP acuity at 6, 17, and 52 weeks to the infants in the human milk group, and significantly better than the VEP acuity of the standard formula group (LA:ALA = 15:1.5). These results suggest that in terms of visual development, the level of DHA in the diet has to be higher than 0.1% to have a beneficial impact.

This theme is continued as regards the effects of supplemented formulas on cognitive function. Only a handful of studies to date have found a positive impact of added n-PUFAs, and these were with DHA at the levels of 0.35% or 0.36% of total fatty acids (Birch et al. 2000; Birch et al. 2007; Drover et al. 2009). Other studies that have used supplemented formulas where the level of DHA added to the formula was around 0.1% DHA (e.g., Lucas et al. 1999; Makrides et al. 2000; Auestad et al. 2001) have failed to show any differences in cognitive or motor development between infants fed a supplemented formula over the standard one. Interestingly, the level of DHA in American mother's milk is estimated at around 0.13% DHA, whereas only higher levels of DHA in the formula have been shown to be effective at producing improvements over placebo in these studies.

It is a logical progression to investigate the developmental impact of supplementing the maternal diet with DHA and other n-3 PUFAs (in the absence of a similar n-6 PUFA shortage in the maternal diet). Indeed, the results of a large ($N = 11,875$) prospective epidemiological study, Hibbeln et al. (2007) reported that consumption of less than 340 g of seafood per week was associated with increased risk for suboptimal outcomes for prosocial behavior and fine motor, communication, and social development scores and increased risk for being the lowest quartile for verbal intelligence. Helland et al. (2003) recruited 341 women at 17–19 weeks of their pregnancy and randomly allocated them to a daily regimen of 10 mL of either corn or cod liver oil (1180 mg DHA + 803 mg EPA) until three months after delivery. Plasma levels of DHA were significantly higher in both the infants and the mothers of the cod liver arm compared to the placebo group, demonstrating that maternal dietary supplementation with n-3 PUFAs is reflected in a simultaneous increase in plasma lipid levels of the infant. Fish oil supplementation during pregnancy in this way has been shown to have a positive impact on infant development. The same authors assessed these children at 4 years using the Kaufman Assessment Battery for Children (K-ABC) as an outcome for intelligence and achievement. Infants whose mothers were in the cod liver oil treatment group scored higher on the Mental Processing Composite of the K-ABC, and in a multiple regression model, maternal intake of DHA was the only variable to significantly predict this difference in mental processing at age 4 (Helland et al. 2003); however, these differences disappeared at the 7 year follow-up (Helland et al. 2008). In another randomized double-blind trial study, children whose mothers had been given a fish oil supplement (2.2 g DHA + 1.1 g EPA; $N = 33$) had better hand–eye coordination at 2.5 years of age than those whose mothers had been given olive oil ($N = 39$) during pregnancy (Dunstan et al. 2008). There were, however, no significant differences between groups on measures of receptive language or behavior. It is worth noting that maternal supplementation with 2.82 g/day ALA from week 14 of pregnancy to 32 weeks following delivery had no impact on either the infant's DHA status as measured by plasma lipid levels or on their cognitive function compared to the control group suggesting that the infant requires preformed DHA to meet requirements (de Groot et al. 2004).

NORMALLY DEVELOPING CHILDREN AND HEALTHY ADULTS

To date, only a limited number of intervention trials have assessed the effects of dietary n-3 PUFAs on behavioral measures in healthy children and adults. Dalton et al. (2009) found beneficial effects of a daily fish flour spread (335 mg ALA + 82 mg EPA + 192 mg DHA) given to healthy children aged 7–9 years ($N = 183$) for 6 months in a single-blind placebo-controlled trial. Following the treatment, participant's plasma and erythrocyte concentrations of EPA and DHA significantly increased in the active treatment group and AA significantly decreased. The children were also better than placebo on the Hopkins Verbal Learning Recognition and Discrimination outcomes (Immediate and Delayed Word Recall, Word Recognition), as well as on a spelling test. Marginally significant results were also seen on a reading test, compared to placebo. On the other hand, three other studies in similar age groups did not find any effects of n-3 PUFA supplements (EPA + DHA or DHA in isolation) on a range of behavioral outcomes including computerized cognitive tasks, IQ, and reading and spelling (Osendarp et al. 2007; Kennedy et al. 2009; Kirby et al. 2010). Similarly, another study in a slightly younger age group (4 years, $N = 175$) also found no effect of 400 mg/day algal DHA following 4 months of supplementation on any of the outcomes that included measures of memory, attention, vocabulary acquisition, listening comprehension and impulsivity; although the authors did find evidence of a positive association between blood concentrations of DHA and performance on a measure of vocabulary acquisition and listening comprehension (Ryan and Nelson 2008).

n-3 PUFA intervention trials in healthy adults have yielded similar results. Fontani et al. (2005) reported significant within-group reductions on participants' reaction times on attention tasks following supplementation with fish oil (1.60 g EPA + 0.80 g DHA) for 35 days ($N = 33$, mean age 33 years); however, no other studies in this population have found any effects on cognitive function (Hamazaki et al. 1996; Rogers et al. 2008; Antypa et al. 2009). Overall, the available evidence suggests that n-3 PUFA supplementation in normally developing children and healthy adults, has no observable effects on behavior.

FUTURE DIRECTIONS

More recently, there has been increased interest in the effects of n-3 PUFA supplementation on brain physiology. An earlier study in aged rhesus monkeys provided evidence supporting a role for DHA in cerebrovascular function. In this study, supplementation with DHA (150 mg/kg/day) for 4 weeks restored age-related impairment of the rCBF response to tactile stimulation (as measured by PET), compared to placebo (Tsukada et al. 2000). The effects of DHA supplementation on functional activation of the brain have also been demonstrated in humans. Using fMRI, McNamara et al. (2010) revealed that compared to placebo, DHA supplementation for 8 weeks resulted in increased activation of the dorsolateral prefrontal cortex during a sustained attention task in 33 school children (males, 8–10 years). Interestingly, these changes were not accompanied by any effects of the intervention on task performance. Jackson et al. (In press) also found evidence of increased task-related cerebral blood flow in the prefrontal cortex using Near Infrared Spectroscopy following 12 weeks' supplementation with 1 g DHA-rich fish oil (450 mg DHA + 90 mg EPA), compared to placebo in healthy adults aged 18–35 years. While the evidence presented in the previous section could not be taken as encouragement for further investigations of the direct effect of n-3 PUFA supplementation on cognitive function, the use of neuroimaging techniques, even in the absence of behavioral modification, may well be useful in delineating the effects that n-3 PUFAs are exerting on overall brain function and physiology, and provide many interesting avenues for future investigation.

SUMMARY

DHA and EPA are involved in a number of fundamental functions at the cellular level. In the brain, DHA is heavily enriched in the cerebral cortex where its incorporation into the phospholipid bilayer of neural cell membranes confers optimal membrane fluidity, resulting in improved membrane function as regards signal transduction and neurotransmission. Further, there is evidence to suggest that the expression of a number of genes and the production of various neurotransmitters are sensitive to dietary intake of n-3 PUFAs, suggesting a role for n-3 PUFAs in these processes. In addition, DHA- and EPA-derived secondary signaling molecules are essential in the resolution of the immune response and may also be neuroprotective. Finally, dietary n-3 PUFAs modulate a number of cardiovascular parameters, which may contribute to reduced risk of cardiovascular events. Given the fundamental nature of n-3 PUFAs and DHA and EPA in particular, it is plausible that alterations in dietary intake could potentially impact upon brain function and hence behavior.

Using models of n-3 PUFA deficiency and subsequent repletion, research that has investigated the effects of n-3 PUFAs on behavioral outcomes in animals has demonstrated that brain depletion of n-3 PUFAs occurs in the complete absence of dietary n-3, and is associated with cognitive costs which can be ameliorated once n-3 PUFAs are reintroduced into the diet. Human studies have been far less conclusive. Low n-3 PUFA status is associated with poorer behavioral outcomes, but the evidence provided by intervention studies in the treatment of conditions such as depression, schizophrenia, ADHD, and dementia has been mixed and inconclusive as a whole, although results from a few positive studies have been compelling enough to pursue further research in the area. Overall, the benefit of providing n-3 PUFAs to infants on behavioral outcomes appears to be transient, although the majority of studies have only evaluated the effects of relatively low amounts of DHA, and those providing more than 0.3% DHA in the formula have been more effective. The issue of cognitive enhancement via n-3 PUFA supplementation in normally developing children and healthy younger and older adults suggests that supplementation with n-3 PUFAs has little observable effect on behavioral outcomes, even when dietary intake of n-3 PUFAs is low. On the other hand, emerging evidence from neuroimaging studies suggests that supplementation with n-3 PUFAs may be exerting an effect on cerebrovascular parameters in healthy populations. Future investigations using a variety of imaging techniques to assess the causal relationship between n-3 PUFA intake and brain function in physiological terms are, therefore, warranted.

REFERENCES

Aid, S., S. Vancassel et al. (2003). Effect of a diet-induced n-3 PUFA depletion on cholinergic parameters in the rat hippocampus. *Journal of Lipid Research* **44**(8): 1545–1551.

Amminger, G. P., G. E. Berger et al. (2007). Omega-3 fatty acids supplementation in children with autism: A double-blind randomized, placebo-controlled pilot study. *Biological Psychiatry* **61**(4): 551–553.

Anderson, G. J., M. Neuringer et al. (2005). Can prenatal n-3 fatty acid deficiency be completely reversed after birth? Effects on retinal and brain biochemistry and visual function in rhesus monkeys. *Pediatric Research* **58**(5): 865–872.

Antypa, N., A. J. W. Van der Does et al. (2009). Omega-3 fatty acids (fish-oil) and depression-related cognition in healthy volunteers. *Journal of Psychopharmacology* **23**(7): 831–840.

Appleton, K. M., R. C. Hayward et al. (2006). Effects of n-3 long-chain polyunsaturated fatty acids on depressed mood: Systematic review of published trials. *American Journal of Clinical Nutrition* **84**(6): 1308–1316.

Armah, C. K., K. G. Jackson et al. (2008). Fish oil fatty acids improve postprandial vascular reactivity in healthy men. *Clinical Science* **114**(11–12): 679–686.

Assies, J., R. Lieverse et al. (2001). Significantly reduced docosahexaenoic and docosapentaenoic acid concentrations in erythrocyte membranes from schizophrenic patients compared with a carefully matched control group. *Biological Psychiatry* **49**(6): 510–522.

Auestad, N., R. Halter et al. (2001). Growth and development in term infants fed long-chain polyunsaturated fatty acids: A double-masked, parallel, prospective, multivariate study. *Pediatrics* **108**(2): 372–381.

Auestad, N., M. B. Montalto et al. (1997). Visual acuity, erythrocyte fatty acid composition, and growth in term infants fed formulas with long chain polyunsaturated fatty acids for one year. *Pediatric Research* **41**(1): 1–10.

Barberger-Gateau, P., L. Letenneur et al. (2002). Fish, meat, and risk of dementia: Cohort study. *British Medical Journal* **325**(7370): 932–933.

Barberger-Gateau, P., C. Raffaitin et al. (2007). Dietary patterns and risk of dementia: The Three-City cohort study. *Neurology* **69**(20): 1921–1930.

Barcelo-Coblijn, G., K. Kitajka et al. (2003). Gene expression and molecular composition of phospholipids in rat brain in relation to dietary n-6 to n-3 fatty acid ratio. *Biochimica et Biophysica Acta* **1632**(1–3): 72–79.

Bazan, N. G. (2006). The onset of brain injury and neurodegeneration triggers the synthesis of docosanoid neuroprotective signaling. *Cellular and Molecular Neurobiology* **26**(4–6): 901–913.

Bekaroglu, M., Y. Aslan et al. (1996). Relationships between serum free fatty acids and zinc, and attention deficit hyperactivity disorder: A research note. *Journal of Child Psychology and Psychiatry and Allied Disciplines* **37**(2): 225–227.

Bell, J. G., J. R. Sargent et al. (2000). Red blood cell fatty acid compositions in a patient with autistic spectrum disorder: A characteristic abnormality in neurodevelopmental disorders? *Prostaglandins Leukotrienes and Essential Fatty Acids* **63**(1–2): 21–25.

Belluzzi, A., S. Boschi et al. (2000). Polyunsaturated fatty acids and inflammatory bowel disease. *American Journal of Clinical Nutrition* **71**(1): 339S–342S.

Birch, E. E., S. Garfield et al. (2000). A randomized controlled trial of early dietary supply of long-chain polyunsaturated fatty acids and mental development in term infants. *Developmental Medicine and Child Neurology* **42**(3): 174–181.

Birch, E. E., S. Garfield et al. (2007). Visual acuity and cognitive outcomes at 4 years of age in a double-blind, randomized trial of long-chain polyunsaturated fatty acid-supplemented infant formula. *Early Human Development* **83**(5): 279–284.

Birch, E. E., D. R. Hoffman et al. (1998). Visual acuity and the essentiality of docosahexaenoic acid and arachidonic acid in the diet of term infants. *Pediatric Research* **44**(2): 201–209.

Borja-Hart, N. L. and J. Marino (2010). Role of omega-3 fatty acids for prevention or treatment of perinatal depression. *Pharmacotherapy* **30**(2): 210–216.

Boudrault, C., R. P. Bazinet et al. (2009). Experimental models and mechanisms underlying the protective effects of n-3 polyunsaturated fatty acids in Alzheimer's disease. *Journal of Nutritional Biochemistry* **20**(1): 1–10.

Bountziouka, V., E. Polychronopoulos et al. (2009). Long-term fish intake is associated with less severe depressive symptoms among elderly men and women: The MEDIS (MEDiterranean ISlands Elderly) epidemiological study. *Journal of Aging and Health* **21**(6): 864–880.

Bourre, J. M. (2005). Omega-3 fatty acids in psychiatry. *Medical Science (Paris)* **21**(2): 216–221.

Bourre, J. M., L. Dinh et al. (1997). Possible role of the choroid plexus in the supply of brain tissue with polyunsaturated fatty acids. *Neuroscience Letters* **224**(1): 1–4.

Browne, J. C., K. M. Scott et al. (2006). Fish consumption in pregnancy and omega-3 status after birth are not associated with postnatal depression. *Journal of Affective Disorders* **90**(2–3): 131–139.

Burdge, G. C. and P. C. Calder (2005). Conversion of alpha-linolenic acid to longer-chain polyunsaturated fatty acids in human adults. *Reproduction Nutrition Development* **45**(5): 581–597.

Burdge, G. C., Y. E. Finnegan et al. (2003). Effect of altered dietary n-3 fatty acid intake upon plasma lipid fatty acid composition, conversion of [C-13]alpha-linolenic acid to longer-chain fatty acids and partitioning towards beta-oxidation in older men. *British Journal of Nutrition* **90**(2): 311–321.

Burdge, G. C., A. E. Jones et al. (2002). Eicosapentaenoic and docosapentaenoic acids are the principal products of alpha-linolenic acid metabolism in young men. *British Journal of Nutrition* **88**(4): 355–363.

Burgess, J. R., L. Stevens et al. (2000). Long-chain polyunsaturated fatty acids in children with attention-deficit hyperactivity disorder. *American Journal of Clinical Nutrition* **71**(Suppl 1): 327S–330S.

Calder, P. C. (2006a). n-3 polyunsaturated fatty acids, inflammation, and inflammatory diseases. *American Journal of Clinical Nutrition* **83**(6 Suppl): 1505S–1519S.

Calder, P. C. (2006b). Polyunsaturated fatty acids and inflammation. *Prostaglandins Leukotrienes and Essential Fatty Acids* **75**(3): 197–202.

Calder, P. C. (2007). Immunomodulation by omega-3 fatty acids. *Prostaglandins Leukotrienes and Essential Fatty Acids* **77**(5–6): 327–335.

Carlson, S. E., A. J. Ford et al. (1996). Visual acuity and fatty acid status of term infants fed human milk and formulas with and without docosahexaenoate and arachidonate from egg yolk lecithin. *Pediatric Research* **39**(5): 882–888.

Carrie, I., M. Clement et al. (2000). Phospholipid supplementation reverses behavioral and biochemical alterations induced by n-3 polyunsaturated fatty acid deficiency in mice. *Journal of Lipid Research* **41**(3): 473–480.

Colcombe, S. J., K. I. Erickson et al. (2003). Aerobic fitness reduces brain tissue loss in aging humans. *Journals of Gerontology Series A: Biological Sciences and Medical Sciences* **58**(2): 176–180.

Colquhoun, I. and S. Bunday (1981). A lack of essential fatty-acids as a possible cause of hyperactivity in children. *Medical Hypotheses* **7**(5): 673–679.

Connor, W. E., M. Neuringer et al. (1990). Dietary-effects on brain fatty-acid composition— The reversibility of n-3 fatty-acid deficiency and turnover of docosahexaenoic acid in the brain, erythrocytes, and plasma of rhesus-monkeys. *Journal of Lipid Research* **31**(2): 237–247.

Conquer, J. A., M. C. Tierney et al. (2000). Fatty acid analysis of blood plasma of patients with Alzheimer's disease, other types of dementia, and cognitive impairment. *Lipids* **35**(12): 1305–1312.

Cyhlarova, E., J. G. Bell et al. (2007). Membrane fatty acids, reading and spelling in dyslexic and non-dyslexic adults. *European Neuropsychopharmacology* **17**(2): 116–121.

Dalton, A., P. Wolmarans et al. (2009). A randomised control trial in schoolchildren showed improvement in cognitive function after consuming a bread spread, containing fish flour from a marine source. *Prostaglandins Leukotrienes and Essential Fatty Acids* **80**(2–3): 143–149.

Das, U. N. (2007). Is depression a low-grade systemic inflammatory condition? *American Journal of Clinical Nutrition* **85**(6): 1665–1666; author reply 1666.

Davis, K. L., R. S. Kahn et al. (1991). Dopamine in schizophrenia—A review and reconceptualization. *American Journal of Psychiatry* **148**(11): 1474–1486.

De Caterina, R. and G. Basta (2001). n-3 fatty acids and the inflammatory response— Biological background. *European Heart Journal Supplements* **3**(D): D42–D49.

De Caterina, R., J. K. Liao et al. (2000). Fatty acid modulation of endothelial activation. *American Journal of Clinical Nutrition* **71**(1): 213S–223S.

Del Parigi, A., F. Panza et al. (2006). Nutritional factors, cognitive decline, and dementia. *Brain Research Bulletin* **69**(1): 1–19.

Delion, S., S. Chalon et al. (1994). Chronic dietary alpha-linolenic acid deficiency alters dopaminergic and serotoninergic neurotransmission in rats. *Journal of Nutrition* **124**(12): 2466–2476.

Drover, J., D. R. Hoffman et al. (2009). Three randomized controlled trials of early long-chain polyunsaturated fatty acid supplementation on means-end problem solving in 9-month-olds. *Child Development* **80**(5): 1376–1384.

Dullemeijer, C., J. Durga et al. (2007). n-3 Fatty acid proportions in plasma and cognitive performance in older adults. *American Journal of Clinical Nutrition* **86**(5): 1479–1485.

Dunstan, J. A., K. Simmer et al. (2008). Cognitive assessment of children at age 2(1/2) years after maternal fish oil supplementation in pregnancy: A randomised controlled trial. *Archives of Disease in Childhood. Fetal and Neonatal Edition* **93**(1): F45–F50.

Edwards, R., M. Peet et al. (1998). Omega-3 polyunsaturated fatty acid levels in the diet and in red blood cell membranes of depressed patients. *Journal of Affective Disorders* **48**(2–3): 149–155.

Engelhart, M. J., M. I. Geerlings et al. (2002). Diet and risk of dementia: Does fat matter? The Rotterdam study. *Neurology* **59**(12): 1915–1921.

Farooqui, A. A., H. C. Yang et al. (1997). Phospholipase A(2) and its role in brain tissue. *Journal of Neurochemistry* **69**(3): 889–901.

Farquharson, J., F. Cockburn et al. (1993). Effect of diet on infant subcutaneous tissue triglyceride fatty-acids. *Archives of Disease in Childhood* **69**(5): 589–593.

Fedorova, I. and N. Salem, Jr. (2006). Omega-3 fatty acids and rodent behavior. *Prostaglandins Leukotrienes and Essential Fatty Acids* **75**(4–5): 271–289.

Feller, S. E., K. Gawrisch et al. (2002). Polyunsaturated fatty acids in lipid bilayers: Intrinsic and environmental contributions to their unique physical properties. *Journal of the American Chemical Society* **124**(2): 318–326.

Fenton, W. S., F. Dickerson et al. (2001). A placebo-controlled trial of omega-3 fatty acid (ethyl eicosapentaenoic acid) supplementation for residual symptoms and cognitive impairment in schizophrenia. *American Journal of Psychiatry* **158**(12): 2071–2074.

Ferri, C. P., M. Prince et al. (2005). Global prevalence of dementia: A Delphi consensus study. *Lancet* **366**(9503): 2112–2117.

Fontani, G., A. Corradeschi et al. (2005). Cognitive and physiological effects of Omega-3 polyunsaturated fatty acid supplementation in healthy subjects. *European Journal of Clinical Investigation* **35**: 691–699.

Fotuhi, M., P. Mohassel et al. (2009). Fish consumption, long-chain omega-3 fatty acids and risk of cognitive decline or Alzheimer disease: A complex association. *Nature Clinical Practice Neurology* **5**(3): 140–152.

Frangou, S., M. Lewis et al. (2006). Efficacy of ethyl-eicosapentaenoic acid in bipolar depression: Randomised double-blind placebo-controlled study. *British Journal of Psychiatry* **188**: 46–50.

Fraser, T., H. Taylor et al. (2009). Fatty acid composition of frontal, temporal and parietal neocortex in the normal human brain and in Alzheimer's disease. *Neurochemical Research* **35**(3): 503–513.

Freeman, M. P., M. Davis et al. (2008). Omega-3 fatty acids and supportive psychotherapy for perinatal depression: A randomized placebo-controlled study. *Journal of Affective Disorders* **110**(1–2): 142–148.

Freeman, M. P., J. R. Hibbeln et al. (2006a). Randomized dose-ranging pilot trial of omega-3 fatty acids for postpartum depression. *Acta Psychiatrica Scandinavica* **113**(1): 31–35.

Freeman, M. P., J. R. Hibbeln et al. (2006b). Omega-3 fatty acids: Evidence basis for treatment and future research in psychiatry. *Journal of Clinical Psychiatry* **67**(12): 1954–1967.

Freund-Levi, Y., H. Basun et al. (2007). Omega-3 supplementation in mild to moderate Alzheimer's disease: Effects on neuropsychiatric symptoms. *International Journal of Geriatric Psychiatry* **23**(2): 161–169.

Freund-Levi, Y., M. Eriksdotter-Jonhagen et al. (2006). Omega-3 fatty acid treatment in 174 patients with mild to moderate Alzheimer disease: OmegAD study—A randomized double-blind trial. *Archives of Neurology* **63**(10): 1402–1408.

Gamoh, S., M. Hashimoto et al. (2001). Chronic administration of docosahexaenoic acid improves the performance of radial arm maze task in aged rats. *Clinical and Experimental Pharmacology and Physiology* **28**(4): 266–270.

Garcia-Calatayud, S., C. Redondo et al. (2005). Brain docosahexaenoic acid status and learning in young rats submitted to dietary long-chain polyunsaturated fatty acid deficiency and supplementation limited to lactation. *Pediatric Research* **57**(5 Pt 1): 719–723.

van Gelder, B. M., M. Tijhuis et al. (2007). Fish consumption, n-3 fatty acids, and subsequent 5-y cognitive decline in elderly men: The Zutphen Elderly study. *American Journal of Clinical Nutrition* **85**(4): 1142–1147.

Geleijnse, J. M., E. J. Giltay et al. (2002). Blood pressure response to fish oil supplementation: Metaregression analysis of randomized trials. *Journal of Hypertension* **20**(8): 1493–1499.

Gerster, H. (1998). Can adults adequately convert alpha-linolenic acid (18: 3n-3) to eicosapentaenoic acid (20: 5n-3) and docosahexaenoic acid (22: 6n-3)? *International Journal for Vitamin and Nutrition Research* **68**(3): 159–173.

Gibney, M. J. and B. Hunter (1993). The effects of short-term and long-term supplementation with fish oil on the incorporation of n-3 polyunsaturated fatty-acids into cells of the immune-system in healthy-volunteers. *European Journal of Clinical Nutrition* **47**(4): 255–259.

Greiner, R. S., T. Moriguchi et al. (2001). Olfactory discrimination deficits in n-3 fatty acid-deficient rats. *Physiology & Behavior* **72**(3): 379–385.

Gronn, M., E. Christensen et al. (1991). Peroxisomal retroconversion of docosahexaenoic acid (22–6(n-3)) to eicosapentaenoic acid (20–5(n-3)) studied in isolated rat-liver cells. *Biochimica et Biophysica Acta* **1081**(1): 85–91.

de Groot, R. H. M., J. Adam et al. (2004). Alpha-linolenic acid supplementation during human pregnancy does not effect cognitive functioning. *Prostaglandins Leukotrienes and Essential Fatty Acids* **70**: 41–47.

Haggarty, P. (2004). Effect of placental function on fatty acid requirements during pregnancy. *European Journal of Clinical Nutrition* **58**(12): 1559–1570.

Hamazaki, T. and S. Hirayama (2004). The effect of docosahexaenoic acid-containing food administration on symptoms of attention-deficit/hyperactivity disorder— A placebo-controlled double-blind study. *European Journal of Clinical Nutrition* **58**(5): 838–838.

Hamazaki, T., S. Sawazaki et al. (1996). The effect of docosahexaenoic acid on aggression in young adults. A placebo-controlled double-blind study. *Journal of Clinical Investigation* **97**(4): 1129–1133.

Harris, W. S., G. S. Rambjor et al. (1997). n-3 fatty acids and urinary excretion of nitric oxide metabolites in humans. *American Journal of Clinical Nutrition* **65**(2): 459–464.

Hashimoto, M., S. Hossain et al. (2002). Docosahexaenoic acid provides protection from impairment of learning ability in Alzheimer's disease model rats. *Journal of Neurochemistry* **81**(5): 1084–1091.

Hashimoto, M., S. Hossain et al. (2005a). Docosahexaenoic acid-induced amelioration on impairment of memory learning in amyloid beta-infused rats relates to the decreases of amyloid beta and cholesterol levels in detergent-insoluble membrane fractions. *Biochimica et Biophysica Acta—Molecular and Cell Biology of Lipids* **1738**(1–3): 91–98.

Hashimoto, M., S. Hossain et al. (2008). The protective effect of dietary eicosapentaenoic acid against impairment of spatial cognition learning ability in rats infused with amyloid beta((1–40)). *Journal of Nutritional Biochemistry*.

Hashimoto, M., Y. Tanabe et al. (2005b). Chronic administration of docosahexaenoic acid ameliorates the impairment of spatial cognition learning ability in amyloid beta-infused rats. *Journal of Nutrition* **135**(3): 549–555.

Helland, I. B., L. Smith et al. (2003). Maternal supplementation with very-long-chain n-3 fatty acids during pregnancy and lactation augments children's IQ at 4 years of age. *Pediatrics* **111**(1): e39–e44.

Helland, I. B., L. Smith et al. (2008). Effect of supplementing pregnant and lactating mothers with n-3 very-long-chain fatty acids on children's IQ and body mass index at 7 years of age. *Pediatrics* **122**(2): e472–e479.

Heude, B., P. Ducimetiere et al. (2003). Cognitive decline and fatty acid composition of erythrocyte membranes—The EVA Study. *American Journal of Clinical Nutrition* **77**(4): 803–808.

Hibbeln, J. R. (1998). Fish consumption and major depression. *Lancet* **351**(9110): 1213.

Hibbeln, J. R. (2002). Seafood consumption, the DHA content of mothers' milk and prevalence rates of postpartum depression: A cross-national, ecological analysis. *Journal of Affective Disorders* **69**(1–3): 15–29.

Hibbeln, J. R., J. M. Davis et al. (2007). Maternal seafood consumption in pregnancy and neurodevelopmental outcomes in childhood (ALSPAC study): An observational cohort study. *Lancet* **369**(9561): 578–585.

Hibbeln, J. R., L. R. Nieminen et al. (2006). Healthy intakes of n-3 and n-6 fatty acids: Estimations considering worldwide diversity. *American Journal Clinical Nutrition* **83**(6 Suppl): 1483S–1493S.

Holman, R. T., S. B. Johnson et al. (1982). A case of human linolenic acid deficiency involving neurological abnormalities. *American Journal of Clinical Nutrition* **35**(3): 617–623.

Holman, R. T., S. B. Johnson et al. (1991). Deficiency of essential fatty-acids and membrane fluidity during pregnancy and lactation. *Proceedings of the National Academy of Sciences of the United States of America* **88**(11): 4835–4839.

Huang, T. L., P. P. Zandi et al. (2005). Benefits of fatty fish on dementia risk are stronger for those without APOE epsilon 4. *Neurology* **65**(9): 1409–1414.

Innis, S. M. (1991). Essential fatty-acids in growth and development. *Progress in Lipid Research* **30**(1): 39–103.

Jackson, P. A., J. L. Reay et al. (2012). DHA-rich fish oil modulates cerebral haemodynamic response to cognitive tasks in healthy young adults: A Near Infrared Spectroscopy pilot study. *British Journal of Nutrition* **107**: 1093–1098.

Joy, C. B., R. Mumby-Croft et al. (2006). Polyunsaturated fatty acid supplementation for schizophrenia. *Cochrane Database of Systematic Reviews* (3).

Kalmijn, S., M. P. van Boxtel et al. (2004). Dietary intake of fatty acids and fish in relation to cognitive performance at middle age. *Neurology* **62**(2): 275–280.

Kalmijn, S., L. J. Launer et al. (1997). Dietary fat intake and the risk of incident dementia in the Rotterdam Study. *Annals of Neurology* **42**(5): 776–782.

Kang, J. X. and A. Leaf (1996). Evidence that free polyunsaturated fatty acids modify Na+ channels by directly binding to the channel proteins. *Proceedings of the National Academy of Sciences of the United States of America* **93**(8): 3542–3546.

Keck, P. E., J. Mintz et al. (2006). Double-blind, randomized, placebo-controlled trials of ethyl-eicosapentanoate in the treatment of bipolar depression and rapid cycling bipolar disorder. *Biological Psychiatry* **60**(9): 1020–1022.

Kennedy, D. O., P. A. Jackson et al. (2009). Cognitive and mood effects of 8 weeks' supplementation with 400 mg or 1000 mg of the omega-3 essential fatty acid docosahexaenoic acid (DHA) in healthy children aged 10–12 years. *Nutritional Neuroscience* **12**(2): 48–56.

Kidd, P. M. (2007). Omega-3 DHA and EPA for cognition, behavior, and mood: Clinical findings and structural-functional synergies with cell membrane phospholipids. *Alternative Medicine Review* **12**(3): 207–227.

Kim, H. Y., M. Akbar et al. (2000). Inhibition of neuronal apoptosis by docosahexaenoic acid (22: 6n-3)—Role of phosphatidylserine in antiapoptotic effect. *Journal of Biological Chemistry* **275**(45): 35215–35223.

Kirby, A., A. Woodward et al. (2010). A double-blind, placebo-controlled study investigating the effects of omega-3 supplementation in children aged 8–10 years from a mainstream school population. *Research in Developmental Disabilities* **31**(3): 718–730.

Kitajka, K., L. G. Puskas et al. (2002). The role of n-3 polyunsaturated fatty acids in brain: Modulation of rat brain gene expression by dietary n-3 fatty acids. *Proceedings of the National Academy of Sciences of the United States of America* **99**(5): 2619–2624.

Kremer, J. M. (2000). n-3 Fatty acid supplements in rheumatoid arthritis. *American Journal of Clinical Nutrition* **71**(1): 349S–351S.

Kroger, E., R. Verreault et al. (2009). Omega-3 fatty acids and risk of dementia: The Canadian Study of Health and Aging. *American Journal of Clinical Nutrition* **90**(1): 184–192.

Landfield, P. W., J. L. McGaugh et al. (1978). Impaired synaptic potentiation processes in hippocampus of aged, memory-deficient rats. *Brain Research* **150**(1): 85–101.

Levant, B., J. D. Radel et al. (2006). Reduced brain DHA content after a single reproductive cycle in female rats fed a diet deficient in n-3 polyunsaturated fatty acids. *Biological Psychiatry* **60**(9): 987–990.

Lim, W. S., J. K. Gammack et al. (2006). Omega 3 fatty acid for the prevention of dementia. *Cochrane Database of Systematic Reviews* (1): CD005379.

Lucas, A., M. Stafford et al. (1999). Efficacy and safety of long-chain polyunsaturated fatty acid supplementation of infant-formula milk: A randomised trial. *Lancet* **354**(9194): 1948–1954.

Makrides, M., M. A. Neumann et al. (1994). Fatty-acid composition of brain, retina, and erythrocytes in breast-fed and formula-fed infants. *American Journal of Clinical Nutrition* **60**(2): 189–194.

Makrides, M., M. Neumann et al. (1995). Are long-chain polyunsaturated fatty acids essential nutrients in infancy? *Lancet* **345**(8963): 1463–1468.

Makrides, M., M. A. Neumann et al. (2000). A critical appraisal of the role of dietary long-chain polyunsaturated fatty acids on neural indices of term infants: A randomized, controlled trial. *Pediatrics* **105**(1): 32–38.

Mamalakis, G., E. Jansen et al. (2006). Depression and adipose and serum cholesteryl ester polyunsaturated fatty acids in the survivors of the seven countries study population of Crete. *European Journal of Clinical Nutrition* **60**(8): 1016–1023.

Mamalakis, G., M. Kiriakakis et al. (2004a). Depression and adipose polyunsaturated fatty acids in an adolescent group. *Prostaglandins Leukotrienes and Essential Fatty Acids* **71**(5): 289–294.

Mamalakis, G., M. Kiriakakis et al. (2004b). Depression and adipose polyunsaturated fatty acids in the survivors of the Seven Countries Study population of Crete. *Prostaglandins Leukotrienes and Essential Fatty Acids* **70**(6): 495–501.

Mamalakis, G., M. Tornaritis et al. (2002). Depression and adipose essential polyunsaturated fatty acids. *Prostaglandins Leukotrienes and Essential Fatty Acids* **67**(5): 311–318.

Marangell, L. B., J. M. Martinez et al. (2003). A double-blind, placebo-controlled study of the omega-3 fatty acid docosahexaenoic acid in the treatment of major depression. *American Journal of Psychiatry* **160**: 996–998.

Marangell, L. B., J. M. Martinez et al. (2004). Omega-3 fatty acids for the prevention of postpartum depression: Negative data from a preliminary, open-label pilot study. *Depression and Anxiety* **19**(1): 20–23.

McGahon, B. M., D. S. Martin et al. (1999). Age-related changes in synaptic function: Analysis of the effect of dietary supplementation with omega-3 fatty acids. *Neuroscience* **94**(1): 305–314.

McNamara, R. K., J. Able et al. (2010). Docosahexaenoic acid supplementation increases prefrontal cortex activation during sustained attention in healthy boys: A placebo-controlled, dose-ranging, functional magnetic resonance imaging study. *American Journal of Clinical Nutrition*.

McNamara, R. K. and S. E. Carlson (2006). Role of omega-3 fatty acids in brain development and function: Potential implications for the pathogenesis and prevention of psychopathology. *Prostaglandins Leukotrienes and Essential Fatty Acids* **75**(4–5): 329–349.

McNamara, R. K., C. G. Hahn et al. (2007). Selective deficits in the omega-3 fatty acid doco-sahexaenoic acid in the postmortem orbitofrontal cortex of patients with major depressive disorder. *Biological Psychiatry* **62**(1): 17–24.

Menon, N. K. and G. A. Dhopeshwarkar (1983). Essential fatty-acid deficiency and brain-development. *Progress in Lipid Research* **21**: 309–326.

Moore, S. A. (2001). Polyunsaturated fatty acid synthesis and release by brain-derived cells in vitro. *Journal of Molecular Neuroscience* **16**(2–3): 195–200.

Mori, T. A., L. J. Beilin et al. (1997). Interactions between dietary fat, fish, and fish oils and their effects on platelet function in men at risk of cardiovascular disease. *Arteriosclerosis Thrombosis and Vascular Biology* **17**(2): 279–286.

Mori, T. A., I. B. Puddey et al. (2000). Effect of omega 3 fatty acids on oxidative stress in humans: GC-MS measurement of urinary F-2-isoprostane excretion. *Redox Report* **5**(1): 45–46.

Mori, T. A. and R. J. Woodman (2006). The independent effects of eicosapentaenoic acid and docosahexaenoic acid on cardiovascular risk factors in humans. *Current Opinion in Clinical Nutrition and Metabolic Care* **9**(2): 95–104.

Moriguchi, T., R. S. Greiner et al. (2000). Behavioral deficits associated with dietary induction of decreased brain docosahexaenoic acid concentration. *Journal of Neurochemistry* **75**(6): 2563–2573.

Morris, M. C., D. A. Evans et al. (2003). Consumption of fish and n-3 fatty acids and risk of incident Alzheimer disease. *Archives of Neurology* **60**(7): 940–946.

Morris, M. C., D. A. Evans et al. (2005). Fish consumption and cognitive decline with age in a large community study. *Archives of Neurology* **62**(12): 1849–1853.

Morris, M. C., F. Sacks et al. (1993). Does fish-oil lower blood-pressure—A metaanalysis of controlled trials. *Circulation* **88**(2): 523–533.

Nestel, P. J. (2000). Fish oil and cardiovascular disease: Lipids and arterial function. *American Journal of Clinical Nutrition* **71**(1): 228S–231S.

Ng, K. H., B. J. Meyer et al. (2009). Dietary PUFA intakes in children with attention-deficit/hyperactivity disorder symptoms. *British Journal of Nutrition* **102**(11): 1635–1641.

Noaghiul, S. and J. R. Hibbeln (2003). Cross-national comparisons of seafood consumption and rates of bipolar disorders. *American Journal of Psychiatry* **160**(12): 2222–2227.

Ohara, K. (2007). The n-3 polyunsaturated fatty acid/dopamine hypothesis of schizophrenia. *Progress Neuro-Psychopharmacology & Biological Psychiatry* **31**(2): 469–474.

Osendarp, S. J. M., K. I. Baghurst et al. (2007). Effect of a 12-mo micronutrient intervention on learning and memory in well-nourished and marginally nourished school-aged children: 2 parallel, randomized, placebo-controlled studies in Australia and Indonesia. *American Journal of Clinical Nutrition* **86**(4): 1082–1093.

Osher, Y., Y. Bersudsky et al. (2005). Omega-3 eicosapentaenoic acid in bipolar depression: Report of a small open-label study. *Journal of Clinical Psychiatry* **66**(6): 726–729.

Peet, M. and D. F. Horrobin (2002). A dose-ranging exploratory study of the effects of ethyl-eicosapentaenoate in patients with persistent schizophrenic symptoms. *Journal of Psychiatric Research* **36**(1): 7–18.

Peet, M., B. Murphy et al. (1998). Depletion of omega-3 fatty acid levels in red blood cell membranes of depressive patients. *Biological Psychiatry* **43**(5): 315–319.

Peters, R., J. Peters et al. (2008). Alcohol, dementia and cognitive decline in the elderly: A systematic review. *Age and Ageing* **37**(5): 505–512.

Pratico, D. and J. Q. Trojanowski (2000). Inflammatory hypotheses: Novel mechanisms of Alzheimer's neurodegeneration and new therapeutic targets? Commentary. *Neurobiology of Aging* **21**(3): 441–445.

de la Presa Owens, S. D. and S. M. Innis (1999). Docosahexaenoic and arachidonic acid prevent a decrease in dopaminergic and serotoninergic neurotransmitters in frontal cortex caused by a linoleic and alpha-linolenic acid deficient diet in formula-fed piglets. *Journal of Nutrition* **129**(11): 2088–2093.

Quinn, J. (2007). A randomized double-blind placebo-controlled trial of the effects of doco-sahexaenoic acid (DHA) in slowing the progression of Alzheimer's disease. Retrieved November 18, 2009, from http://clinicaltrials.gov/ct2/show/NCT00440050.

van de Rest, O., J. M. Geleijnse et al. (2008). Effect of fish oil on cognitive performance in older subjects: A randomized, controlled trial. *Neurology* **71**(6): 430–438.

Richardson, A. J. (2006). Omega-3 fatty acids in ADHD and related neurodevelopmental disorders. *International Review of Psychiatry* **18**(2): 155–172.

Richardson, A. J. (2008). n-3 Fatty acids and mood: The devil is in the detail. *British Journal of Nutrition* **99**(2): 221–223.

Richardson, A. J., C. M. Calvin et al. (2000). Fatty acid deficiency signs predict the severity of reading and related difficulties in dyslexic children. *Prostaglandins Leukotrienes and Essential Fatty Acids* **63**(1–2): 69–74.

Richardson, A. J. and P. Montgomery (2005). The Oxford-Durham study: A randomized, controlled trial of dietary supplementation with fatty acids in children with developmental coordination disorder. *Pediatrics* **115**(5): 1360–1366.

Richardson, A. J. and B. K. Puri (2002). A randomized double-blind, placebo-controlled study of the effects of supplementation with highly unsaturated fatty acids on ADHD-related symptoms in children with specific learning difficulties. *Progress in Neuro-Psychopharmacology Biological Psychiatry* **26**: 233–239.

Richardson, A. J. and M. A. Ross (2000). Fatty acid metabolism in neurodevelopmental disorder: A new perspective on associations between attention-deficit hyperactivity disorder, dyslexia, dyspraxia and the autistic spectrum. *Prostaglandins Leukotrienes and Essential Fatty Acids* **63**(1–2): 1–9.

Rogers, P. J., K. M. Appleton et al. (2008). No effect of n-3 long-chain polyunsaturated fatty acid (EPA and DHA) supplementation on depressed mood and cognitive function: A randomised controlled trial. *British Journal of Nutrition* **99**(2): 421–431.

Ryan, A. S. and E. B. Nelson (2008). Assessing the effect of docosahexaenoic acid on cognitive functions in healthy, preschool children: A randomized, placebo-controlled, double-blind study. *Clinical Pediatrics (Philadelphia)* **47**(4): 355–362.

Sagduyu, K., M. E. Dokucu et al. (2005). Omega-3 fatty acids decreased irritability of patients with bipolar disorder in an add-on, open label study. *Nutrition Journal* **4**: 6.

Sanders, T. and P. Emery (2003). *Molecular Basis of Human Nutrition*. London, U.K.: Taylor & Francis Group.

von Schacky, C. (2000). n-3 Fatty acids and the prevention of coronary atherosclerosis. *American Journal of Clinical Nutrition* **71**(1): 224S–227S.

Schaefer, E. J., V. Bongard et al. (2006). Plasma phosphatidylcholine docosahexaenoic acid content and risk of dementia and Alzheimer disease—The Framingham heart study. *Archives of Neurology* **63**(11): 1545–1550.

Schmitz, G. and J. Ecker (2008). The opposing effects of n-3 and n-6 fatty acids. *Progress in Lipid Research* **47**(2): 147–155.

Schuchardt, J. P., M. Huss et al. (2009). Significance of long-chain polyunsaturated fatty acids (PUFAs) for the development and behaviour of children. *European Journal of Pediatrics* **162**(2): 149–164.

Seebungkert, B. and J. W. Lynch (2002). Effects of polyunsaturated fatty acids on voltage-gated K+ and Na+ channels in rat olfactory receptor neurons. *European Journal of Neuroscience* **16**(11): 2085–2094.

Shukitt-Hale, B., G. Mouzakis et al. (1998). Psychomotor and spatial memory performance in aging male Fischer 344 rats. *Experimental Gerontology* **33**(6): 615–624.

Silvers, K. M. and K. M. Scott (2002). Fish consumption and self-reported physical and mental health status. *Public Health Nutrition* **5**(3): 427–431.

Simopoulos, A. P. (1991). Omega-3 fatty acids in health and disease and in growth and development. *American Journal of Clinical Nutrition* **54**(3): 438–463.

Simopoulos, A. P. (2000). Human requirement for n-3 polyunsaturated fatty acids. *Poultry Science* **79**(7): 961–970.

Simopoulos, A. P. (2008). The importance of the omega-6/omega-3 fatty acid ratio in cardiovascular disease and other chronic diseases. *Experimental Biology and Medicine* **233**(6): 674–688.

Slater, S. J., M. Kelly et al. (1995). Polyunsaturation in cell membranes and lipid bilayers and its effects on membrane proteins. *2nd Congress of the International Society for the Study of Fatty Acids and Lipids*, Bethesda, MD, American Oil Chemists' Society.

Sliwinski, S., J. Croonenberghs et al. (2006). Polyunsaturated fatty acids: Do they have a role in the pathophysiology of autism? *Neuroendocrinology Letters* **27**(4): 465–471.

Stevens, L., W. Zhang et al. (2003). EFA supplementation in children with inattention, hyperactivity, and other disruptive behaviors. *Lipids* **38**(10): 1007–1021.

Stevens, L. J., S. S. Zentall et al. (1995). Essential fatty-acid metabolism in boys with attention-deficit hyperactivity disorder. *American Journal of Clinical Nutrition* **62**(4): 761–768.

Stevens, L. J., S. S. Zentall et al. (1996). Omega-3 fatty acids in boys with behavior, learning, and health problems. *Physiology & Behavior* **59**(4–5): 915–920.

Stillwell, W. and S. R. Wassall (2003). Docosahexaenoic acid: Membrane properties of a unique fatty acid. *Chemistry and Physics of Lipids* **126**(1): 1–27.

Stoll, A. L., W. E. Severus et al. (1999). Omega 3 fatty acids in bipolar disorder: A preliminary double-blind, placebo-controlled trial. *Archives of General Psychiatry* **56**(5): 407–412.

Stordy, B. J. (2000). Dark adaptation, motor skills, docosahexaenoic acid, and dyslexia. *American Journal of Clinical Nutrition* **71**(1): 323S–326S.

Strom, M., E. L. Mortensen et al. (2009). Fish and long-chain n-3 polyunsaturated fatty acid intakes during pregnancy and risk of postpartum depression: A prospective study based on a large national birth cohort. *American Journal of Clinical Nutrition* **90**(1): 149–155.

Swan, G. E. and C. N. Lessov-Schlaggar (2007). The effects of tobacco smoke and nicotine on cognition and the brain. *Neuropsychology Review* **17**(3): 259–273.

Tanskanen, A., J. R. Hibbeln et al. (2001). Fish consumption and depressive symptoms in the general population in Finland. *Psychiatric Services* **52**(4): 529–531.

Tassoni, D., G. Kaur et al. (2008). The role of eicosanoids in the brain. *Asia Pacific Journal of Clinical Nutrition* **17**(Suppl 1): 220–228.

Taylor, K. E., C. J. Higgins et al. (2000). Dyslexia in adults is associated with clinical signs of fatty acid deficiency. *Prostaglandins Leukotrienes and Essential Fatty Acids* **63**(1–2): 75–78.

Tsukada, H., K. Sato et al. (2000). Age-related impairment of coupling mechanism between neuronal activation and functional cerebral blood flow response was restored by cholinesterase inhibition: PET study with microdialysis in the awake monkey brain. *Brain Research* **857**(1–2): 158–164.

Turnbull, T., M. Cullen-Drill et al. (2008). Efficacy of omega-3 fatty acid supplementation on improvement of bipolar symptoms: A systematic review. *Archives of Psychiatric Nursing* **22**(5): 305–311.

Turner, N., P. L. Else et al. (2003). Docosahexaenoic acid (DHA) content of membranes determines molecular activity of the sodium pump: Implications for disease states and metabolism. *Naturwissenschaften* **90**(11): 521–523.

Ulmann, L., V. Mimouni et al. (2001). Brain and hippocampus fatty acid composition in phospholipid classes of aged-relative cognitive deficit rats. *Prostaglandins Leukotrienes and Essential Fatty Acids* **64**(3): 189–195.

Vancassel, S., G. Durand et al. (2001). Plasma fatty acid levels in autistic children. *Prostaglandins Leukotrienes and Essential Fatty Acids* **65**(1): 1–7.

Voigt, R. G., A. M. Llorente et al. (2001). A randomized, double-blind, placebo-controlled trial of docosahexaenoic acid supplementation in children with attention-deficit/hyperactivity disorder. *Journal of Pediatrics* **139**(2): 189–196.

Willis, L. M., B. Shukitt-Hale et al. (2009). Dose-dependent effects of walnuts on motor and cognitive function in aged rats. *British Journal of Nutrition* **101**(8): 1140–1144.

Xiao, Y. F., A. M. Gomez et al. (1997). Suppression of voltage-gated L-type Ca2+ currents by polyunsaturated fatty acids in adult and neonatal rat ventricular myocytes. *Proceedings of the National Academy of Sciences of the United States of America* **94**(8): 4182–4187.

Yao, J. K. and D. P. van Kammen (2004). Membrane phospholipids and cytokine interaction in schizophrenia. *Disorders of Synaptic Plasticity and Schizophrenia.* **59**: 297–326.

Yehuda, S., S. Rabinovitz et al. (1999). Essential fatty acids are mediators of brain biochemistry and cognitive functions. *Journal of Neuroscience Research* **56**(6): 565–570.

Yurko-Mauro, K., D. McCarthy et al. (2010). Beneficial effects of docosahexaenoic acid on cognition in age-related cognitive decline. *Alzheimers Dement* 6(6): 456–464.

Zimmer, L., S. Delion-Vancassel et al. (2000a). Modification of dopamine neurotransmission in the nucleus accumbens of rats deficient in n-3 polyunsaturated fatty acids. *Journal of Lipid Research* **41**(1): 32–40.

Zimmer, L., S. Delpal et al. (2000b). Chronic n-3 polyunsaturated fatty acid deficiency alters dopamine vesicle density in the rat frontal cortex. *Neuroscience Letters* **284**(1–2): 25–28.

9 Omega-3 Fatty Acids
From Ancient Nutrients to Modern Nutraceuticals

Joanne Bradbury

CONTENTS

ABBREVIATIONS

AA   arachidonic acid (20:5n-6)
ALA   alpha-linolenic acid or linolenic acid (18:3n-3)
C   carbon atom
COX   cyclooxygenase
D5D   delta-5 desaturase
D6D   delta-6 desaturase
DHA   docosahexaenoic acid (22:6n-3)
DPA   docosapentaenoic acid (22:5n-3)
EFA   essential fatty acid
EPA   eicosapentaenoic acid (20:5n-3)
EPO   evening primrose oil
FA   fatty acids
LA   linoleic acid (18:2n-6)
LC   longer chain ($\geq$20 carbon atoms)
LOX   lipoxygenase
LT   leukotriene
LX   lipoxin
n-3   omega-3
n-6   omega-6
PD   protectin
PG   prostaglandin
PL   phospholipids
PUFA   polyunsaturated fatty acid
Rv   resolvin
TX   thromboxane

INTRODUCTION TO FATTY ACIDS

The word "fat" can be traced back to the Sanskrit word *payate* meaning "swells, grows, fattens." In Latin it was *pinguis*, in Greek *pakus*, and an old Anglo-Saxon word *faet*, meant "to fatten." Its first formal appearance in the English language was in 1539, when it was defined as the "oily concrete substance of which the fat parts of animal bodies are chiefly composed" [1].

The ancient Greeks had different words for the different types of fats, such that the noun for animal and vegetable fats was *lipos*, the derivative of the modern English term "lipid." Interestingly, olive oil had its own word, *elaion*, which was carried into Latin as *oleium*, as oil or olive oil, and the derivation of the term "oleic" acid, the predominant fatty acid (76%) in olive oil [2].

It has been suggested that the "real beginning" of lipid research was marked by the publication in 1884 of the book *A Treatise on the Chemical Constitution of the Brain* by Johann Ludwig Wilhelm Thudichum (1829–1901). In the preface of his book, Thudichum emphasizes the central role of phospholipids (PLs) to life:

> Phosphotides are the centre, life and chemical soul of all life bioplasm whatsoever, that of plants as well as of animals. Their chemical stability is greatly due to the fact that

their fundamental radicle is a mineral acid of strong and manifold dynamacities. Their varied functions are the result of a collusion of radicles of strongly contrasting properties. Their physical properties are, viewed from the teleological point of standing, eminently adapted to their functions. Amongst these properties none of these are more deserving of further inquiry as those that may be described as their power of colloidation. Without this power, no brain as an organ would be possible, as indeed the existence of all bioplasm is dependent on the colloid state. (Thudichum (1884) in Ref. [1])

DEFINITIONS

A useful current working definition for the term "lipids" is provided by the American Oil Chemists' Society (AOCS) comprehensive website, Lipid Library [3]: "Lipids are fatty acids and their derivatives, and substances related biosynthetically or functionally to these compounds." The most common classes of lipids consist of fatty acids bonded to glycerol, such as triglycerides, or other alcohols such as cholesterol, and to sphingoid bases. Further, lipids may contain an additional alkyl group such as phosphate or carbohydrate.

Lipids can be further classified as "simple" or "complex" based on the presence of the additional alkyl group. This classification is sometimes also referred to as "neutral" or "polar" lipids. Triglycerides, sterols, waxes, and free fatty acids are examples of simple lipids. Complex lipids can be further divided into (1) PLs (more correctly, glycerophospholipids) which consist of a glycerol backbone linking two fatty acids and a phosphate polar head, and (2) glycolipids (both glycoglycerolipids and glycosphingolipids) containing a carbohydrate polar head [3].

Fatty acids are naturally occurring compounds that have been synthesized by a multienzyme complex called *fatty acid synthase*. Fatty acids generally consist of an even number of carbon atoms (C) in the range of 14–24C, in a straight chain with a carboxylic acid group (COOH) at the start and a methyl group (CH_3) at the end of the molecule. They are classified according to whether or not they contain a double bond between carbon atoms (C=C). Fatty acids that do not contain a double bond are known as saturated, and those that contain at least one double bond are unsaturated. Saturated fatty acids are the predominant fatty acids found in animal tissue while unsaturated fatty acids are predominately found in plant tissue.

Unsaturated fatty acids are further classified according to the number and position of the double bond. Fatty acids with one double bond are known as monounsaturated fatty acids and those with more than one are polyunsaturated fatty acids (PUFAs). Double bonds in plant oils are generally in the *cis*-isomer configuration, which induces a "kink" in the carbon side chain that prevents them from stacking tightly together, as with saturated fatty acids, which have no double bonds and form straight chains (see Figure 9.1). Partial hydrogenation, as in commercial food technology, changes the *cis*-isomer to a *trans*-isomer, which has the effect of "straightening" the side chain so that it behaves more like a saturated fatty acid than an unsaturated fatty acid.

NOMENCLATURE OF UNSATURATED FATTY ACIDS

In the biochemical nomenclature of the unsaturated fatty acids, omega (the last letter in the Greek alphabet) is used to denote that the counting of carbon atoms begins from the methyl terminus. Holman [4] introduced the omega nomenclature in 1964

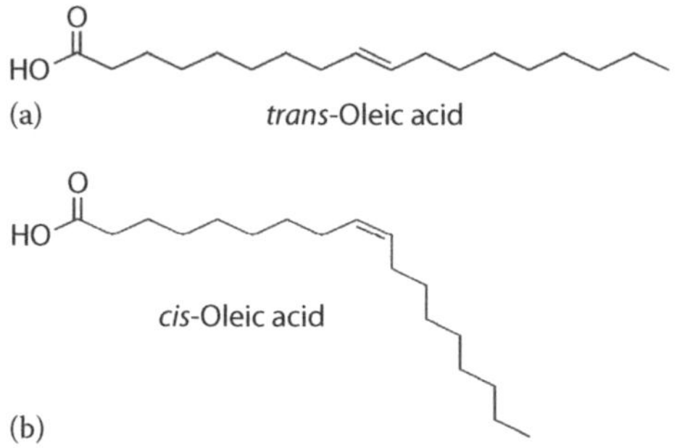

FIGURE 9.1   Molecular structure of oleic acid (18:1n-9) in the *trans*-isomer (a) and *cis*-isomer (b).

because the terminal group remains unaltered during desaturation. In this system, the number of carbon atoms is given, followed by a colon; after the colon the number of double bonds is given, followed by the position of the first double bond from the methyl group. Thus, oleic acid is 18:1n-9 to denote 18C and one double bond, found on the ninth carbon atom from the methyl end.

Linoleic acid (LA) is known as 18:2n-6 to denote 18C and two double bonds with the first one found on the sixth carbon from the methyl terminus while alpha-linolenic acid (ALA) is 18:3n-3, denoting three double bonds; that is, a higher degree of unsaturation. The first double bond is found on the third carbon from the methyl terminus for ALA and indeed all the omega-3 fatty acids. Moreover, arachidonic acid (AA) is denoted as 20:4n-6, eicosapentaenoic acid (EPA) as 20:5n-3, and docosahexaenoic acid (DHA) as 22:6n-3. With its six double bonds, DHA has the highest degree of unsaturation of the fatty acids [5]. Whether the molecule belongs to the omega-6 or the omega-3 family of fatty acids is thus denoted by n-6 or n-3, respectively.

ESSENTIAL FATTY ACIDS

LA (18:2n-6) and ALA (18:3n-3 or simply "linolenic" acid) are readily synthesized from oleic acid in plants but mammals lack the specific enzymes for the synthesis of double bonds beyond the ninth carbon atom from the carboxyl group [6]. Because they are required for vital biological functions in humans and the body shows deficiency symptoms with inadequate intakes, both LA and linolenic acids meet the requirements of nutritional essentiality [7].

The ultimate dietary source of LA and ALA are thus from plants. As such ALA is sometimes known as the "plant omega-3." An important function of these fatty acids, in particular ALA, in plants involves the production of hormonelike signaling molecules that act as defense mechanisms in response to stressors, such as physical damage by insects and animals or pathogenic invasions. In plants, ALA is the precursor to a family of oxygenated fatty acids, fatty acids which have been generated by *cyclooxygenase (COX)* or *lipoxygenase (LOX)* enzymes. Such oxygenated fatty acid signaling molecules in plants are collectively known as oxylipins or, less frequently, octadecanoids [8].

In humans, LA itself is used in the protective barrier in the outer layer of the skin, but a putative role for ALA per se is not as apparent. Metabolically, LA and ALA are both precursors for the generation of longer chain (LC, >20C) fatty acids of the omega-6 and omega-3 families, respectively. The essential function of ALA in humans has long been held as limited to this metabolic role. Dietary ALA is transported to the liver where it is sequentially desaturated and elongated into the 20C EPA (20:5n-3) and the 22C DHA (22:6n-3). Dietary LA is similarly metabolized to the 20C AA (20:4n-6), competing for the same *desaturase* and *elongase* enzymes.

The LC (≥20C) PUFAs are precursors for the oxygenation of hormonelike signaling molecules that are produced and secreted in response to tissue stress such as infection, inflammation, and other threats to homeostasis. Those produced from the 20C fatty acids are known as eicosanoids, and include the prostanoids, which are similar in structure to the jasmonates, a class of plant oxylipins. Oxygenated fatty acid signaling molecules that use DHA as a substrate are known as docosanoids and include the resolvins (Rvs) and protectins (PDs).

Sources of EFA

Dietary Sources of LA

LA (18:2n-6) is synthesized from oleic acid (18:1n-9) in plants and is the main fatty acid in the oil of most vegetable seeds such as sunflower (*Helianthus annuus*) (78% LA, <1% ALA), safflower (*Carthamus tinctorius*) (69% LA, <1% ALA), and corn (*Zea mays*) (61% LA, 1% ALA) [2]. Deficiency is extremely rare in the West, due to the large-scale use of these omega-6 rich seed oils as generic vegetable oils in the modern Western diet and the use of soybean (*Glycine max*) (54% LA, 7% ALA) meal as a protein enrichment supplement for domestic animals. The major sources of LA in the diet are through ingestion of fats and oils (22%), meat and poultry (15%), and cereal products (30%). LA-rich cereal products include but are not limited to bread, rice, pasta, pastry, biscuits, and cakes [9].

Dietary Sources of ALA

ALA (18:3n-3), also increasingly known simply as linolenic acid but not to be confused with the omega-6 gamma-linolenic acid (18:2n-6), is synthesized from LA by the chloroplast membranes in higher plants and is thus available for human consumption directly through green, leafy vegetables and indirectly through the consumption of animals which consume green leafy vegetables [10]. It has been shown that wild animals have higher levels of ALA in their meat than domestic animals, including free range, due to a higher dietary diversity of chloroplast-rich plants [11].

Generally, ALA concentrations increase with increased chloroplast activity such as that found in purslane, spinach, rye grass, clover leaves, and marine algae [12]. Recently it was discovered that chloroplasts synthesize more ALA in colder temperatures. This was explained as a defense mechanism used by the plant. The higher degree of unsaturation provided by ALA to the PLs increases membrane fluidity which offsets the suppression to photosynthesis that occurs in low temperatures in the presence of light [13].

The absolute levels of ALA intake from leafy greens, however, are low relative to seed oils as the total lipid content from leafy plants is less than 1% [14]. ALA also accumulates as oil in the seed of certain plants. It was first isolated from hemp seed oil in 1887 (Deul, 1951, as cited by Ref. [15]), but the richest known edible sources from seeds are chia (*Salviahispanica*) (61% ALA and 21% LA) and linseed or flax (*Linum usitatissimum*) (46% ALA and 18% LA) [2,16]. Certain tree nuts, notably the walnut (*Juglans regia*), also concentrate ALA, although walnuts also contain high levels of LA (15% ALA and 58% LA) [17].

Canola oil (*Brassica napus*) is also commonly used as a dietary source of ALA, and it has the added advantage of containing high concentrations of monounsaturated fatty acids (59% oleic acid, 29% LA, and 10% ALA). The issue with canola oil, from the human health perspective, is that it is a mutation of a low-erucic acid variety of rapeseed oil. Erucic acid is a compound that is very useful in industrial applications, having many of the same applications as mineral oil, but may be harmful to human health if consumed in large quantities [2]. The erucic acid content of canola oil is regulated to be less than 5% in the EU [18].

Sources of Arachidonic Acid

AA is rarely found in plant oils but is commonly found in the flesh and organs of animals and in eggs. For instance, 17.2% of the total fatty acids in pork liver is AA. Similarly 1.7% of the total fat content of lean chicken flesh and 1.0% of chicken egg yolks is AA. Human milk contains 0.7% AA [19]. The average intake of AA in the Western diet, largely from meat, poultry, and game, has been estimated at 54 mg/day in sample of 10,851 Australian adults [9], seemingly adequate to replace the membrane AA used up in the generation of various eicosanoids, estimated to be about 1 mg/day [20].

The main source of AA, however, in tissue PLs is derived from its biosynthesis from LA [21]. Sinclair et al. [22] established that dietary LA was a moderating factor for the incorporation of omega-6 fatty acids into plasma PLs. When LA in the background diet was high, plasma phospholipids incorporated LA and AA from cooked lean red meat. However, when LA was low in the background diet, a much more balanced profile of fatty acids were incorporated into PLs; dihomo-gamma-linolenic acid (dGLA, 20:3n-6) and EPA (20:5n-3) were also incorporated along with AA. Both dGLA and EPA are the 20C precursors for two different families of anti-inflammatory and anti-thrombotic eicosanoids, which effectively balance and regulate the potent pro-inflammatory, pro-thrombotic eicosanoids produced by AA, as illustrated in Figure 9.2.

Dietary Sources of EPA and DHA

Marine algae are a class of photosynthetic phytoplankton that produce ALA but also synthesize EPA and DHA de novo [6,23]. Fish feeding on phytoplankton rich in EPA and DHA accumulate these nutrients, which are stored either in the liver (as in cod fish) or flesh (as in mackerel or herring) [24]. Access to EPA and DHA sources has been directly correlated with developmental growth rates in fish larvae and crustacean eggs [25,26]. Larger, cold water fish in particular, such as salmon and tuna, bioaccumulate these nutrients and are thus the richest source of preformed EPA and DHA for humans [16]. Fish are also capable of synthesizing EPA and DHA by desaturation and elongation from ALA. However, as in humans, these conversions

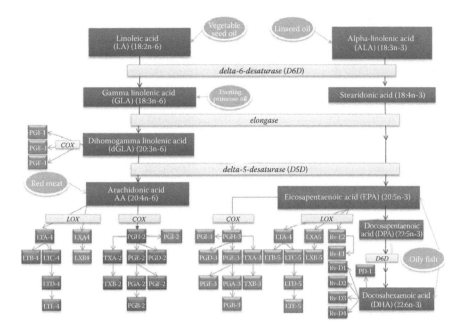

FIGURE 9.2 (**See color insert.**) Metabolic pathways of the essential fatty acids. 'Linseed oil' is also known as flaxseed or flax oil; 'vegetable seed oil' denotes generic linoleic acid (LA) rich vegetable oils as used in industrial food production, including sunflower, safflower oil or corn oils. Dietary LA and alpha linoleic acid (ALA) respectively are absorbed through the gastrointestinal tract and transported to the liver where they are desaturated and elongated into arachidonic acid (AA) and eicosapentaenoic acid (EPA), respectively. AA and EPA circulate in the blood stream until they are taken up into the cell membrane phospholipids (PLs). Upon perturbation of the cell membrane, *phospholipase A$_2$ (PLA$_2$)* releases AA, dihomogamma linolenic acid (dGLA) and/or EPA, which are subsequently converted through *cyclooxygenase (COX)* to the prostaglandins (PGs) and thromboxanes (TXs), and through *lipoxygenases (LOX)* into the leukotrienes (LTs) and lipoxins (LXs). Blue colored mediators denote anti-inflammatory and anti-thrombotic mediators; red colored mediators denote pro-inflammatory and pro-thrombotic mediators.

are limited and believed not to meet the metabolic demands of the organisms, which rely on phytoplankton to provide these essential nutrients [27]. EPA and DHA are thus often referred to as marine omega-3s.

Biosynthesis of EPA and DHA in Humans

Dietary ALA is absorbed through the gastrointestinal tract and transported to the liver where it is taken into cells and serves as a substrate for the synthesis of the LC omega-3 fatty acids. EPA is synthesized from ALA in three steps, as illustrated in Figure 9.2; a double bond is first inserted via *D6D*, the chain is then lengthened with *elongase,* and another double bond is finally inserted with *D5D*.

The metabolic fate of EPA is twofold, as indicated in Figure 9.2. It can serve as a substrate in either of two metabolic pathways. It may be oxidized via *COX* or *LOX* into anti-inflammatory, anti-thrombotic signaling molecules or it may be further elongated to docosapentaenoic acid (DPA).

In humans, DHA is synthesized from DPA in three relatively complex and less completely understood steps. Indeed, the old texts report the use of *delta-4 desaturase*, but current thinking holds that, in humans at least, *D6D* is used again in the insertion of the final double bond [24]. It was recently proposed that there may exist two isomers of *D6D*, such that it is a different isomer that desaturates DHA than used between LA and GLA [37].

The DPA chain is first elongated, then it is desaturated via *D6D* to form the intermediate molecule 24:6n-3. It must then undergo chain shortening, which occurs after it has been transported into the peroxisomes where two carbon atoms are removed in a process of retro-conversion first proposed by "Sprecher," and sometimes referred to as the "Sprecher pathway" [6,24]. Patients with Zellweger's syndrome and generalized peroxisomal disorders are unable to synthesize DHA [28]. It is interesting to note that marine algae bypass the last three steps required by humans and simply metabolize DHA in one step from DPA with *delta-4 desaturase* [6].

Delta-6 Desaturase: The Rate-Limiting Step

D6D and, to a lesser extent, *D5D* are rate-limiting steps in the conversions from ALA to EPA and from DPA to DHA and also from LA to AA. Omega-3 and omega-6 fatty acids compete for these same enzymes for metabolism to the LC 20C fatty acids, which are the eicosanoid precursors. Many dietary and hormonal factors have been identified which influence the activity of the enzyme, thus the rate of desaturation of fatty acids. The presence of large amounts of LA in the diet shifts the enzymatic preference from omega-3- to omega-6-dominated metabolism. In adults, most *D6D* activity occurs in liver cells, but it is doubtful whether the liver supply of DHA is adequate to provide optimal conditions for the development and maintenance of the human brain.

As shown in Figure 9.2, both families of PUFAs compete for the same *desaturase* enzymes, giving rise to antagonistic regulation between the n-6 and n-3 fatty acid metabolism through competitive inhibition [29]. Under normal physiological conditions, enzymatic preference of the desaturases is given to omega-3 metabolism, and has been demonstrated in vitro and in vivo as descending in the preferential order: n-3 > n-6 > n-9 [30].

Enzyme activity is regulated by many factors. For instance, a negative feedback loop by raised levels of the end products of metabolism (i.e., the LC-PUFAs) inhibits the enzyme, effectively blocking further metabolism. Further, when LC n-3 fatty acids (EPA and DHA) are deficient, soy oil (with its n-6/n-3 ratio of 7:1) increases the levels of the n-3 metabolites before those of the n-6 family.

Similarly, high levels of DHA can inhibit the metabolism of LA into AA [31]. That fish oil supplementation decreases desaturase activity, suggests the possibility of a negative feedback loop for further n-3 metabolism and, importantly, antagonistic inhibition of n-6 metabolism [26]. Conversely, high levels of AA reduces the conversions of ALA into its LC derivatives, EPA and DHA [32].

Perhaps not surprisingly, the enzyme *D6D* becomes "overwhelmed" in the presence of a predominance of LA. It responds by shifting its preference to favor n-6 fatty acid metabolism [33]. This implies that high levels of AA and LA in the diet will inhibit enzyme, effectively blocking omega-3 metabolism [32], a point that is often overlooked in the design of research intervention and prevention studies involving human participants. High dietary intakes of LA have been reported to diminish the conversions from ALA to EPA + DHA by 40%–50% [34].

Desaturase activity is also influenced by the type and amount of other dietary fats consumed. Saturated fats and *trans* fatty acids inhibit the desaturases and thereby inhibit essential fatty acid metabolism [35,36]. Many other dietary and hormonal factors influence the activity of the enzymes. For instance, essential fatty acid deficiency and protein are activators of *D6D* while fasting, low protein, glucose, fructose, LA, and AA are depressors. Important cofactors for the enzymes are vitamin A, niacin (vitamin B3), pyridoxine (vitamin B6), zinc, calcium, magnesium [37].

Hormonal controls include insulin as an activator of the enzyme while glucagon and the stress hormones adrenaline and cortisol are depressors [32]. Interestingly, insulin resistance has been associated with altered fatty acid metabolism including high saturated fat conversion to monounsaturated fatty acids (by the enzyme *stearoyl CoA desaturase*) and high *D6D* activity but low *D5D* activity [38], reflecting low conversions to LC omega-3 fatty acids. Interestingly, when a sample of (n = 1062) children were randomized into a longitudinal study at 7 months of age, those for whom the parents had been given advice to reduce total fat and replace saturated fatty acids with unsaturated fatty acids had reduced insulin sensitivity when retested at the age of 9 years [39].

Psychological stress suppresses *D6D* and *D5D* activity in humans [40] and rats [41–43]. In study involving 144 Greek lawyers, a positive correlation was found between the ratio between parent essential fatty acids to the LC derivatives (LA + ALA)/(AA+ DHA) in adipose tissue and trait anxiety on the Spielberger scale (r = 0.25, p = 0.0002). The relationship between anxiety and lower levels of long chain essential fatty acids was accounted for by the inhibition of *D6D* and *D5D* activity by the catecholamines [40]. An impairment of desaturase activity has also been associated with aging, diabetes, high alcohol intake, high cholesterol levels, hepatic disorders, atopic eczema, the premenstrual syndrome, and viral infections [44].

The desaturases have recently been found to be under genetic control. The genes which encode for the desaturases are found within the FAD cluster on chromosome 11, close to those known for common atopic diseases, such as asthma. Common variants of FAD cluster genes have recently been associated with low levels of LC fatty acids and high levels of the 18C fatty acids, implying a functional inhibition of the desaturases in certain genotypes in a large Caucasian sample from Germany (n = 727) [45].

An important clinical implication of such genetic haplotypes that increase *D6D* and *D5D* activity in the context of the Western diet is that more LA is metabolized to AA, the precursor for the strongest, most potent, pro-inflammatory, pro-thrombotic, vasoconstrictive signaling molecules. This combination of the modern Western diet, with its ratio of LA to ALA in the range of 10–50:1, in those with a genetic predisposition to increased *D6D* and *D5D* activity predisposes to atherosclerotic disease. Martinelli et al. [46] have described these nutrient/genetic/metabolic interactions under the banner of a "Desaturase Hypothesis" for atherosclerosis.

Genetic expression and activity after stimulation by a signaling molecule is regulated at the point of gene transcription. That is, genetic control is regulated by factors which influence the selection of specific genes, or segments of nuclear DNA to be transcribed to the RNA molecule, which results in production of a specific protein [47]. The two important nuclear transcription factors that regulate *D6D* expression are SREBP (sterol regulatory element binding protein), which downregulates D6D expression and PPARs (peroxisome proliferator-activated receptor), which upregulates *D6D* expression.

N-3 and n-6 PUFAs inhibit *D6D* gene transcription by reducing the availability of the nuclear content of the SREBP-1 protein, by either delaying its proteolytic release or hastening its decay. Conversely, intake of fibrates, a pharmacological inducer of PPAR-α activator WY 14,643, was shown to induce transcription of *D6D* expression in the liver by some 500% [48]. These observations indicate that *D6D* is under the regulatory control of specific transcription factors. Influencing these transcription factors provide mechanisms for the omega-3 fatty acids and certain pharmacological agents to influence *D6D* expression and activity.

Finally, *D6D* activity was shown 20 years ago to be approximately 60–100 times higher in the liver than the brain of the adult human [49]. It was also understood that human brain *D6D* activity peaks during neurodevelopment, but rapidly declines with age [49]. It was therefore generally accepted that the liver supplies most of the *D6D* conversions required by the brain and body [50]. Moreover, it was widely assumed as highly unlikely that the supply of DHA from endogenous liver synthesis from ALA was adequate to provide the levels of DHA required for optimal human brain function, given that >30% of the brain lipids are ≥20C [51]. Thus, a strong argument exists that preformed DHA must be included in the diet for optimal brain function.

New recent evidence from Cho et al. [37], however, has challenged these long-held assumptions. It was widely assumed that the enzyme consists of three components, including cytochrome *b*5. However, Cho et al. have demonstrated that the enzyme actually functions in the absence of cytochrome *b*5 but is instead dependent on the presence of a cytochrome *b*5-like domain. Cho et al. argue that this shift in understanding of the structure of enzyme may now require a reassessment of previously held assumptions of the distribution and activity of the enzyme. Further, Cho et al. used Northern analysis of mRNA to show that *D6D* is indeed distributed in heart and lung tissue in levels comparable to hepatic tissue. Moreover, they demonstrated that brain *D6D* mRNA was twice that in of the liver. They suggest further research with *D6D* cloning and RNA expression will further our understanding of the constitution, isoforms, and behavior of the enzyme.

D6D is a rate-limiting step in the biosynthesis of long chain fatty acids from their 18C precursors. This appears to be particularly evident for the endogenous synthesis of DHA from ALA. Many factors, both known and unknown, appear to influence this pathway. *D6D* comes under genetic, hormonal and dietary controls. An interaction between genetics and a Western diet, with its characteristic high LA intake, could result in a phenotype with a predisposition to atherosclerosis.

EFA DEFICIENCY

LA was the first, and for a great many years the only, fatty acid to be considered nutritionally essential, having been recognized some 80 years ago. The essentiality of ALA was finally recognized during the 1980s. The issue of whether DHA should be regarded as nutritionally essential is quite controversial and is currently the subject of debate in the literature.

LA Deficiency Symptoms

Essential fatty acid deficiency in animals was first described in 1929 by Burr and Burr [52]. In weaned rats reared on a fat-free diet, they observed an abnormal scaling of the skin between 70 and 90 days of life. When prolonged, there was inflammation commencing in the tip of the tail which later became necrotic, followed by a similar process in the feet. Pruritus was indicated by the constant scratching of the scaly face with the inflamed feet. A decline in weight commenced around the 5th month of life with internal abnormalities such as fatty liver, kidney lesions, lung changes, suppressed reproductive capacity, and ultimately death 3–4 months thereafter. This process was immediately and fully reversed at any stage prior to death with the administration of 2% fatty acids from lard. The particular lard used in the study was not analyzed but published approximations of lard were provided: palmitic acid (16:0) 25%, stearic acid (18:0) 15%, oleic acid (18:1n-9) 50%, LA (18:3n-6) 10%, and traces of linolenic acid and AA.

The following year Burr and Burr set out to discover which fatty acids were curative in the fatty acid deficiency disease [53]. They demonstrated that linseed oil and corn oil were equivalent in reversing the loss of growth and scaly skin disease processes. They studied isolated forms of stearic acid, oleic acid, and LA but did not include a study of linolenic acid (omega-3), concluding that LA (omega-6) was the essential fatty acid.

In the late 1970s, LA was identified as a structural component of glycosphingolipids and ceramides in the *stratum corneum* of pig and human epidermis [54,55]. Gray and White [55] attribute the high melting point of LA as providing the conditions for increased resilience of the skin to withstand wide temperature changes, atmospheric oxidation, and ultraviolet radiation. Ziboh [56] demonstrated that epidermal LA is spared from metabolism to AA, further reinforcing the structural importance of the LA molecule for the skin. Further animal studies by Ziboh et al. [57] demonstrated that (LA-rich) safflower oil could reverse the severe scaly dermatitis, epidermal hyper-proliferation, and percutaneous water loss in EFA-deficient rats. By comparison, a marine oil preparation (rich in EPA + DHA) was not effective at reversing the cutaneous symptoms of EFA deficiency, strengthening the evidence for the essential requirement of LA for the protective barrier in the outer layer of the skin.

EFA Deficiency Diagnosis

In the research literature, essential fatty acid deficiency is determined by levels of eicosatrienoic acid (20:3n-9), often referred to simply as trienoic acid. This molecule is normally present in trace levels in the PLs but its synthesis is increased from oleic acid during EFA deficiency in an attempt to maintain a degree of unsaturation in cellular membranes, required to provide a certain molecular order or fluidity in the membrane [58].

Piefer and Holman [59] demonstrated that a high dietary saturated fatty acid to LA ratio shifts the lipid profile of heart PLs in a manner similar to that observed in frank EFA deficiency. That is, in both EFA and high saturated fatty acid diets, the heart PLs selectively increased the long chain PUFA concentrations (AA, EPA, and DHA) in an apparent "sparing" effect of PUFAs.

Moreover, Piefer and Holman observed that trienoic acid was raised in high saturated fatty acid diets, even when LA was included in the diet, raising the possibility of a "relative deficiency." Holman [60] argues that an increased intake of saturated fatty acids in the diet therefore increases the requirement of PUFAs. In addition, Holman argues that a more accurate indicator of EFA deficiency is therefore the triene/tetraene (AA) ratio, the 20:3n-9/20:4n-6 ratio. Holman demonstrated that when this ratio in blood or erythrocyte PLs is less than 0.4, the minimum requirement of LA (approximately 1% of caloric intake) has been met.

ALA Deficiency Symptoms

The first recorded instance of ALA deficiency in humans was published by Holman et al. [60] in 1982 as a case report of a hospitalized 6-year-old girl that was maintained on total parental nutrition after surgery in which she had lost part of her intestine as a result of a gunshot wound. After 5 months she presented with "numbness, paresthesia, weakness, inability to walk, pain in the legs, and blurring of vision." As part of the investigation, the nutritional formula was investigated and its PUFA content was found to be sourced from safflower oil, which is rich in LA but contains negligible ALA. When the oil was changed from safflower to soybean oil, which includes ALA, the neurological disturbances were reversed. Holman et al. established the minimal dose for the prevention of the observed deficiency symptoms was 44 mg/kg weight/day or 0.5%–0.6% total energy.

Soon after, Nueringer et al. [61] studied ALA dietary deficiency in Rhesus monkeys, demonstrating a progressive loss of visual acuity associated with reduced tissue levels of DHA. Pregnant monkeys were fed ALA-deficient (safflower oil, n = 7) diets during pregnancy and the neonate was continued on the diet and compared with controls. The control diet was fed soybean oil (n = 8). At birth, the ALA-deficient diet group had 42% less DHA in plasma PLs. After 4 weeks on the diet, blood lipid DHA levels had dropped to 79% of the control values, then 91% at 8 weeks and 94% by 12 weeks after birth. This depletion of DHA was associated with reduced visual acuity, as assessed by the preferred looking method, by one quarter at 4 weeks ($p < 0.05$), and one half at 8 and 12 weeks ($p < 0.0005$) compared with controls.

In addition, Neuringer et al. discussed an interesting but unanticipated finding. In the Rhesus mothers, tissue DHA levels dropped in the control (soybean oil) group to a similar level to that of the safflower oil mothers' group. This was explained by the presence of DHA in the stock diet that the monkeys were consuming prior to the study. When this diet was replaced by the experimental diet of soybean or safflower oil, there was no longer a source of dietary preformed DHA. Thus it appears that there was a very limited capacity of the monkeys to endogenously synthesize DHA from its precursor ALA. Further, the stock diet had contained only 0.4% DHA to total fatty acids but the plasma PLs accumulated >4% DHA to total fatty acids, indicating a selective preference for the retention of DHA by the PLs.

In the late 1980s, Bjerve et al. [58] observed four cases of ALA deficiency in institutionalized elderly patients fed entirely by gastric tubes. The supplemental nutrient provided was a formula, registered for the use in gastric feeding, with its PUFA component provided by corn oil. Because of the age of the patients, central nervous system function was not assessed. However, the patients presented with cutaneous symptoms, notably a light scaly dermatitis ("much like dandruff"), a pronounced skin atrophy, and low blood and erythrocyte PL levels of ALA, EPA, and DHA but normal levels of LA and AA.

Both the clinical symptoms and blood and erythrocyte omega-3 levels were completely reversed within 4 weeks of omega-3 fatty acid supplementation (10–20 mL cod liver oil + 20 mL linseed oil). This corresponded with a ratio of 0.8% ALA to 6.5% LA of total energy. Of note was that blood and erythrocyte PLs of LA and AA levels were not affected by omega-3 supplementation. Based on these observations, Bjerve et al. recommend that adults require 0.2%–0.3% of ALA and 0.1%–0.2% EPA + DHA in the diet for the prevention of omega-3 deficiency syndromes.

Bjerge et al. [62] followed these observational case studies with an interventional case study of the treatment of ALA deficiency symptoms in an elderly patient. Through gastric tube feeding, the patient had been fed a diet consisting of 0.024% total calories as ALA for 2 years and presented with scaly dermatitis. She also had signs of seriously deteriorating health. When 0.1 ethyl linolenate was added directly to the gastric tube nutritional formula, increasing ALA to 0.5% of total energy, and the scaly dermatitis cleared up within 5 days. Within 14 days of treatment, erythrocyte PL DPA and DHA, but interestingly not EPA, had increased four- to fivefold, and the abnormal blood triglycerides and cholesterol levels were normalized. On the basis of this observation, Bjerge et al. speculated that conversion to DHA may be selectively enhanced during EFA deficiency. However, with just one case study it is difficult to generalize, especially as the patient was seriously ill and died soon after.

Until recently, a putative role for ALA per se was unknown. Sinclair [63] argues, however, that there is good evidence to suggest such a role for ALA in the sebaceous gland secretions of the skin into the fur of mammals, protecting the skin and fur from the harmful effects of water and light. Furthermore, ALA acts as a carrier for the fat-soluble antioxidant, vitamin E, into skin and fur providing protection from oxidation. In their earlier tracing studies in guinea pigs, Fu and Sinclair [64] demonstrated that almost half (46%) of a single dose of ALA was deposited in the skin and fur lipids (most in the fur) and 39% was exhaled as CO_2 in the first 12 h. From this they surmise that as almost 90% of its intake is used up in these pathways, very little ALA remains for the use as a substrate for metabolism to LC fatty acids.

FUNCTIONS OF OMEGA-3 FATTY ACIDS

At the 1989 NATO Advanced Research Workshop on the biological effects and nutritional essentiality of dietary n-3 and n-6 fatty acids, it was agreed that n-3 fatty acids generally: (1) have anti-inflammatory properties; (2) lower serum triglycerides and cholesterol; and (3) decrease thrombosis and platelet aggregation. Therefore administration was recommended as beneficial in cardiovascular disease, hypertension and rheumatoid arthritis [65].

In the past 20 years, there has been a wealth of evidence to support the notion that the omega-3 fatty acids are not bioequivalent and that the LC EPA and DHA are much more important than ALA [66,67]. The effects of EPA are thought to be mediated via synthesis to its eicosanoids, as shown in Figure 9.2, including the recently identified lipoxins (LXs) and Rvs. DHA is selectively maintained in cell membrane PLs, especially in the retina and brain, and recently new families of molecules have been identified, including Rvs and PDs, which use DHA as the substrate for their synthesis.

Physiological Effects of ALA

Consumption of just four walnuts per day for 3 weeks in 10 volunteers doubled ALA (from 0.23 ± 0.07 to 0.47 ± 0.13) and more than tripled EPA (from 0.23 ± 0.37 to 0.82 ± 0.41) fractions in plasma PLs, but interestingly had no effect on DHA [68]. This was consistent with earlier studies which found that 20 mL of linseed oil daily for 2 weeks increased EPA in plasma PLs from 1.3% to 2.6% and in blood platelets from 0.2% to 1.2% but did not change DHA [69]. This seemingly small increase in EPA in platelet PLs is associated with a significant functional change in platelet behavior.

In their letter to the Lancet in 1983 entitled "Small is Beautiful," Renaud and Nordoy [70] demonstrated that small increases in platelet membrane EPA are associated with "striking" alterations in platelet activity. They use data from their study of 25 French farmers, in which butter was replaced with a margarine enriched with erucic acid-free rapeseed oil, providing oleic acid and, most importantly, ALA.

At the conclusion of the 1 year study, ALA and EPA concentrations in blood PLs had significantly increased. ALA was not taken into platelet membranes, but EPA had significantly increased in platelets (from 0.36% to 0.52%, $p < 0.02$). DHA was not significantly increased in either blood lipids or platelets. The concentrations of LA and its metabolite, AA, were not significantly altered in plasma PLs but both were reduced in platelet PLs (LA from 6.11% to 5.58%, $p < 0.02$, and AA from 29.10 to 26.76, $p < 0.001$). The seemingly small changes to the lipid profile of the platelet membranes were associated with highly significant reductions in platelet aggregation activity across a number of potent physiological inducers (thrombin, adenosine diphosphate [ADP]) and collagen, all $p < 0.001$.

Whether ALA itself is cardioprotective or its effects are mediated via the synthesis of EPA is incompletely understood [71]. There is evidence that ALA itself has antiarrhythmic effects on cardiac cells by directly blocking potassium voltage-gated channels (Kv1.5 channels) in a manner similar to EPA and DHA. However, the mechanisms of EPA and DHA are much more effective as they also downregulate the expression of the atria-specific Kv1.5 channels [72].

One of the best known outcomes of increased ALA in the diet, particularly for heart disease and other inflammatory diseases, is that it competes with the metabolism of LA, the precursor to AA, which is, in turn, the precursor to the most potent form of pro-inflammatory mediators, such as the platelet-aggregating thromboxane (TX) (2 series) and leukotrienes (LTs) (4 series), as illustrated in Figure 9.2. Indeed, Renaud [73] argues that even diets high in plant and marine omega-3 fatty acids are not effective in reducing ADP-induced platelet aggregation against a background diet where the polyunsaturated to saturated fatty acid ratio is higher than 0.7. Renaud cites the failure of his own study (1989) in secondary prevention and a study in primary prevention conducted by Frantz et al. (1989) [74], in both of which this ratio was >1.5, and both of which failed to demonstrate the efficacy of omega-3 fatty acids in reducing mortality. Renaud suggests that in order to reduce the potency of ADP to stimulate platelet aggregation, saturated fatty acids should not exceed 10% of total caloric intake, and thus total dietary PUFAs, of which LA is by far the most dominant, should not exceed 7% of total energy intake.

The widespread assumption that ALA is an essential nutrient for humans due to its ability to maintain tissue levels of its LC metabolites, EPA and DHA, is currently being debated in the literature. Barceló-Coblijn and Murphy [75] argue that ALA is the only essential omega-3 fatty acid required by humans as it is a substrate for the synthesis of the LC fatty acids. However, there is growing support for the argument that that these conversions are not efficient as the evidence mounts that intake of high doses of ALA does not necessarily result in a significant increase in tissue levels of DHA.

While conversions to EPA and the intermediary molecule DPA (22:5n-3) are more consistently demonstrated, those to DHA are much lower and more limited than previously assumed. For instance, when stable isotope-labeled ALA was administered in a series of British studies by Burdge et al., there was very little conversion right through to DHA in healthy male volunteers (EPA 7.9%, DPA 8.1%, and DHA 0%–0.04%) [76] but almost 10% conversion to DHA was found in females (EPA 21%, DPA 6%, and DHA 9%) [77]. Moreover, in a population of pregnant women, supplementation with ALA significantly increased maternal and neonatal EPA and DPA but not DHA [78].

There is no doubt that EPA and DHA are critical nutrients for the human body; both are precursors for families of oxygenated fatty acids that function as beneficial signaling molecules, which maintain tissue homeostasis during inflammation. Further, EPA is a critical molecule for the regulation of platelet aggregation and the DHA molecule itself is an essential structural component of every cellular membrane in the body, and is particularly required for the development and maintenance of mammalian the brain and retina [79].

The essence of the current debate is whether ALA alone is a sufficient dietary source for the biosynthesis of DHA in the quantities required for the optimal maintenance of the human body and brain [63]. There is now a body of strong evidence that plant omega-3 fatty acids do not provide optimal levels of DHA in humans in all circumstances [26] and a call for recognition of DHA as "conditionally essential" [80].

EPA-Specific Functions

Toward the late 1970s, it was noted that myocardial infarction was rare in Greenland Eskimos where plasma PL erythrocyte EPA were astonishingly high and archidonic acid was extremely low. For instance, in blood PLs of a sample of Eskimos, EPA was 7.1% of total fatty acids, compared with just 0.2% in Danes. For AA, the Eskimos had only 0.8%, where the Danes had 8% of AA to total fatty acids in blood PLs [81]. The Eskimos' blood PL profile coincided with low plasma total cholesterol and triglycerides, and low levels of low-density lipoproteins (LDLs), very low levels of very low lipoproteins (VLDLs) and high levels of high-density lipoproteins (HDLs); the blood lipid profile that health authorities in the Western world promote as protective against heart disease. Of note, as there is very little ALA in the diet of the Eskimo, EPA is consumed directly from the diet, predominately from cold water oily fish [82].

Deryberg et al. [82] noted that the longer bleeding times in the Eskimos indicated a role for the powerful clotting agent, thromboxane (TXA_2), an AA metabolite from platelet PLs in the mediation of thrombosis. They deduced that the discovery of the synthesis of a powerful anti-aggregating mediator by blood vessel wall cells, also from AA, prostacyclin (PGI_2), indicated that some kind of net balance of pro- and anti-aggregating forces between these two AA mediators could be a key determining factor in thrombosis. In their groundbreaking study, Dyerberg et al. demonstrated that not only EPA was the precursor to a non-aggregating thromboxane (TXA_3) but also a powerful anti-aggregating prostacyclin (PGI_3) synthesized from EPA by the vessel wall cells. Therefore, displacement of even a small proportion of AA in platelet membranes by EPA could shift the TX-PGI balance toward an anti-aggregatory state with a low thrombotic tendency.

When an attempt was made to generalize these findings for EPA to ALA from linseed oil, it was quickly rebutted. Hornstra et al. [83] suggested that EPA need not be sourced directly from the diet and that it could sourced from linseed oil, via liver biosynthesis from ALA. In their rebuttal, Dyerberg et al. [81] retorted that the literature did not support such an assumption and they further reported on a case study with a volunteer that took 15 mL cod liver oil (containing 10% EPA) three times per day for 8 days. Then, 2 months later, the same volunteer was given the same dose of linseed oil (containing 60% ALA) for the same period of time. With cod liver oil, EPA plasma lipids increased "strikingly," compared with nonsignificant EPA changes, even after the dose of linseed oil had been increased fivefold. From this, Dyerberg et al. concluded that the desaturations and elongations from ALA to EPA are inefficient in humans. However, they did not report on the presence of LA in the background diet of the volunteer. As ALA and LA compete for *D6D* for desaturation, high levels of LA could effectively block the metabolism of ALA to EPA.

Eicosanoids

The term "eicosa" comes from the Greek, meaning "twenty." The term "eicosanoid" refers to the oxygenated products of the 20C fatty acids. These are AA (20:4n-6), EPA (20:5n-3), and to a lesser extent, dGLA (20:3n-6). These 20C fatty acids are the substrates for a variety of powerful hormonelike signaling molecules, collectively known as eicosanoids [84]. Eicosanoids are characterized by their potent, short-range action on neighboring tissue, low tissue levels, and rapid metabolic turnover. The important classes of eicosanoids are the PGIs, prostaglandins (PGs), TXs, LTs, and LXs [85].

The PGs, TXs, and LTs have the potential to be extremely strong pro-inflammatory, pro-thrombotic signaling molecules when generated from AA. Conversely, when they are generated from EPA, PGs, TXs, and LTs are much lighter inflammatory mediators, often competitive and antagonistic of AA-derived eicosanoids and thus effectively anti-inflammatory in nature. PGIs and LXs, from both AA and EPA, are anti-inflammatory and anti-thrombotic. The LXs from both AA and EPA are pro-resolution of inflammation. These mediators, from either AA or EPA, attempt to balance the actions of the pro-inflammatory eicosanoids during inflammatory processes and as such are involved in tissue homeostasis.

Eicosanoids are not stored in the cell but are produced de novo upon the slightest provocation. In response to any cell membrane disturbance, whether physiological as in the stretching of smooth muscle or pathological as in inflammation, *phospholipase A_2 (PLA$_2$)* signals the cell membrane to release the 20C fatty acids. PLA_2 itself may be activated in response to local tissue damage, phagocytic particles, microorganisms, the pro-inflammatory cytokine, IL-8, or platelet activating factor (PAF) [86].

The release of PGs from cells in response to any change in the membrane may be understood as defense mechanisms, aimed at reducing the stimulus of cell membrane change or minimizing its disruption and, in some cases, such as the stomach, bladder, and uterus, facilitating the adaptation to change by promoting relaxation of the stretched muscle [87]. In the cytosol of the cell after its release from the membrane PLs, the 20C fatty acid substrates are oxidized into various potent signaling molecules, depending upon the fatty acid substrate and the enzymatic pathways involved in the conversions.

A major determinant of which series of eicosanoids are produced is the balance of fatty acids in the membrane PL, which in turn is determined by dietary intake and fatty acid metabolism [88,89]. Eicosanoid synthesis appears to respond rapidly to dietary changes in fatty acid intakes [90]. Individuals with low intakes of omega-3 fatty acids have been shown to release high levels of AA from cell membranes during inflammation [85]. Dietary EPA displaces AA from the cell membranes and shifts the metabolic state of the organism toward an anti-inflammatory, anti-thrombotic state.

Prostanoids and Leukotrienes

PGs and TXs, collectively known as prostanoids, are formed by the *COX* enzymes while the LTs and LXs are generated by *LOX* enzymes. The *COX* enzyme has two isoforms, *COX-1* and *COX-2*. While *COX-1* is expressed constitutionally in most cells, for instance in the gastric mucosa, *COX-2* is not normally present but is inducible at sites of inflammation [91].

After its release from the PL membrane, the *COX-1* and/or *COX-2* converts AA to the intermediate molecule PGG$_2$, then to PGH$_2$. PGH$_2$ is then the substrate for a variety of prostanoids, PGE$_2$, PGF-α, PGI$_2$, and TXA$_2$. The types of prostanoids produced depends upon several factors, most notably the specific profile of the synthases enzymes that may be expressed in the particular cell type but also the presence of *COX-2*. For instance, macrophages predominantly produce PGE$_2$ and TXA$_2$. Interestingly, macrophages produce different profiles of prostanoids depending upon activation state. During activation, for example, PGE$_2$ secretion from macrophages is enhanced. This change in prostanoid synthesis during activation may result from inducement of the expression of *PGE synthase* by the pro-inflammatory cytokine, IL-1β [92]. It may also be influenced by the presence of *COX-2*, which favors secretion of PGE$_2$ and PGI$_2$ as distinct from the *COX-1* induced prostanoid profile, which favors a balanced array of prostanoids with more diverse effects [93].

Prostanoids have a wide range of biological actions, the detailed description of which has been the subject of voluminous scholarly attention since their discovery in semen in the 1930s [84]. In general, they are inflammatory mediators and their clinical relevance can be measured by the prolific use of drugs which target their biosynthesis, the nonsteroidal anti-inflammatory drugs (NSAIDs).

TXs are generally involved with platelet aggregation and the LTs with vascular constriction, but the actions of the PGs are quite complex. For instance, PGE_2 is involved with both pro-inflammatory and anti-inflammatory activity. It would appear that in the early stages of inflammation, PGE_2 is pro-inflammatory, but at a certain point it switches the transcription of the *LOX* enzymes away from LT synthesis and toward LX synthesis. The generation of LX coincides with the spontaneous resolution of the inflammatory process [94].

Competitive Inhibition

The PGs, TXs, and LTs produced from n-3 fatty acids are antagonistic to those produced by n-6 fatty acids by competing for the same enzyme systems. For instance, PGE_3 inhibits PLA_2, effectively blocking the release of AA from cell membranes. In this way, the presence and activities of n-3 derived PGs play a critical role in the regulation of the potent n-6 derived PGs.

PGs from the n-6 dGLA, such as PGE_1, have physiological effects remarkably similar to the n-3 derived PGE_3. For instance, PGE_1 is a vasodilator, inhibits platelet aggregation, inhibits collagen deposition and has been shown to have a direct effect on the regulation of cellular energy [95]. Further, 15-OH-dGLA, a metabolite of dGLA, inhibits *LOX*, which effectively blocks the conversion of AA into the LTs. It is important to note that although dGLA and AA are both n-6 fatty acids, they are precursors for different series of eicosanoids, which are often antagonistic with each other. The series-1 PGs are critical in the regulation of the pro-inflammatory series-2 AA-derived eicosanoids, although they are present in much smaller quantities [95].

In summary, eicosanoids derived from AA (series-2 PGs and TXs and series-4 LTs) instigate pro-inflammatory cascades. For instance, PGE_2 and LTB_4 induce the secretion of many other pro-inflammatory mediators, such as pro-inflammatory cytokines (including the interleukins [IL], interferons [IFN] and tumor necrosis factors [TNF]). Conversely, omega-3 fatty acids (series-3 PGs and TXs and series-5 LTs) are generally much milder inflammatory mediators or anti-inflammatory and anti-thrombotic in nature.

Lipoxins

The most exciting breakthrough in PUFA research in the recent past has been the discovery of new families of molecules, derived from both 20C and 22C PUFAs, which actively mediate the resolution and termination of the inflammatory response. The first to be discovered were the LXs (trihydroxy-eicosatetraenoic acids) formed endogenously from AA and EPA substrates, followed by the Rvs formed from EPA and DHA and the PDs, formed from DHA.

LXs and aspirin-triggered lipoxins (ATL) are now known as important "braking" signal molecules during the inflammatory response. It is known that aspirin has its anti-inflammatory effects through the inhibition of the actions of the *COX* enzymes by causing the acetylation of the enzymes. What was recently discovered is that in epithelial and endothelial cells, the aspirin-induced acetylated *COX-2* enzyme subsequently uses AA as its substrate to form 15*R* hydroxyeicosatetraenoic acid (15*R*-HETE). In monocytes and leukocytes, 15*R*-HETE is rapidly converted to 15-epi-lipoxins (also known as epi-LX or ATL) by *5-LOX*. Two additional pathways are now known whereby LXs such as LXA_4 and LXB_4 are formed from AA, independently of the aspirin-*COX-2* interaction, via other *LOX* enzymes in airway epithelia, leukocytes, and platelets.

One of the *LOX* pathways starts with *5-LOX* converting AA into 5-HPETEs (5-hydroperoxyeicosatetraenoic acid), an intermediate hydroperoxide molecule that is rapidly converted into 5-HETE. 5-HETE is then the substrate for LTs, such as LTB_4, and LXs, such as LXA_4 and LXB_4. In a second pathway, *15-LOX* converts AA to 15-HPETE and then *5-LOX* converts this to 5-HETE and then to the LXs, LXA_4 and LXB_4 [8]. Interestingly, in the process of producing LXs, less LTs are produced.

LXs are secreted in the inflammatory milieu in response to activation of *LOX* enzymes by IL-4 and IL-13. LXs are anti-inflammatory in nature, counteracting the initial pro-inflammatory response phase and restoring homeostasis for the tissues after the activation of the acute inflammatory response. Impairment of LX pathways has been suggested as a predisposition to chronic inflammatory diseases. LXs and ALTs suppress leukocyte trafficking to sites of inflammation, particularly eosinophil and neutrophil recruitment and activation. They promote resolution of the inflammatory process by directly activating endogenous anti-inflammatory pathways. For instance, LXs downregulate the highly inflammatory nuclear transcription factor NF-kB and upregulate the transcriptional co-suppressor factor NAB1, which works in the endogenous anti-inflammatory pathways with the glucocorticoids [96]. Importantly, LXs promote clearance of apoptotic cells by stimulating macrophages to phagocytosis and reprogramming cytokine-primed macrophages from pro-inflammatory activity to phagocytic activity [97].

The discovery of the pro-resolving nature of LXs and the recognition that the termination of inflammation was not a passive process but a concerted active response led to a search for further understanding of the conditions that facilitate the resolution phase of inflammation. The subsequent discovery of very small but extremely potent bioactive molecules, known as "resolvins" and "protectins," derived from long chain omega-3 fatty acids, represents a paradigm shift in understanding the scope of potential mechanisms that these fatty acids use to terminate inflammatory processes and maintain tissue homeostasis [98].

DHA-Specific Functions

The major functions of DHA appears to be the maintenance of the structure and function of cellular membranes, a role which is critical for the gray matter of the brain as it is rich in membranes. The rate of incorporation of DHA into cell membranes depends upon the cell and tissue type and dietary intake [99]. Cells in the retina and brain selectively concentrate DHA. DHA constitutes 95% of the omega-3 fatty acids in the retina and 97% in neurons [100]. Recently, DHA has been identified as the substrate for family of Rvs and a molecule with neuroprotective actions, called neuroprotectin when found in the brain.

Cell Membrane Phospholipids

All cells are contained within a cell membrane, which consists of thin bilayer of PLs, within which proteins are suspended. The cell membranes are not static and passive but are highly dynamic, fluid structures. Perhaps their most fundamental role is to maintain a fluid environment in order to carry out vital functions for the cell [101].

Critical functions of the cell membrane include maintenance of the membrane potential, regulation of ion channels, receptor formation, ligand binding to receptors, intracellular and intercellular signaling, gene expression and cell differentiation, and the formation of eicosanoids [102] and the newly discovered docosanoids [103].

Cell membrane fluidity generally increases with increasing unsaturation of the fatty acids with are contained within its PLs [104]. When compared with saturated fatty acids, unsaturated fatty acids in PLs are associated with thinner, more permeable membranes with increased water permeation that are less sensitive to mechanical stress [105]. DHA is a unique fatty acid, as it has the highest degree of unsaturation with its six *cis*-isomer double bonds.

It has been shown that the loss of a single double bond from the hydrocarbon chain in DHA significantly alters the properties of the cell membrane [79]. Computerized 3D structures comparing DHA (22:6n-3) with DPA (22:5n-6) demonstrated that the final double bond in DHA, not present in DPA, enables the molecule to fold itself into a slightly spiral structure [106]. Recent molecular dynamic modeling of PLs bilayers have demonstrated hundreds of likely configurations for DHA in PLs [107]. In fluid, the molecule is in constant motion, extending itself to take a twisted, helical configuration, and folding itself back in a "hairpin" configuration where the terminus end is close to the surface of the bilayer.

The degree of unsaturation of the fatty acids incorporated into the PL bilayers determines the membrane fluidity and function. Known modifiers of membrane fluidity in the periphery are dietary fatty acids and cholesterol and drugs such as alcohol, cannabinoids, barbiturates, steroids, and antidepressants. However, in the brain, membrane fluidity is determined by the types and ratios of the dietary PUFAs [108].

Brain Phospholipids

The brain contains the highest concentration of lipids of all body tissue, with the exception of adipose tissue [109]. Gray matter consists of 40% lipids, while white matter is 50%–70% lipids, as it includes myelin. The most abundant lipids in the brain are PLs, constituting 60% of its dry weight. DHA constitutes between 20% and 40% of the PLs in the brain, retina, and testes [110]. In the brain, DHA is the predominant omega-3 fatty acid and AA is the predominant omega-6 fatty acid.

Glycerophospholipids and sphingolipids are critical types of PLs and are contained in all cellular membranes. They are found particularly concentrated in the gray matter of the brain because gray matter is dense with synapses and is therefore membrane rich [111]. The two fatty acids attached to the glycerol backbone in glycerophospholipids are generally a saturated fatty acid in the Sn-1 position, and an unsaturated fatty acid in the Sn-2 position. The Sn-3 position attaches the phosphate moiety, bonded to a defining polar head group, such as choline, serine, ethanolamine or inositol [111]. In the brain, the PLs with the highest content of DHA are the ethanolamine (PE) and serine (PS) glycerophospholipids (also known as phosphoglycerines or simply but less accurately PLs) [106].

The specific type of unsaturated fatty acid in the Sn-2 position of the glycerophospholipids is determined by the tissue type; monounsaturated FAs predominate in white matter higher while polyunsaturated FAs predominate in gray matter. The proportion of DHA to AA is generally higher in gray matter, but in white matter this is reversed [111].

The unsaturated fatty acid composition of normal brain tissue appears to be age-specific. Svennerholm [112] demonstrated that the glycerophospholipids in the cerebral cortex show steady increases in the concentration of 22:6n-3 DHA with corresponding reductions of 20:4n-6 AA in the normal aging brain. Using tissue samples from 11 brains from fetal to 82 years, the percentages of AA and DHA in cerebral cortex ethanolamine glycerophospholipids was roughly equal or 1:1 (16.5% AA and 16.1% DHA) in the 1-month-old human infant, but by the 82nd year the percentage of DHA had more than doubled, while AA had reduced, resulting in an increased proportion of AA to DHA close to approximately 1:4 (10.3% AA and 33.9% DHA).

It is generally well accepted that neuronal membrane PLs preferentially retain DHA and do not release it readily. Support cells such as astrocytes, on the other hand, readily release DHA into the chemical environment around neurons [78]. There are no appreciable levels of the precursors ALA or LA in the brain even when present in large amounts in serum [49]. PUFAs easily cross the blood-brain barrier by simple diffusion, but only the longer chain fatty acids are selectively incorporated into brain PLs. Dietary supplementation with DHA substantially influences brain PL DHA content [113,114].

Resolvins and Protectins

Very recently novel families of oxygenated lipid mediators from 20C and 22C omega-3 fatty acids (EPA + DHA) were identified which, like LXs, promote the resolution of the acute phase of an inflammation response. The Rvs have also been identified from the omega-3 DPA (22:5n-3) and the omega-6 DPA (22:5n-6). In addition, a family of distinct docosanoids have also been recently discovered. The first molecule identified was initially called neuroprotectin D1 (NP-D1), to designate that it was found in neural tissue and was a product of DHA. However, as NP-D1 has since been found in a wide array of tissue outside the brain, it is simply referred to as protectin (PD-1) when found in peripheral tissue. A third distinct family of oxygenated docosanoids have also been identified called maresins. At this stage very little is known about maresin mediators, although initial findings suggest similar actions as those for Rvs and PDs, that is counter-inflammatory and pro-resolutionary [86].

In conditions of tissue stress, Rvs (derived from "resolution phase interaction products") counteract inflammation, reduce neutrophil trafficking, regulate pro-inflammatory cytokines and reactive oxygen species, and promote resolution and clearance of apoptotic cells in the area. Rvs are generated from EPA or DHA and the naming of each series takes the first letter from the fatty acid substrate. That is, E-series Rvs are generated from EPA and D-series Rvs from DHA. Rvs are generated in pathways similar to LXs. That is, in the presence of aspirin, *COX-2* converts EPA to 18*R*-HEPE (18*R*-hydroxyeicosapentanoic acid) by endothelial cells lining the blood vessel walls. 18*R*-HEPE can be released to neighboring circulating leukocytes for conversion to RvE1 and RvE2 by *5-LOX*.

The DHA side chain is similarly converted to one of its four distinct bioactive Rvs (RvD1–RvD4) with or without the presence of aspirin through *LOX* pathways. If aspirin is present, DHA is first converted to 17*R*-HDHA (17*R*-hydroxydocosahexaenoic acid) by acetylated *COX-2*, but if aspirin is not present, DHA is converted to 17*S*-HDHA by *15-LOX* in vascular cells, then released to neutrophils for further conversion with *5-LOX*.

DHA is also the substrate for *15-LOX* conversion to the PDs via an intermediate molecule, 17S-hydroperoxy-DHA. As yet, only one PD has been identified (PD-1), which is synthesized from peripheral blood mononuclear cells and Th2 CD4 cells and has been found in various tissues including neuronal, microglial, peripheral blood and lung tissue. It is generated in response to oxidative stress and/or neutrophil activation. Maresins are product of the further oxidation of the 17S-hydroperoxy-DHA substrate [86].

Among other brain activities, NP-D1 has been found to induce nerve regeneration, reduce leukocyte infiltration, and maintain homeostasis through aging by reducing pro-apoptotic and pro-inflammatory signaling [115]. It has also been shown to suppress the proinflammatory cytokine (IL-1β) stimulation of *COX* [116]. NP-DI is induced by oxidative stress and protects retinal and neuronal cells from oxidative stress-induced apoptosis. The discovery of NP-D1 and its multiple mechanisms offer new therapeutic opportunities for a wide range of conditions from neurodegenerative diseases, such as Alzheimer's Disease, to the cognitive decline during aging [117].

EVOLUTION OF THE HUMAN BRAIN

There is a compelling argument that the human brain would not have evolved its extensive cerebral hemispheres without a rich and consistent source of preformed DHA in the diet [23,118]. Crawford et al. [106,119] argue that marine blue-green algae, which have populated the oceans for over 2.5 billion years, supplied the DHA that was used in the development of the first visual systems, which eventually evolved into a nervous system and ultimately a brain.

Briefly, Crawford et al. argue that the first visual systems were developed, around 600 million years ago, using DHA as structural support for a photon-sensitive primordial vitamin A molecule. Then, around 300 million years ago, this structure began to utilize solar energy to convert carbon dioxide and water into chemical energy, in the form of glucose. The oxygen released by the newly acquired photosynthesis together with the eventual mitochondrial symbiosis facilitated the development of oxidative metabolism and the electron transport chain, which generated electrical potential [106]. Electrical impulses were initially conducted along electron-sensitive molecules that would develop into primordial nerves and eventually complex nervous systems and brain. Throughout the ensuing hundreds of millions of years, the photon-sensitive vitamin A molecule was supported by DHA as the major structural component of its PL membrane [23].

Crawford et al. [120] studied the brains of 42 land-based mammals, which were found to be similar in chemistry, particularly with regard to the predominant use of DHA in membrane-rich neural tissue, such as that in the retina and at synapses. The human brain was clearly distinguished by its size in proportion to the body. In every other mammalian species they observed, brain size decreased as body size increased.

Together with evidence of early human population of areas concentrated around land/water interfaces, Crawford et al. [106] argue that these trends are consistent with the theory that brain growth was rate-limited by insufficient endogenous synthesis of DHA. As the body grew and its cells required more of the liver *D6D*, there was less DHA available for the membrane-rich synaptic tissue required for higher cerebral development. Further, in the absence of a significant dietary source of preformed DHA, the brains of land-based species did not appear to use the omega-6 DPA (22:5n-6) as a substitute, despite its relative abundance in the food supply. Liver biosynthesis produced adequate DHA to supply the primitive brain and retina, but the size of the evolving brain was limited by dietary supply of preformed DHA.

Crawford et al. [120] point out that over 3 million years of evolution did not appear to remarkably affect the brain capacity of *Australopithecus*. In contrast, brain capacity doubled in the million years between *Homo erectus* and *Homo sapiens*. Further, over the past 200,000 years, there was an exponential encephalization growth rate [23,106]. This evidence supports their theory that *Homo erectus* migrated down rivers to estuaries where the land/water interface supplied a rich source of easily obtainable dietary preformed DHA, seafood that have eaten fish which have fed off the algae. Since DHA is between 10 and 100 times more concentrated in seafood than in terrestrial meats [121], multigenerational exploitation of this food source may account, in part at least, for the extensive cerebral expansion unique to humans.

This theory is consistent with the findings of a groundbreaking analysis conducted by Richards et al. [122]. When a new database of the fossil record was created, about 10 years ago, cataloguing individual collections of fossils, Richards et al. were able to demonstrated that a turning point in human evolution coincided with the inclusion of seafood in the diet [122,123]. They did this by comparing stable isotope values of bone collagen samples from the database of early modern humans with those of Neanderthal populations in similar regions, sourced from the published literature. Stable isotope values of the bone collagen from human remains enabled an estimation of the proportion of protein obtained in the diet from terrestrial versus aquatic sources in the 10 years prior to death. Stable isotopes can additionally distinguish between protein from marine coasts and freshwater wetlands.

Richards et al. [122] analyzed bone of specimens that had lived during the Mid-Upper Paleolithic period (20–28,000 years ago) in the regions now known as Britain, Russia, and the Czech Republic with published carbon isotope values from five Neanderthal specimens that had lived in similar geographic regions, but were dated from an earlier time (from 28–130,000 years ago). There was no evidence of aquatic or marine sources of protein in the bone collagen of the Neanderthal samples. They sourced protein from the red meat of herbivorous prey, most likely wolves, large felids, and hyenas. However, seafood appeared to be a "staple" of the diet of the early modern humans (Mid-Upper Paleolithic period).

Marine or freshwater sources of protein constituted between 10% and 50% of the diet for the early modern humans, depending upon geographical region. Freshwater sources along rivers included fish and/or water fowl while marine sources were along coastlines and included fish, shellfish, and probably small slow-moving animals such as turtles/tortoises. When the radio isotope values were averaged for each group, the difference in the group means was statistically significant (p = 0.005).

Of note was that the exploitation of food from marine and other such aquatic sources apparently coincides with a rise of more elaborate enrichment in material culture, such as personal ornamentation, decoration of burials, pottery figurines, and knotted textiles. Richards et al. [122] speculated that the expansion of the diet to include seafood is likely to have "...rendered humans more resilient to natural pressures and the increasingly packed social environments of Late Pleistocene Europe."

DIETARY IMBALANCE OF EFAs

Humans have evolved consuming an almost equal balance of omega-3 to omega-6 fatty acids in the diet [65]. While some argue this balance was 1:1, others maintain that the Paleolithic diet had a proportion of omega-6 to omega-3 of 5:1. In Japan and Greenland, this balance is approximately 3:1. However, in Europe this ratio is estimated in the range of 10–50:1, while in the United States, this range is estimated in the range 30–50:1 [124].

After the agricultural and industrial revolutions the balance was dramatically shifted to a predominance of omega-6 fatty acids. Inventions such as the seed drill dramatically increased the availability of seed oils in the food supply; multiplied by the feeding of an omega-6 rich seed diet to domesticated animals. The PUFA composition of the meat reflects the changes in feeding: Omega-6 is the main PUFA in domesticated meat and eggs, but omega-3 fatty acids are predominant in free-range cattle and poultry allowed to forage for grass, the terrestrial relative of marine algae [15].

The substantial rise in Western consumption of meat from domesticated livestock in the last century was linked, through observational studies in the 1960s, to the Western epidemic of coronary heart disease. Blood cholesterol was subsequently identified as major risk factor for heart disease, resulting in public health policy recommendations to reduce dietary cholesterol, particularly by replacing dietary saturated fatty acids with PUFAs [125].

As the omega-6 rich PUFA sources, such as sunflower oil, were the most readily available, these recommendations effectively worsened the conditions required for optimal omega-3 metabolism. Further, as PUFAs are readily oxidized, manufacturers developed products by partial hydrogenation to increase shelf life, such as margarines, despite the fact that commercial hydrogenation produced *trans*-isomers, rarely found in nature.

In addition to saturated fats, domesticated meat also contains higher proportions of omega-6 fatty acids in their PUFA fractions, which are passed through consumption to humans [5]. Hibbeln et al. [126] argue that in fact saturated fatty acids and cholesterol were not significantly increased during the twentieth century at all, but the amount of the omega-6 LA consumption was increased threefold in the U.S. diet from an average of 11.7–34.3 g/person/day from 1909 to 1999.

In the same period, estimated PL levels of AA levels rose from 64% to 75%. This change in resulted in a displacement of the long chain omega-3 fatty acids (EPA + DHA) from the cell membranes [29]. This new PL profile has been associated through observational and clinical intervention studies with many chronic inflammatory and autoimmune diseases, such as coronary vascular disease, rheumatoid arthritis, mental health and psychiatric problems, and neurodegenerative diseases [127,128].

In a remarkable multinational study published in the *Lancet*, Hibbeln [127] demonstrated an extremely high inverse correlation between the prevalence of depression in a country and the fish consumption per capita for that country ($r = -0.84$, $p < 0.005$).

Moreover, in a sample of 20 moderately to severely depressed patients, Adams [129] correlated the severity of the symptoms of depression with the imbalance of omega-3 to omega-6 fatty acids in red blood cell PLs ($r = 0.729$, $p < 0.01$). Although the correlations are strong, causality cannot be inferred from observational studies alone. However, these data contribute to the mounting evidence demonstrating a protective effect of the preformed dietary long chain omega-3 fatty acids in mental health.

Omega-3 fatty acids were recently shown in a prospective, intervention study to help prevent psychosis in at-risk young people. In a clinical trial of a 12 week supplementation period of 1.2 g/day of EPA + DHA in 81 young people (13–25 years old) with ultrahigh risk of developing schizophrenia, the supplement was significantly better than placebo at reducing risk of transitioning to psychosis. Where 27.5% of the control group transitioned to psychosis, only 4.9% of the omega-3 group ($p = 0.007$) had developed psychosis within the 40 week follow-up period [130].

A high ratio of omega-6 to omega-3 fatty acids has been associated with violent and aggressive behavior. Hibbeln et al. [131] directly correlated intakes of omega-6 fatty acids with homicide rates in five Western countries, including Australia ($r = 0.94$, $p < 0.00001$). In a sample of habitually violent and impulsive male offenders with antisocial personality, plasma PL DHA were significantly lower than controls ($p < 0.05$), while the omega-6 fatty acid, AA, metabolites PGE_2, and TXB_2 levels were elevated ($p < 0.01$) [132]. In a randomized, placebo-controlled double blind trial involving 231 young adult male prisoners, supplementation of a combination of fish oil with gamma-linolenic acid and multivitamins led to a 23% ($p = 0.03$) reduction in violent and antisocial behavior compared with those taking the placebo. After 2 weeks of supplementation, violence and antisocial behavior were reduced by 35% ($p < 0.001$) in those taking the active supplement where the placebo group was unchanged [133].

In summary, humans have evolved with a rich dietary source of dietary preformed EPA and DHA. Consumed as seafood or as capsules, these nutrients are readily incorporated into lipid transport mechanisms and distributed around the body where they are taken up into the PL membranes of cells in a dose-dependent manner [134]. Brain and retinal tissue rely on DHA, in particular, for optimal functioning and are sensitive to dietary fluctuations [135]. DHA can theoretically be converted from ALA in the liver, but the conversions are rate-limited by many known and unknown factors.

RECOMMENDED DAILY INTAKES

There is no longer any disparity in the literature with regards to the essentiality of PUFAs. Widespread discussions, however, are canvassing the optimal levels of essential fatty acid intakes required for health and well-being. There are considerable differences of opinion regarding the practicalities of providing recommendations on a dietary ratio of omega-6 to omega-3 fatty acids.

It is argued that a recommendation regarding the ratio of n-6 to n-3 fatty acids would not be useful as the different omega-3 fatty acids have different biological effects and a reduction in omega-6 fatty acids is not equivalent to an increase in omega-3 fatty acids. Therefore most health authorities focus on absolute intakes of the individual fatty acids [136]. Nevertheless, the argument is maintained that because of competitive inhibition, there is an imperative to keep the ratio from exceeding 10:1 although some argue that a far more desirable goal should be 4–6:1 [34].

A joint expert committee between the Food and Agriculture Organisation for the United Nations and the World Health Organisation recommended back in 1993 that when intakes of the n-3 to n-6 ratio exceeds 1:10 then omega-3 rich foods should be increased where possible. They also proposed that the goals for the total intake of PUFAs should be between 6% and 10% of total energy intake, with 1%–2% from omega-3 fatty acids and 5%–8% from the omega-6 fatty acids [137]. In France, 6% of energy intake from LA, 1% from ALA, and 0.4% from EPA + DHA is considered acceptable [124].

Recommended daily intakes are difficult to establish due to the lack of data indicating dose-response relationships. In Australia, the National Health and Research Medical Council (NHRMC) gives estimates for the average intakes that are based on data provided by the 1995 Australian National Nutrition Survey. However, there is an acknowledgment that these values are not necessarily indicative of optimal intakes, simply those "found in a population with little apparent essential fatty acid deficiency" [138].

This raises an important point about difficulty in diagnosing essential fatty acid deficiency. Severe or frank essential fatty acid deficiency, as observed in the gastric feeding case studies, is not nearly as common as the more subtle subclinical deficiency, perhaps better described as insufficiency [139]. For instance, low-fat diets in the population may lead to essential fatty acid "insufficiency" after several months, where it may take years to manifest as a complete depletion or "deficiency" of essential fatty acids. Thus, diets high in processed foods containing large amounts of LA and *trans* fatty acids may create an insufficiency of omega-3 fatty acids that is not the same as the neurological deficiency syndrome seen in patients on long-term parental nutrition devoid of essential fatty acids.

Low blood PL and erythrocyte DHA concentrations are widespread in the Western diet [128,140]. Low PL levels of DHA have been associated with many diseases and conditions, particularly related to psychiatric and neurodegenerative diseases, for instance, depression [141,142]; learning and behavioral deficits [143] and ADHD in children [144]; Parkinson's disease and Alzheimer's disease [145]; impulsivity and suicide [146]; violence, hostility, and aggression [147,148]; and age-related cognitive alterations [149,150].

Australia is currently facing mental health issues of epidemic proportions. Approximately 20% of Australians aged between 18 and 65 years had a mental disorder in 2007 [151]. Most (14.4%) were anxiety disorders, with posttraumatic stress disorder the most predominate (6.4%) of the anxiety disorders. Mental health disorders are the leading cause of the nonfatal burden of disease and injury in Australia [152]. Chronic exposure to stress in the workplace at least doubles the risk of cardiovascular disease and depression and "translates to a large, preventable chronic disease burden" [153].

The NHMRC have now established Strategic Dietary Targets for the Prevention of Chronic Diseases for the long chain omega-3 fatty acids at 610 mg/day for men and 430 mg/d for women. In the 1995 National Nutrition Survey, the average intake from fish and seafood sources was estimated to be 189 mg/d while the median was just 30 mg/day [9]. These estimates were recently revised to account for the contribution of the omega-3 fatty acid DPA from red meat. The Australian average daily intake was reestimated to be 246 mg but the median daily intake was 121 mg [154]. The average was believed to be skewed to the right because a subgroup of people supplement with high doses of fish oil. Therefore, most people in Australian are consuming around a quarter (25%) of the daily intake recommended to prevent chronic diseases.

For the prevention of coronary heart disease, most Western countries around the word recommend that individuals should consume an average of 500 mg/day from EPA + DHA, the equivalent of approximately two oily fish meals per week [155]. France is the only country which specifically recommends preformed DHA at 120 mg/day. Curiously, the average intake of DHA is 448 mg/day in France, which is much higher than other Western countries. Compare this with approximately 70 mg/day in the United States, 106 mg/day in Australia, and 170 mg/day in Germany [9,156]. During pregnancy and breastfeeding, there is an increased requirement; some recommend 200 mg/day DHA, while intakes up to 3 g/day of DHA have been shown to be safe [157].

The current Australian NHMRC recommendation AIs (adequate intakes) have been estimated on the median population intakes as estimated from the 1995 National Nutrition Survey. These are for men 13 g/day LA, 1.3 g/day ALA, and 160 mg/day combined EPA + DPA + DHA. For women, they are 8 g/day LA, 0.8 g/day ALA, and 90 mg/day combined LC omega-3 fatty acids. These values for the omega-3 fatty acids are comparable with those given in the United States by the Institute of Medicine; 1.6 g/day ALA for men and 1.1 g/day for women, 10% of which can be consumed as EPA + DHA, which represents the current mean intake of EPA + DHA in the American diet, approximately 100 mg/day [155].

Many in the population require increase levels of long chain omega-3 fatty acids, such as during pregnancy and lactation. In addition, requirements are increased during inflammation, as more EPA and DHA are used up in the conversions to Rvs. The optimal dose for treating inflammatory symptoms in Rheumatoid Arthritis has been estimated through meta-analysis to be 3 g/day [158]. A recent review found that the amount required to reduce blood pressure in people with hypertension is in the range of 3–9 g/day of EPA and DHA [124]. A large sample (n = 11,324) open label, randomized controlled Italian study, known as the GISSI-P trial, demonstrated that 875 mg (0.88 g) EPA + DHA per day was sufficient to reduce the risk of all-cause mortality in survivors of myocardial infarction [159].

MODERN NUTRACEUTICALS

When nutrients are consumed, as nutritional supplements or fortified foods, in order to treat or prevent diseases, they come under the relatively new banner of the so-called nutraceuticals [160]. Omega-3 fatty acids, particularly DHA, are increasingly being added back to the food supply in the form of individual supplementation, direct fortification of foods and fortification of animal feed in order to confer beneficial health benefits. Driven by consumer demand, DHA and EPA are increasingly incorporated directly into industrial food production, despite increased production costs. ALA-rich foods, such as linseeds, are being added to animal feed and are increasing the EPA and DHA levels of the eggs, dairy products, and meat produce.

Milk, bread, and eggs are frequently being fortified with algal oil and fish oil [161]. These new functional foods have the potential to positively benefit both individuals and whole societies without any change to dietary habits. For instance, a recent review of the fortification of milks established that inclusion of an average of 300 mg of EPA + DHA increased plasma PL levels of the nutrients by 25%–50% within 6 weeks accompanied by beneficial health effects such as the reduction of LDL cholesterol and serum triglycerides [162].

Domestic animal feed that has been enriched with EPA + DHA has been shown to enrich both red meat and eggs [163]. Similarly feeding ALA to farmed animals produces health benefits in human consumers of the produce. In livestock where extruded linseed replaced 5% of other oil sources, the butter had reduced n-6/n-3 ratio by 54%, the meat by 60%, and the eggs by 86% [164].

The beneficial ratio in animals fed linseed supplementation transferred to consumers, where those that ate the linseed-fed animal produce had a reduction of in blood lipids to an n-6/n-3 ratio of 10.2 compared with 14.3 in controls. These beneficial blood lipid alterations coincided with increased EPA and DHA in erythrocyte PLs and reduced LA and AA, bringing the PL n-6/n-3 ratio down from 4.2 to 3.8 [164]. Thus, without changing dietary habits, supplementation of animal fodder has the potential to effect health benefits in the population. Novel GM (genetically modified) oilseeds are also being patented for future use in domestic animal feed [165].

The meat and eggs from pasture-fed animals naturally contain significantly higher omega-3 fatty acids than those from farm intensive methods that incorporate omega-6-rich feeding of animals [166,167]. However, wild animals have substantially higher omega-3 fatty acids in their produce than free-range or organic animals [168]. In a remarkable recent study, Crawford et al. [168] bought 12 samples of meat from a Western supermarket and compared it with 12 matched samples of African buffalo, using meat from skeletal muscle (the semitendinosis muscle).

In the Western meat, 97.8% of the total fatty acids were saturated and monounsaturated, so that just 2.2% on the fatty acid content consisted of essential PUFAs. In African meat, this ratio was 78.7% saturated and monounsaturated to 21.9% PUFAs. Interestingly, when the buffalo were further categorized into in parkland buffalo (n = 5) or woodland buffalo (n = 7), there was a huge difference in PUFA content of the PLs in the meat of the skeletal muscle; parkland consisted of 10% PUFA content, while woodland consisted of 30% PUFA. The difference between parkland and woodland buffalo was compared by Crawford et al. with the difference in available grazing diversity between the first type of historic animal enclosures for domesticated animals and the modern pasture-limited monoculture available to so-called "free-range" livestock.

Crawford et al. also present data on the fatty acid content in a range of modern meats, including organic and wild game, and consistently show a dramatic loss of EPA and DHA from the meat of domesticated livestock. For instance, the EPA fraction (percentage per total fatty acids) = 0.14 in a beef sirloin steak, 0.32 in organic beef steak, 2.52 in venison and 3.8 in wetlands buffalo. Similarly DHA = trace (<0.1%) in the conventionally and organically reared beef, 0.7 in venison, and 0.9 in buffalo. Although the organic meat had twice the EPA content as conventionally farmed meat, neither had any DHA, and the differences were negligible when compared with the EPA and DHA content of wild meats, the type of meat with which humans have evolved and adapted.

EPA and DHA are unique nutrients that should be regularly consumed as oily fish or supplemented as fish oil or algal supplements for the prevention and treatment of chronic disease. Consumers of the Western diet are meeting just 25% of the dietary recommendation for the prevention of chronic disease. Consumers can address this nutritional deficit by product replacement of milk and bread with those fortified in omega-3s, by taking fish oil or algae supplements, or by changing the diet to include the regular consumption of foods naturally high in omega-3 fatty acids, such as seafood, wild game, and green leafy vegetables.

CONCLUDING REMARKS

It is clear that the majority of Western societies are consuming diets that are characterized by suboptimal levels of omega-3 fatty acids, but excessive consumption of LA. Through a myriad of both direct and indirect mechanisms that are currently being further elucidated, EPA and DHA mediate anti-inflammatory, pro-resolving actions that maintain homeostasis during local tissue stress. In addition, DHA provides a unique structure for all cellular membranes and has recently been discovered as the precursor to a family of PDs that have widespread protective effects against the accumulation of oxidative stress, particularly in the brain.

Modern humans have adapted to a dietary intake of preformed EPA and DHA. The loss of omega-3 fatty acids from the diet together with the overconsumption of omega-6 fatty acids has led to the depletion of EPA and DHA and corresponding increase of LA and AA in cell membrane PLs. This fatty acid PL profile is characteristic of the Western diet and coincides with modern epidemics in heart disease and brain disorders. Western consumers and health bodies alike are becoming increasingly aware of the enormous range of health benefits offered by the reinstatement of these ancient nutrients as a modern nutritional staple.

ACKNOWLEDGMENT

The author thanks Holly Muggleston, dietician and senior lecturer of nutrition at Southern Cross University, for critically reviewing the manuscript in its final drafts.

REFERENCES

1. Gidez LI. The lore of lipids. *Journal of Lipid Research* 1984, **25**(13):1430–1436.
2. Hildebrand D. Plant lipid biochemistry: Production of unusual fatty acids in plants. November 15, 2010. Retrieved January 5, 2011, from http://lipidlibrary.aocs.org/plantbio/unusualfa/index.htm
3. Christie WW. What is a lipid? January 15, 2010. Retrieved January 12, 2011, from http://lipidlibrary.aocs.org/Lipids/whatlip/index.htm
4. Holman RT. Nutritional and metabolic interrelationships between fatty acids. *Federation Proceedings* 1964, **23**:1062–1067.
5. Holman RT. The slow discovery of the importance of omega 3 essential fatty acids in human health. *Journal of Nutrition* 1998, **128**(2S):427S–433S.
6. Christie WW. Fatty acids: Methylene-interrupted double bonds: Structures, occurrence and biochemistry. *Lipid Library* November 25, 2010. Retrieved December 16, 2010, from http://lipidlibrary.aocs.org/Lipids/fa_poly/index.htm

7. Shils ME, Olson JA, Shike M, Ross AC (eds.). *Modern Nutrition in Health and Disease*, 9th edn. Philadelphia, PA: Lippincott Williams & Wilkins, 1999.

8. Christie WW. Fatty acids and eicosanoids. November 15, 2010. Retrieved January 5, 2011, from http://lipidlibrary.aocs.org/Lipids/fa_eic.html

9. Meyer BJ, Mann NJ, Lewis JL, Milligan GC, Sinclair AJ, Howe PRC. Dietary intakes and food sources of omega-6 and omega-3 polyunsaturated fatty acids. *Lipids* 2003, **38**(4):391–398.

10. Crawford MA. Fatty acid ratios in free living and domestic animals: Possible implications for atheroma. *Lancet* 1968, **291**(7556):1329–1333.

11. Watson RR, De Meester F, Zibadi S, Crawford MA, Wang Y, Lehane C, Ghebremeskel K. Fatty acid ratios in free-living and domestic animals. In: F. De Meester, S. Zibadi and RR. Watsons (Eds.). *Modern Dietary Fat Intakes in Disease Promotion*. New York: Humana Press, 2010, pp. 95–108.

12. Christie WW. Lipid compositions of plants and microorganisms. December 17, 2009. Retrieved January 5, 2011, from http://lipidlibrary.aocs.org/Lipids/comp_plant/index.htm

13. Takami T, Shibata M, Kobayashi Y, Shikanai T. De novo biosynthesis of fatty acids plays critical roles in the response of the photosynthetic machinery to low temperature in Arabidopsis. *Plant and Cell Physiology* 2010, **51**(8):1265–1275.

14. Nettleton JA. Omega-3 fatty acids: Comparison of plant and seafood sources in human nutrition. *Journal of American Dietetics Association* 1991, **91**(3):331–337.

15. Simopoulos AP. Omega-3 fatty acids and antioxidants in edible wild plants. *Biological Research* 2004, **37**(2):263–277.

16. Tinoco J. Dietary requirements and functions of alpha-linolenic acid in animals. *Progress in Lipid Research* 1982, **21**(1):1–45.

17. Pereira JA, Oliveira I, Sousa A, Ferreira ICFR, Bento A, Estevinho L. Bioactive properties and chemical composition of six walnut (*Juglans regia* L.) cultivars. *Food and Chemical Toxicology* 2008, **46**(6):2103–2111.

18. Commission Directive 80/891/EEC of July 25, 1980 relating to the Community method of analysis for determining the erucic acid content in oils and fats intended to be used as such for human consumption and foodstuffs containing added oils or fats. *EurLex Official Journal* 1980, **254**. Retrieved February 4, 2011, from http://eur-lex.europa.eu/LexUriServ/LexUriServ.do?uri=CELEX:31980L0891:EN:HTML

19. Beare-Rogers J, Dieffenbacher A, Holm JV. Lexicon of lipid nutrition (IUPAC Technical Report). *Pure Applied Chemistry* 2001, **73**(4):685–744.

20. Das UN. Essential fatty acids: Biochemistry, physiology and pathology. *Biotechnology Journal* 2006, **1**(4):420–439.

21. Mann NJ, Johnson LG, Warrick GE, Sinclair AJ. The arachidonic acid content of the Australian diet is lower than previously estimated. *Journal of Nutrition* 1995, **125**(10):2528–2535.

22. Sinclair AJ, Johnson L, O'Dea K, Holman RT. Diets rich in lean beef increase arachidonic acid and long-chain omega 3 polyunsaturated fatty acid levels in plasma phospholipids. *Lipids* 1994, **29**(5):337–343.

23. Broadhurst CL, Wang Y, Crawford MA, Cunnane SC, Parkington JE, Schmidt WF. Brain-specific lipids from marine, lacustrine, or terrestrial food resources: Potential impact on early African *Homo sapiens*. *Comparative Biochemistry and Physiology Part B, Biochemistry and Molecular Biology (SAUS)* 2002, **131**(4):653–673.

24. Gerr MI. *Lipids in Nutrition and Health: A Reappraisal*. Bridgwater, U.K.: P.J. Barnes & Associates (The Oily Press Ltd), 1999.

25. Dawczynski C, Martin L, Wagner A, Jahreis G. n-3 LC-PUFA-enriched dairy products are able to reduce cardiovascular risk factors: A double-blind, cross-over study. *Clinical Nutrition* 2010, **29**(5):592–599.

26. Brenna JT, Salem N, Jr., Sinclair AJ, Cunnane SC. [Alpha]-linolenic acid supplementation and conversion to n-3 long-chain polyunsaturated fatty acids in humans. *Prostaglandins, Leukotrienes, and Essential Fatty Acids* 2009, **80**(2–3):85–91.

27. Wu CH, Popova EV, Hahn EJ, Paek KY. Linoleic and [alpha]-linolenic fatty acids affect biomass and secondary metabolite production and nutritive properties of *Panax ginseng* adventitious roots cultured in bioreactors. *Biochemical Engineering Journal* 2009, **47**(1–3):109–115.

28. Martinez M. Tissue levels of polyunsaturated fatty acids during early human development. *Journal of Pediatrics* 1992, **120**(4 Pt 2):S129–S138.

29. Whelan J. Antagonistic effects of dietary arachidonic acid and n-3 polyunsaturated fatty acids. *Journal of Nutrition* 1996, **126**(4):1–7.

30. Siguel EN, Maclure M. Relative activity of unsaturated fatty acid metabolic pathways in humans. *Metabolism* 1987, **36**(7):664–669.

31. Brenner RR, Peluffo RO. Inhibitory effect of docosa-4,7,10,13,16,19-hexaenoic acid upon the oxidative desaturation of linoleic into gamma-linolenic and of alpha-linolenic into octadeca-6,9,12,15-tetraenoic acid. *Biochimica et Biophysica Acta* 1967, **137**:184–186.

32. Brenner R. Nutritional and hormonal factors influencing desaturation of essential fatty acids. *Progress in Lipid Research* 1981, **20**:41–47.

33. Huang YS, Nassar BA. Modulation of tissue fatty acid composition, prostaglandin production and cholesterol levels by dietary manipulation of n-3 and n-6 essential fatty acid metabolites. In: *Omega 6 Essential Fatty Acids: Pathophysiology and Roles in Clinical Medicine*. Ed. Horrobin D. New York: Wiley-Liss, 1990, pp. 127–144.

34. Gerster H. Can adults adequately convert alpha-linolenic acid (18:3n-3) to eicosapentaenoic acid (20:5n-3) and docosahexaenoic acid (22:6n-3)? *International Journal for Vitamin and Nutrition Research* 1998, **68**(3):159–173.

35. Decsi T, Koletzko B. Do trans fatty acids impair linoleic acid metabolism in children? *Annals of Nutrition & Metabolism* 1995, **39**(1):36–41.

36. Kummerow FA, Zhou Q, Mahfouz MM, Smiricky MR, Grieshop CM, Schaeffer DJ. Trans fatty acids in hydrogenated fat inhibited the synthesis of the polyunsaturated fatty acids in the phospholipid of arterial cells. *Life Sciences* 2004, **74**(22):2707–2723.

37. Cho HP, Nakamura MT, Clarke SD. Cloning, expression, and nutritional regulation of the mammalian Delta-6 desaturase. *Journal of Biological Chemistry* 1999, **274**(1):471–477.

38. Riserus U. Fatty acids and insulin sensitivity. *Current Opinion in Clinical Nutrition & Metabolic Care* 2008, **11**(2):100–105.

39. Kaitosaari T, Ronnemaa T, Viikari J, Raitakari O, Arffman M, Marniemi J, Kallio K, Pahkala K, Jokinen E, Simell O. Low-saturated fat dietary counseling starting in infancy improves insulin sensitivity in 9-year-old healthy children: The special turku coronary risk factor intervention project for children (STRIP) study. *Diabetes Care* 2006, **29**(4):781–785.

40. Mamalakis G, Kafatos A, Tornaritis M, Alevizos B. Anxiety and adipose essential fatty acid precursors for prostaglandin E1 and E2. *Journal of the American College of Nutrition* 1998, **17**(3):239–243.

41. Mills DE, Ward RP. Effects of eicosapentaenoic acid (20:5 omega 3) on stress reactivity in rats. *Proceedings of the Society for Experimental Biology and Medicine* 1986, **182**(1):127–131.

42. Mills DE, Ward RP. Effects of essential fatty acid administration on cardiovascular responses to stress in the rat. *Lipids* 1986, **21**(2):139–142.

43. Mills DE, Huang YS, Narce M, Poisson JP. Psychosocial stress, catecholamines, and essential fatty acid metabolism in rats. *Proceedings of the Society for Experimental Biology and Medicine* 1994, **205**(1):56–61.

44. Horrobin D (ed.). *Omega 6 Essential Fatty Acids: Pathophysiology and Roles in Clinical Medicine*. New York: Wiley-Liss, 1990.
45. Schaeffer L, Gohlke H, Muller M, Heid IM, Palmer LJ, Kompauer I, Demmelmair H, Illig T, Koletzko B, Heinrich J. Common genetic variants of the FADS1 FADS2 gene cluster and their reconstructed haplotypes are associated with the fatty acid composition in phospholipids. *Human Molecular Genetics* 2006, **15**(11):1745–1756.
46. Martinelli N, Consoli L, Olivieri O. A 'Desaturase hypothesis' for atherosclerosis: Janus-Faced enzymes in n-6 and n-3 polyunsaturated fatty acid metabolism. *Journal of Nutrigenetics and Nutrigenomics* 2009, **2**(3):129–139.
47. Latchman DS. Transcription factors: An overview. *International Journal of Biochemistry & Cell Biology* 1997, **29**(12):1305–1312.
48. Tang C, Cho HP, Nakamura MT, Clarke SD. Regulation of human delta-6 desaturase gene transcription: Identification of a functional direct repeat-1 element. *Journal of Lipid Research* 2003, **44**(4):686–695.
49. Cook HW. Brain metabolism of alpha-linolenic acid during development. *Nutrition* 1991, **7**(6):440–446.
50. Scott BL, Bazan NG. Membrane docosahexaenoate is supplied to the developing brain and retina by the liver. *Proceedings of the National Academy of Sciences of the United States of America* 1989, **86**(8):2903–2907.
51. Crawford MA, Costeloe K, Ghebremeskel K, Phylactos A, Skirvin L, Stacey F. Are deficits of arachidonic and docosahexaenoic acids responsible for the neural and vascular complications of preterm babies? *American Journal of Clinical Nutrition* 1997, **66**(4 Suppl):1032S–1041S.
52. Burr GO, Burr MM. A new deficiency disease produced by the rigid exclusion of fat from the diet. *Journal of Biological Chemistry* 1929, **82**(2):345–367.
53. Burr GO, Burr MM. On the nature and role of the fatty acids essential in nutrition. *Journal of Biological Chemistry* 1930, **86**(2):587–621.
54. Gray GM, King IA, Yardley HJ. The plasma membrane of granular cells from pig epidermis: Isolation and lipid and protein composition. *Journal of Investigative Dermatology* 1978, **71**(2):131–135.
55. Gray GM, White RJ. Glycosphingolipids and ceramides in human and pig epidermis. *Journal of Investigative Dermatology* 1978, **70**(6):336–341.
56. Ziboh VA, Chapkin RS. Metabolism and function of skin lipids. *Progress in Lipid Research* 1988, **27**(2):81–105.
57. Ziboh V, Chapkin R. Biologic significance of polyunsaturated fatty acids in the skin. *Archives of Dermatology* 1987, **123**:1686–1690.
58. Bjerve K, Mostad I, Thoresen L. Alpha-linolenic acid deficiency in patients on long-term gastric-tube feeding: Estimation of linolenic acid and long-chain unsaturated n-3 fatty acid requirement in man. *American Journal of Clinical Nutrition* 1987, **45**(1):66–77.
59. Peifer JJ, Holman RT. Effect of saturated fat upon essential fatty acid metabolism of the rat. *Journal of Nutrition* 1959, **68**(1):155–168.
60. Holman RT. The ratio of trienoic: Tetraenoic acids in tissue lipids as a measure of essential fatty acid requirement. *Journal of Nutrition* 1960, **70**:405–410.
61. Neuringer M, Connor WE, Van Petten C, Barstad L. Dietary omega-3 fatty acid deficiency and visual loss in infant rhesus monkeys. *Journal of Clinical Investigation* 1984, **73**(1):272–276.
62. Bjerve K, Fischer S, Alme K. Alpha-linolenic acid deficiency in man: Effect of ethyl linolenate on plasma and erythrocyte fatty acid composition and biosynthesis of prostanoids. *American Journal of Clinical Nutrition* 1987, **46**(4):570–576.
63. Sinclair AJ, Attar-Bashi NM, Li D. What is the role of alpha-linolenic acid for mammals? *Lipids* 2002, **37**(12):1113–1123.

64. Fu Z, Sinclair AJ. Novel pathway of metabolism of alpha-linolenic acid in the guinea pig. *Pediatric Research* 2000, **47**(3):414–417.
65. Simopoulos AP. Summary of the NATO advanced research workshop on dietary omega 3 and omega 6 fatty acids: Biological effects and nutritional essentiality. *Journal of Nutrition* 1989, **119**(4):521–528.
66. Gebauer SK, Psota TL, Harris WS, Kris-Etherton PM. n–3 Fatty acid dietary recommendations and food sources to achieve essentiality and cardiovascular benefits. *American Journal of Clinical Nutrition* 2006, **83**(6):S1526–S1535.
67. Le HD, Meisel JA, de Meijer VE, Gura KM, Puder M. The essentiality of arachidonic acid and docosahexaenoic acid. *Prostaglandins, Leukotrienes, and Essential Fatty Acids* 2009, **81**(2–3):165–170.
68. Marangoni F, Colombo C, Martiello A, Poli A, Paoletti R, Galli C. Levels of the n-3 fatty acid eicosapentaenoic acid in addition to those of alpha linolenic acid are significantly raised in blood lipids by the intake of four walnuts a day in humans. *Nutrition, Metabolism and Cardiovascular Diseases* 2007, **17**(6):457–461.
69. Sanders TA, Younger KM. The effect of dietary supplements of omega 3 polyunsaturated fatty acids on the fatty acid composition of platelets and plasma choline phosphoglycerides. *British Journal of Nutrition* 1981, **45**(3):613–616.
70. Renaud S, Nordoy A: "Small is beautiful": Alpha-linolenic acid and eicosapentaenoic acid in man. *Lancet* 1983, **1**(8334):1169.
71. De Lorgeril M, Salen P. Alpha-linolenic acid and coronary heart disease. *Nutrition, Metabolism, and Cardiovascular Diseases* 2004, **14**(3):162–169.
72. Guizy M, David M, Arias C, Zhang L, Cofán M, Ruiz-Gutiérrez V, Ros E, Lillo MP, Martens JR, Valenzuela C. Modulation of the atrial specific Kv1.5 channel by the n 3 polyunsaturated fatty acid, [alpha]-linolenic acid. *Journal of Molecular and Cellular Cardiology* 2008, **44**(2):323–335.
73. Renaud S. Linoleic acid, platelet aggregation and myocardial infarction. *Atherosclerosis* 1990, **80**(3):255–256.
74. Frantz ID et al., Test of effect of lipid lowering by diet on Cardiovascular risk. The Minnesota Coronary Survey. *Arteriosclerosis*, 1989, **9**(1): 129–35.
75. Barceló-Coblijn G, Murphy EJ. Alpha-linolenic acid and its conversion to longer chain n-3 fatty acids: Benefits for human health and a role in maintaining tissue n-3 fatty acid levels. *Progress in Lipid Research* 2009, **48**(6):355–374.
76. Burdge GC, Jones AE, Wootton SA. Eicosapentaenoic and docosapentaenoic acids are the principal products of alpha-linolenic acid metabolism in young men. *British Journal of Nutrition* 2002, **88**(04):355–363.
77. Burdge GC, Wootton SA. Conversion of alpha-linolenic acid to eicosapentaenoic, docosapentaenoic and docosahexaenoic acids in young women. *British Journal of Nutrition* 2002, **88**(04):411–420.
78. De Groot RHM, Hornstra G, Van Houwelingen AC, Roumen F. Effect of α-linolenic acid supplementation during pregnancy on maternal and neonatal polyunsaturated fatty acid status and pregnancy outcome. *American Journal of Clinical Nutrition* 2004, **79**(2):251–260.
79. Salem N, Jr., Litman B, Kim HY, Gawrisch K. Mechanisms of action of docosahexaenoic acid in the nervous system. *Lipids* 2001, **36**(9):945–959.
80. Calder PC, Dangour AD et al. Essential fats for future health. *Proceedings of the 9th Unilever Nutrition Symposium*, May 26–27, 2010. *European Journal of Clinical Nutrition* 2010, **64**(Suppl 4):S1–13.
81. Dyerberg J, Bang HO. Dietary fat and thrombosis. *Lancet* 1978, **1**(8056):152.
82. Dyerberg J, Bang HO, Stoffersen E, Moncada S, Vane JR. Eicosapentaenoic acid and prevention of thrombosis and atherosclerosis? *Lancet* 1978, **2**(8081):117–119.
83. Hornstra G, Haddeman E, ten Hoor F. Fish oils, prostaglandins, and arterial thrombosis. *Lancet* 1979, **2**(8151):1080.

84. Alexander J. Immunonutrition: The role of omega-3 fatty acids. *Nutrition* 1998, **14**:627–633.
85. Heller A, Koch K et al. Lipid mediators in inflammatory disorders. *Drugs* 1998, **55**(4):487–496.
86. Stables MJ, Gilroy DW. Old and new generation lipid mediators in acute inflammation and resolution. *Progress in Lipid Research* 2010, **50**(1):35–51.
87. Piper P, Vane J. The release of prostaglandins from lung and other tissues. *Annals of the New York Academy of Sciences* 1971, **180**:363–385.
88. Mantzioris E, James M, Gibson R, Cleland L. Dietary substitution with an alpha-linolenic acid-rich vegetable oil increase eicosapentaenoic acid concentrations in tissues. *American Journal of Clinical Nutrition* 1994, **59**(6):1304–1312.
89. Sinclair AJ. Short-term diets rich in arachidonic acid influence plasma phospholipid polyunsaturated fatty acid levels and prostacyclin and thromboxane production in humans. *Journal of Nutrition* 1996, **126**(4):1110S–1117S.
90. Kirtland S. Prostaglandin E1: A review. *Prostaglandins, Leukotrienes, and Essential Fatty Acids* 1988, **32**:165–174.
91. Dubois RN, Abramson SB, Crofford L, Gupta RA, Simon LS, Van De Putte LB, Lipsky PE. Cyclooxygenase in biology and disease. *FASEB Journal: Official Publication of the Federation of American Societies for Experimental Biology* 1998, **12**(12):1063–1073.
92. Jakobsson PJ, Thoren S, Morgenstern R, Samuelsson B. Identification of human prostaglandin E synthase: A microsomal, glutathione-dependent, inducible enzyme, constituting a potential novel drug target. *Proceedings of the National Academy of Sciences of the United States of America* 1999, **96**(13):7220–7225.
93. Brock TG, McNish RW, Peters-Golden M. Arachidonic acid is preferentially metabolized by cyclooxygenase-2 to prostacyclin and prostaglandin E2. *Journal of Biological Chemistry* 1999, **274**(17):11660–11666.
94. Levy BD. Resolvins and protectins: Natural pharmacophores for resolution biology. *Prostaglandins Leukotrienes & Essential Fatty Acids* 2010, **82**(4–6):327–332.
95. Horrobin D. Prostaglandin E1: Physiological significance and clinical use. *Wiener klinische wochenschrift* 1988, **100**(14):471–477.
96. Qiu FH, Devchand PR, Wada K, Serhan CN. Aspirin-triggered lipoxin A4 and lipoxin A4 up-regulate transcriptional corepressor NAB1 in human neutrophils. *FASEB Journal* 2001, **15**(14):2736–2738.
97. Maderna P, Godson C. Lipoxins: resolutionary road. [Research Support, Non-US. Govt. Review]. *British Journal of Pharmacology*, **158**(4):947–959.
98. Ariel A, Serhan CN. Resolvins and protectins in the termination program of acute inflammation. *Trends in Immunology* 2007, **28**(4):176–183.
99. Katan MB, Deslypere JP, van Birgelen AP, Penders M, Zegwaard M. Kinetics of the incorporation of dietary fatty acids into serum cholesteryl esters, erythrocyte membranes, and adipose tissue: an 18-month controlled study. *Journal of Lipid Research* 1997, **38**(10):2012–2022.
100. Kuratko CN, Salem N, Jr. Biomarkers of DHA status. *Prostaglandins, Leukotrienes, and Essential Fatty Acids* 2009, **81**(2–3):111–118.
101. Stoll AL, Lock CA, Marangell LB, Severus WE. Omega 3 fatty acids and bipolar disorder: A review. *Prostaglandins, Leukotrienes, and Essential Fatty Acids* 1999, **60**(5–6):329–337.
102. Lachman HM, Papolos DF. Abnormal signal transduction: A hypothetical model for bipolar affective disorder. *Life Sciences* 1989, **45**:1413–1426.
103. Palacios-Pelaez R, Lukiw WJ, Bazan NG. Omega-3 essential fatty acids modulate initiation and progression of neurodegenerative disease. *Molecular Neurobiology* 2010, **41**(2–3):367–374.

104. Stillwell W, Wassall SR. Docosahexaenoic acid: Membrane properties of a unique fatty acid. *Chemistry and Physics of Lipids* 2003, **126**(1):1–27.
105. Olbrich K, Rawicz W, Needham D, Evans E. Water permeability and mechanical strength of polyunsaturated lipid bilayers. *Biophysical Journal* 2000, **79**(1):321–327.
106. Crawford MA, Bloom M, Broadhurst CL, Schmidt WF, Cunnane SC, Galli C, Gehbremeskel K, Linseisen F, Lloyd-Smith J, Parkington J. Evidence for the unique function of docosahexaenoic acid during the evolution of the modern hominid brain. *Lipids* 1999, **34**(Suppl):S39–S47.
107. Feller SE, Gawrisch K, MacKerell AD, Jr. Polyunsaturated fatty acids in lipid bilayers: Intrinsic and environmental contributions to their unique physical properties. *Journal of the American Chemical Society* 2002, **124**(2):318–326.
108. Ryan AS, Astwood JD, Gautier S, Kuratko CN, Nelson EB, Salem N, Jr. Effects of long-chain polyunsaturated fatty acid supplementation on neurodevelopment in childhood: A review of human studies. *Prostaglandins, Leukotrienes, and Essential Fatty Acids* 2010, **82**(4–6):305–314.
109. Sinclair AJ, Begg D, Mathai M, Weisinger RS. Omega 3 fatty acids and the brain: Review of studies in depression. *Asia Pacific Journal of Nutrition* 2007, **16**(Suppl 1):391–397.
110. Tobin A. Fish oil supplementation. *Lancet* 1988, **1**(8593):1046–1047.
111. Tassoni D, Kaur G, Weisinger RS, Sinclair AJ. The role of eicosanoids in the brain. *Asia Pacific Journal of Nutrition* 2008, **17**(Suppl 1):220–228.
112. Svennerholm L. Distribution and fatty acid composition of phosphoglycerides in normal human brain. *Journal of Lipid Research* 1968, **9**(5):570–579.
113. Rapoport SI, Chang MC, Spector AA. Delivery and turnover of plasma-derived essential PUFAs in mammalian brain. *Journal of Lipid Research* 2001, **42**(5):678–685.
114. Calder PC. Immunomodulation by omega-3 fatty acids. *Prostaglandins, Leukotrienes, and Essential Fatty Acids* 2007, **77**(5–6):327–335.
115. Niemoller TD, Bazan NG. Docosahexaenoic acid neurolipidomics. *Prostaglandins & Other Lipid Mediators* 2010, **91**(3–4):85–89.
116. Bazan NG. Neuroprotectin D1-mediated anti-inflammatory and survival signaling in stroke, retinal degenerations, and Alzheimer's disease. *Journal of Lipid Research* 2009, **50**(Suppl):S400–S405.
117. Bazan NG. Neuroprotectin D1 (NPD1): A DHA-derived mediator that protects brain and retina against cell injury-induced oxidative stress. *Brain Pathology (Zurich, Switzerland)* 2005, **15**(2):159–166.
118. Cunnane SC, Harbige LS, Crawford MA. The importance of energy and nutrient supply in human brain evolution. *Nutrition and Health* 1993, **9**(3):219–235.
119. Crawford M. Cerebral evolution. *Nutrition and Health* 2002, **16**:29–34.
120. Abedin L, Lien EL, Vingrys AJ, Sinclair AJ. The effects of dietary alpha-linolenic acid compared with docosahexaenoic acid on brain, retina, liver, and heart in the guinea pig. *Lipids* 1999, **34**(5):475–482.
121. Elvevoll EO, James DG. Potential benefits of fish for maternal, foetal and neonatal nutrition: a review of the literature. *Food, Nutrition and Agriculture* 2000, **27**:28–39.
122. Richards MP, Pettitt PB, Stiner MC, Trinkaus E. Stable isotope evidence for increasing dietary breadth in the European mid-Upper Paleolithic. *Proceedings of the National Academy of Sciences of the United States of America* 2001, **98**(11):6528–6532.
123. Newman M. A new picture of life's history on Earth. *Proceedings of the National Academy of Sciences of the United States of America* 2001, **98**(11):5955–5956.
124. Grynberg A. Hypertension prevention: From nutrients to (fortified) foods to dietary patterns. Focus on fatty acids. *Journal of Human Hypertension* 2005, **19**(Suppl 3):S25–S33.
125. Taubes G. The soft science of dietary fat. *Science* 2001, **291**(March):2536–2545.

126. Hibbeln JR, Lands EM, Lamoreaux ET. Quantitative changes in the availability of fats in the US food supply. In: *ISSFAL—5th International Conference: 2002*. Montreal, Quebec, Canada: International Society for the Study of Fatty Acids and Lipids, 2002. http://archive.issfal.org/index.php/archive-mainmenu-14/issfal-2002-mainmenu-49/friday-may-10th-mainmenu-57/concurrents-k-p-mainmenu-83

127. Hibbeln JR. Fish consumption and major depression *Lancet*, **351**(9110):1213.

128. Newton IS. Long chain fatty acids in health and nutrition. *Journal of Food Lipids* 1996, **3**:233–249.

129. Adams PB, Lawon S, Sanigorski A, Sinclair AJ. Arachidonic acid to eicosapentaenoic acid ratio in blood correlates positively with clinical symptoms of depression. *Lipids* 1996, **31**(Suppl.):S157–S161.

130. Amminger GP, Schafer MR, Papageorgiou K, Klier CM, Cotton SM, Harrigan SM, Mackinnon A, McGorry PD, Berger GE. Long-chain omega-3 fatty acids for indicated prevention of psychotic disorders: A randomized, placebo-controlled trial. *Archives of General Psychiatry* 2010, **67**(2):146–154.

131. Hibbeln JR, Nieminen LR, Lands WE. Increasing homicide rates and linoleic acid consumption among five Western countries, 1961–2000. *Lipids* 2004, **39**(12):1207.

132. Virkkunen MF, Jenkins ME, Horrobin DF, Jenkins DD, Manku MS. Plasma phospholipid essential fatty acids and prostaglandins in alcoholic, habitually violent and impulsive offenders. *Biological Psychology* 1987, **22**:1087–1096.

133. Gesch CB, Hammond SM, Hampson SE, Eves A, Crowder MJ. Influence of supplementary vitamins, minerals and essential fatty acids on the antisocial behaviour of young adult prisoners. Randomised, placebo-controlled trial. *British Journal of Psychiatry* 2002, **181**:22–28.

134. Calder PC, Yaqoob P. Understanding omega-3 polyunsaturated fatty acids. *Postgraduate Medicine* 2009, **121**(6):148–157.

135. Anderson GJ, Connor WE, Corliss JD. Docosahexaenoic acid is the preferred dietary n-3 fatty acid for the development of the brain and retina. *Pediatric Research* 1990, **27**(1):89–97.

136. De Deckere EA, Korver O, Verschuren PM, Katan MB. Health aspects of fish and n-3 polyunsaturated fatty acids from plant and marine origin. *European Journal of Clinical Nutrition* 1998, **52**(10):749–753.

137. FAO/WHO. Report of a Joint FAO/WHO Consultation. Fats and oils in human nutrition, Nutrition Paper No. 57. Rome, Italy: Food and Agriculture Organisation, United Nations and the World Health Organisation, 1994.

138. NHMRC. Nutrition reference values for Australia and New Zealand including recommended dietary intakes. Commonwealth Department of Health and Ageing, Australia, Ministry of Health, New Zealand; National Health and Medical Research Council, 2004. http://www7.health.gov.au/nhmrc/advice/nrv.htm

139. Siguel E. Identification and quantification of fatty acids. *JPEN Journal of Parenteral and Enteral Nutrition* 1998, **22**(6):401–402.

140. Horrocks LA, Yeo YK. Health benefits of docosahexaenoic acid (DHA). *Pharmacological Research* 1999, **40**(3):211–225.

141. Logan AC. Neurobehavioral aspects of omega-3 fatty acids: Possible mechanisms and therapeutic value in major depression. *Alternative Medicine Review* 2003, **8**(4):410–425.

142. Logan AC. Omega-3 fatty acids and major depression: A primer for the mental health professional. *Lipids in Health and Disease* 2004, **3**:25.

143. Stevens LJ, Zentall SS, Abate ML, Thomas K, Burgess JR. Omega-3 fatty acids in boys with behaviour, learning, and health problems. *Physiology and Behavior* 1996, **59**(4/5):915–920.

144. Ross BM, McKenzie I, Glen I, Bennett CP. Increased levels of ethane, a non-invasive marker of n-3 fatty acid oxidation, in breath of children with attention deficit hyperactivity disorder. *Nutritional Neuroscience* 2003, **6**(5):277–281.

145. Youdim KA, Martin A, Joseph JA. Essential fatty acids and the brain: Possible health implications. *International Journal of Developmental Neuroscience* 2000, **18**(4–5):383–399.
146. Brunner J, Parhofer KG, Schwandt P, Bronisch T. Cholesterol, essential fatty acids, and suicide. *Pharmacopsychiatry* 2002, **35**(1):1–5.
147. Rogers PJ. A healthy body, a healthy mind: Long-term impact of diet on mood and cognitive function. *Proceedings of the Nutrition Society* 2001, **60**(1):135–143.
148. Hibbeln JR, Umhau JC, George DT, Salem N, Jr. Do plasma polyunsaturates predict hostility and depression? *World Review of Nutrition and Dietetics* 1997, **82**:175–186.
149. Barcelo-Coblijn G, Hogyes E, Kitajka K, Puskas LG, Zvara A, Hackler L, Jr., Nyakas C, Penke Z, Farkas T. Modification by docosahexaenoic acid of age-induced alterations in gene expression and molecular composition of rat brain phospholipids. *Proceedings of the National Academy of Sciences of the United States of America* 2003, **100**(20):11321–11326.
150. Innis SM. Dietary (n-3) fatty acids and brain development. *Journal of Nutrition* 2007, **137**(4):855–859.
151. National Survey of Mental Health and Wellbeing. Summary of results, 2007 (2008). Retrieved 4 May, 2009, from http://www.abs.gov.au/Ausstats/abs@.nsf/Lookup/3F8A5 DFCBECAD9C0CA2568A900139380
152. Begg S, Vos T, Barker B, Stevenson C, Stanley L, Lopez A. The burden of disease and injury in Australia 2003. Australian Institute of Health and Welfare, Canberra, Australia, 2007.
153. LaMontagne AD, Ostry A, Shaw A, Louie A, Keegel TG. Workplace stress in Victoria: Developing a systems approach. Full Report. Melbourne, Victoria, Australia: VicHealth, The Victorian Health Promotion Foundation, 2006, p. 152.
154. Calder PC. n-3 polyunsaturated fatty acids, inflammation, and inflammatory diseases. *American Journal of Clinical Nutrition* 2006, **83**(6 Suppl):1505S–1519S.
155. Kris-Etherton PM, Grieger JA, Etherton TD. Dietary reference intakes for DHA and EPA. *Prostaglandins, Leukotrienes, and Essential Fatty Acids* 2009, **81**(2–3):99–104.
156. Whelan J, Jahns L, Kavanagh K. Docosahexaenoic acid: Measurements in food and dietary exposure. *Prostaglandins, Leukotrienes, and Essential Fatty Acids* 2009, **81**(2–3):133–136.
157. Makrides M. Is there a dietary requirement for DHA in pregnancy? *Prostaglandins, Leukotrienes, and Essential Fatty Acids* 2009, **81**(2–3):171–174.
158. James MJ, Cleland LG. Dietary n-3 fatty acids and therapy for rheumatoid arthritis. *Seminars in Arthritis and Rheumatism* 1997, **27**(2):85–97.
159. Stone N. The Gruppo Italiano per lo Studio della Sopravvivenza nell'infarto miocardio (GISSI)-Prevenzione trial on fish oil and vitamin E supplementation in myocardial infarction survivors. *Current Cardiology Reports* 2000, **2**(5):445–451.
160. Brower V. Nutraceuticals: Poised for a healthy slice of the healthcare market? *Nature Biotechnology* 1998, **16**(8):728–731.
161. Woods VB, Fearon AM. Dietary sources of unsaturated fatty acids for animals and their transfer into meat, milk and eggs: A review. *Livestock Science* 2009, **126**(1–3):1–20.
162. Lopez-Huertas E. Health effects of oleic acid and long chain omega-3 fatty acids (EPA and DHA) enriched milks. A review of intervention studies. *Pharmacological Research* 2010, **61**(3):200–207.
163. Kalogeropoulos N, Chiou A, Gavala E, Christea M, Andrikopoulos NK. Nutritional evaluation and bioactive microconstituents (carotenoids, tocopherols, sterols and squalene) of raw and roasted chicken fed on DHA-rich microalgae. *Food Research International* 2010, **43**(8):2006–2013.
164. Weill P, Schmitt B, Chesneau G, Daniel N, Safraou F, Legrand P. Effects of introducing linseed in livestock diet on blood fatty acid composition of consumers of animal products. *Annals of Nutrition and Metabolism* 2002, **46**(5):182–191.

165. Cox DN, Evans G, Lease HJ. The influence of product attributes, consumer attitudes and characteristics on the acceptance of: (1) Novel bread and milk, and dietary supplements and (2) fish and novel meats as dietary vehicles of long chain omega 3 fatty acids. *Food Quality and Preference* 2011, **22**(2):205–212.
166. Ponte PIP, Alves SP, Bessa RJB, Ferreira LMA, Gama LT, Bras JLA, Fontes CMGA, Prates JAM. Influence of pasture intake on the fatty acid composition, and cholesterol, tocopherols, and tocotrienols content in meat from free-range broilers. *Poultry Science* 2008, **87**(1):80–88.
167. Muriel E, Ruiz J, Ventanas J, Antequera T. Free-range rearing increases (n-3) polyunsaturated fatty acids of neutral and polar lipids in swine muscles. *Food Chemistry* 2002, **78**(2):219–225.
168. Crawford MA, Wang Y, Lehane C, Ghebremeskel K. Fatty acid ratios in free-living and domestic animals. In: *Modern Dietary Fat Intakes in Disease Promotion.* Ed. Watson RR, De Meester F, Zibadi S. New York: Humana Press, 2010, pp. 95–108.

Part IV

Herbal Medicines, Nutraceuticals and Neurocognition

10 Chinese Medicine Used to Treat Dementia

Dennis Chang, Ben Colagiuri, and Rong Luo

CONTENTS

INTRODUCTION

Dementia is a progressive intellectual and functional impairment that affects memory and learning ability, activities of daily living, and quality of life. Dementia is a leading cause of mental and physical disability in the elderly and carries the highest disability weighting of all illnesses. In Australia, there were around 245,400 dementia patients in 2009, and this figure is estimated to reach 1.13 million by 2050 (Access Economics, 2009). The total cost of dementia to the Australian health system was estimated at $6.6 billion in 2002 representing 1% GDP and it may exceed 3% of GDP by 2050 (Access Economics, 2003).

Alzheimer's disease (AD) is the most common form of dementia accounting for 80%–85% of all dementia cases. The pathogenesis of AD is heterogeneous and may include extracellular amyloid accumulation in the brain, formation of neurofibrillary tangles inside neurons, decreased central nervous system cholinergic function, genetic mutations (e.g., amyloid-precursor-protein gene), allelic variant of apolipoprotein-E (APOE), and lipid abnormalities (Drachman, 2003). Vascular dementia (VaD) that results from cerebrovascular and cardiovascular diseases is a second common dementia accounting for 15%–20% of the dementia population in Western countries (Ogata, 1999). In Asia and some developing countries, the prevalence of VaD is even higher, equaling or exceeding that of AD (Ogata, 1999). Between 1% and 4% of individuals over the age of 65 years old suffer from VaD (Aggarwal and Decarli, 2007), and this figure is doubled with each additional 5–10 years such that in individuals over the age of 85, the prevalence of VaD surpasses AD.

Several classes of pharmaceutical agents are currently used for the management of dementia, especially AD, among which cholinesterase inhibitors and glutamate receptor antagonists are suggested to produce best clinical outcomes (Farlow et al., 2008). However, these improved clinical outcomes are often limited to modest symptomatic relief, meaning that a large proportion of the disease burden lingers. Further, the safety and the long-term therapeutic benefits of these interventions remain uncertain. As such, currently available pharmaceutical appears unsatisfactory for the treatment of dementia.

Complementary and alternative medicine (CAM) has an extensive history of use worldwide, and several interventions have been explored as therapeutic options for the management and prevention of dementia. According to a recent U.S. national survey conducted by the National Centre for Complementary and Alternative Medicine, approximately 38% of U.S. adults aged 18 years and over and approximately 12% of children use some form of CAM (Barnes et al., 2007). The World Health Organization (WHO) estimates that almost 75% of the world's population has therapeutic experience with herbal remedies (Dubey et al., 2004).

Traditional Chinese medicine (TCM) is an ancient, holistic health care system for promoting health and healing from various diseases and is one of the most popular and fast growing CAM therapies. The use of TCM for treatment of aging-related disorders dates back to 5000 years ago in China where herbal remedies were used to boost memory function and increase longevity. Treatment based on pattern identification is the essence of the theory of TCM. Although in recent years there has been an increase in the number of standardized proprietary herbal medicine products being developed for dementia, treatment based on individual pattern discrimination is still the most common method adopted by TCM practitioners. The common forms of Chinese medicine currently used in the management of dementia include Chinese herbal medicine, acupuncture, moxibustion and cupping, Qigong and Tai Chi exercise, Chinese massage therapy (Tuina), and Chinese nutritional or food therapy (Table 10.1).

This chapter will provide an overview of the scientific evidence for the various Chinese medicine interventions used for the management of dementia focusing, in particular, on the herbal medicine and acupuncture, the two most commonly used modalities of TCM.

TABLE 10.1
Overview of Chinese Medicine Interventions
for the Treatment of Dementia

Chinese herbal medicine	*Ginkgo biloba*
	Huperzia serrata (Huperzine A)
	Curcuma longa
	Panax ginseng
	Panax notoginseng
	Salvia miltiorrhiza
	Camellia sinensis
	Multi-herbal formulations
Acupuncture	Electro-acupuncture
	Acupressure
	Ear acupuncture (auricular therapy)
	Moxibustion and cupping
Other Chinese medicine therapies	Qigong exercise
	Tai Chi
	Chinese massage therapy—Tuina
	Chinese nutritional or food therapy

HERBAL MEDICINE FOR DEMENTIA

Herbal medicines have been used in TCM for thousands of years to boost memory and cognitive functions and to manage behavioral and psychological symptoms associated with dementia. Some of the most commonly used and studied herbs include *Ginkgo biloba, Huperzia serrata, Curcuma longa, Panax ginseng, Panax notoginseng, Salvia miltiorrhiza,* and *Melissa officinalis*. Green tea (*Camellia sinensis*) is also considered to be beneficial for cognitive function owing to its antioxidant effects. Table 10.2 summarizes the nomenclature, key bioactive compounds, and the mechanisms of action of these herbs.

INDIVIDUAL HERBS

Gingko biloba

Ginkgo biloba leaf extract (ginkgo) is one of the most studied herbs. The principal active components in ginkgo are flavonol glycosides (e.g., quercetin and kaempferol) and terpenoids (e.g., ginkgolide and bilobalide). Various preclinical studies demonstrate that gingko's neuroprotective effects derive from its ability to decrease oxygen radical discharge and pro-inflammatory functions of macrophages (antioxidant and anti-inflammatory), reduce corticosteroid production (anxiolytic), increase glucose uptake and utilization and ATP production, improve blood flow by increasing red blood cell deformability, decrease red cell aggregation, induce nitric oxide production, and inhibit platelet-activating factor receptors (Chan et al., 2007). In healthy young adults, ginkgo has been shown to improve speed of processing, working memory,

TABLE 10.2

Nomenclature, Key Bioactive Compounds, and Mechanisms of Commonly Used Herbs for Dementia

Botanic Name	Chinese Pinyin Names	Key Bioactive Compounds	Possible Mechanisms of Action
Ginkgo biloba	Yin Xing Ye	Flavonol glycosides (e.g., quercetin and kaempferol) and terpenoids (e.g., ginkgolide and bilobalide)	Decrease oxygen radical discharge and pro-inflammatory functions of macrophages, reduce corticosteroid production, increase glucose uptake and utilization and ATP production Improve blood flow by increasing red blood cell deformability Decrease red cell aggregation Induce nitric oxide production Inhibit platelet-activating factors
Huperzia serrata	She Zu Shi Shan	Lycopodium alkaloids (especially huperzine A)	Anti-cholinesterase Anti-oxidant Anti-β-amyloid peptide fragment Inhibition of oxygen–glucose deprivation Muscarinic receptor antagonism
Curcuma longa	Jiang Huang	Curcumin, demethoxycurcumin, bisdemethoxycurcumin	Antioxidant Anti-inflammatory Cholesterol-lowering properties Block aggregation and fibril formation
Panax ginseng	Ren Shen	Ginsenoside (e.g., ginsenoside Rg3)	Stimulating central cholinergic and dopaminergic receptors Stimulating hypothalamic-pituitary–adrenal axis
Panax notoginseng	San Qi	Saponins (e.g., ginsenoside Rb1, Rd, Re, notoginsenoside R1, R2, R3, etc.)	Enhance blood circulation to CNS Antiplatelet CNS suppressant Enhance blood perfusion

TABLE 10.2 (continued)
Nomenclature, Key Bioactive Compounds, and Mechanisms of Commonly Used Herbs for Dementia

Botanic Name	Chinese Pinyin Names	Key Bioactive Compounds	Possible Mechanisms of Action
Salvia miltiorrhiza	Dan Shen	Tanshinone IIA	Antioxidant Anti-inflammation Anti-cholinesterase CNS sedatives effects
Crocus sativus	Xi Hong Hua		Antioxidant effect
Camellia sinensis (tea)	Cha	Polyphenols (e.g., epigallocatechin)	Antioxidant effect

executive function, and cognitive function (Kennedy et al., 2007). In clinical trials with dementia patients, however, the effectiveness of ginkgo for enhancing memory and cognitive function remains controversial. Several early controlled clinical studies demonstrated various levels of improvement in memory loss, concentration, anxiety, and other symptoms associated with dementia (Howes and Houghton, 2003). For example, a randomized, double-blind, placebo-controlled trial of 216 participants with AD or VaD showed significant improvement in the attention and memory function in the EGb761 (a standard ginkgo preparation)-treated group after 24 weeks treatment (Kanowski et al., 1996). However, the use of self-assessment questionnaires in some of these early studies raises concerns regarding the validity of the results (Howes and Houghton, 2003). Several more recent clinical trials with greater participant numbers and longer intervention periods reported no difference between ginkgo and placebo (Schneider, 2008). As a result, a recently updated Cochrane systematic review on the use of ginkgo for dementia concluded that the existing evidence is inconsistent and, therefore, unconvincing (Birks and Evans, 2009).

Huperzia serrata

Huperzine A (HupA) is an alkaloid derived from the club moss, *Huperzia serrata*. It has been used to treat dementia in China and is sold over the counter as a dietary supplement in the United States for memory loss and mental impairment. Besides HupA's well-known anticholinesterase property, it has been suggested to exert other pharmacological effects including antioxidant, anti-beta-amyloid peptide fragmentation, inhibition of oxygen–glucose deprivation, and NMDA receptor antagonism (Howes and Houghton, 2003; Table 10.2). To date, the clinical trials of HupA have been mainly conducted and published in China. A meta-analysis of HupA for the treatment of AD identified 11 studies (one open-label study, two centre reports, and eight controlled clinical trials) among which four trials involving 474 patients (235 in the HupA treatment group and 239 in the control group) were included in the final analysis (Wang et al., 2009). The results demonstrated that HupA (300–500 µg/day) significantly improved cognitive function (as assessed by the mini-mental state examination [MMSE] and activities of daily living [ADLs]). A recent Cochrane systematic review conducted in China, which included six clinical trials with a total of 454 patients, also suggested that HupA may improve general cognitive function, global clinical status, behavioral disturbance, and functional performance with minimal side effects in AD patients (Li et al., 2009). A similar Cochrane review investigated HupA for VaD, but only identified one small study involving 14 participants in which HupA proved no better than placebo (Hao et al., 2009). However, as noted in these reviews, the lack of quality data, small sample sizes of individual clinical trials, and short intervention periods limit firm conclusions about HupA's clinical efficacy and highlight the need for rigorous randomized controlled trials with large sample sizes.

Curcuma longa

The perennial herb *Curcuma longa* (turmeric) has been applied in therapeutic preparations for centuries in different parts of the world, and is a well-documented treatment for various disease conditions including asthma, bronchial hyperactivity and respiratory allergy, liver disorders, anorexia, rheumatism, diabetic wounds, runny nose, cough, and sinusitis, as well as for neurodegenerative disorders (Goel et al., 2008). Turmeric contains three structurally closely related chemical components—curcumin, demethoxycurcumin, and bisdemethoxycurcumin, which together are commonly referred to as "curcumin" or "curcuminoids" (Goel et al., 2008). Commercially available "curcumin" extracts are often claimed to contain either 70% or 95% curcuminoids.

Data from various animal and/or in vitro studies suggest that curcuminoids possess antioxidant, anti-inflammatory, and cholesterol-lowering properties, all of which are key processes involved in pathogenesis of AD (Ringman et al., 2005). It has also been suggested that curcuminoids directly bind small beta-amyloid species to block aggregation and fibril formation, supporting the rationale for curcuminoids to be used therapeutically for AD (Yang et al., 2005). However, evidence from rigorous clinical trials to support this therapeutic claim is still generally lacking. In a large, population-based study, the relationship between consumption of curry (often contains turmeric) and cognitive function was investigated in 1010 elderly nondemented Asians (Singaporeans). The data demonstrated that consumption of curry containing turmeric was associated with significantly better cognitive performance measured by MMSE than those who "never or rarely" consumed curry containing turmeric (Ng et al., 2006). A randomized, double-blind, placebo-controlled trial was recently undertaken in 34 AD patients in Hong Kong to evaluate curcumin's effect on AD. Six months treatment of one or four grams of curcumin did not significantly change the MMSE scores or other pathological parameters, although a trend toward increase in serum Aβ40 emerged, which may represent an increase in disaggregation of Aβ deposits by curcumin treatment. This study was one of the earliest attempts to evaluate the clinical effectiveness of curcumin for AD, in which several factors including small sample size, relatively short invention period, and lack of cognitive decline in the placebo group significantly limited the generalization of the findings. Several larger-scale clinical trials are currently underway or soon to be completed, and their results will no doubt help determine the therapeutic value of curcumin for the treatment and prevention of age-related dementia.

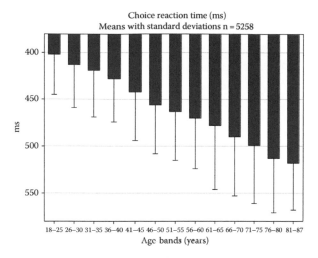

FIGURE 1.1 Declines in normal aging on choice reaction time, a test of attention.

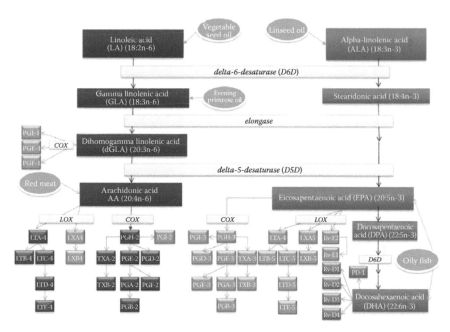

FIGURE 9.2 Metabolic pathways of the essential fatty acids. 'Linseed oil' is also known as flaxseed or flax oil; 'vegetable seed oil' denotes generic linoleic acid (LA) rich vegetable oils as used in industrial food production, including sunflower, safflower oil or corn oils. Dietary LA and alpha linoleic acid (ALA) respectively are absorbed through the gastrointestinal tract and transported to the liver where they are desaturated and elongated into arachidonic acid (AA) and eicosapentaenoic acid (EPA), respectively. AA and EPA circulate in the blood stream until they are taken up into the cell membrane phospholipids (PLs). Upon perturbation of the cell membrane, *phospholipase A_2 (PLA$_2$)* releases AA, dihomogamma linolenic acid (dGLA) and/ or EPA, which are subsequently converted through *cyclooxygenase (COX)* to the prostaglandins (PGs) and thromboxanes (TXs), and through *lipoxygenases (LOX)* into the leukotrienes (LTs) and lipoxins (LXs). Blue colored mediators denote anti-inflammatory and anti-thrombotic mediators; red colored mediators denote pro-inflammatory and pro-thrombotic mediators.

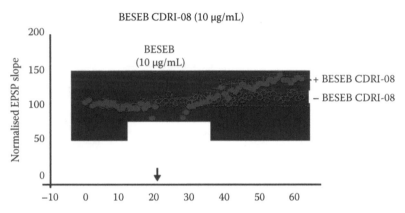

FIGURE 13.14 Intrinsic biological effect of BESEB CDRI-08 on synaptic transmission in the hippocampus. The initial depression reverted to an enduring potentiation when the BESEB CDRI-08 was washed out.

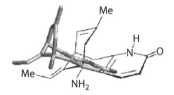

FIGURE 15.1 Chemical structure of HupA and a 3-D view superimposed. (From Zangara, A., *Pharmacol. Biochem. Behav.*, 75(3), 675, 2003.)

Panax ginseng

Panax ginseng (ginseng) root has been used for the management of AD in many Asian countries. Most of the cognition-enhancing effects of ginseng have been studied in animals and healthy individuals. The relevant principal bioactive components of ginseng are ginsenosides, which have been suggested to have antioxidant, anti-inflammatory, and anti-apoptotic effects (Radad et al., 2006). A recent in vitro study also demonstrated that ginsenoside Rg3 promotes beta-amyloid peptide degradation via enhancing gene expression (Yang et al., 2009).

Data from human studies suggest that ginseng modestly improves thinking and secondary and working memory in healthy volunteers (Kennedy et al., 2003; Reay et al., 2006). Two recent small, open-label trials demonstrated the potential therapeutic benefits of ginseng for AD (Heo et al., 2008; Lee et al., 2008). In the former study, 12 week treatment of low-dose (4.5 g/day; $n = 15$) and high-dose (9 g/day; $n = 15$) Korean ginseng showed significant effects on Alzheimer's disease assessment scale-cognitive subscale (ADAS-cog) and clinical dementia rating (CDR) when compared with those in the control group ($n = 31$) (Heo et al., 2008). In the latter study, in which 87 AD patients ($n = 58$ in the ginseng group; $n = 39$ in the control group) were involved, 12 weeks' treatment with ginseng powder (4.5 g/day) produced significant improvements in ADAS-cog and MMSE scores (Lee et al., 2008). Ginseng has also demonstrated clinical benefits when combined with ginkgo in improving cognitive function in humans (Wesnes et al., 2000; Kennedy et al., 2001). Large-scale, long-term studies using standardized extracts are now required to confirm the clinical efficacy of ginseng therapy in AD.

Panax notoginseng

Panax notoginseng (San Qi) is commonly used in Chinese medicine to treat atherosclerosis, hypertension, various thrombosis, external injury, and pain. In addition, San Qi has been suggested to provide therapeutic benefits for dementia. In NG108-15 cells senile dementia model induced by amyloid beta-peptide (Aβ), San Qi significantly increased the survival rate, differentiation, and growth rate of NG108-15 cells, suggesting that the herb can minimize the neural toxic effects of Aβ (Liu et al., 2004). San Qi has also been shown to enhance learning and memory ability as well as to increase ACh content in hippocampus in Aβ and ibotenic acid-induced dementia models in rats (Guo et al., 2004; Sun et al., 2007). Saponins (e.g., ginsenoside Rb1, Rd, Re, notoginsenoside R1, R2, R3, etc.) are the key bioactive components responsible for the neuroprotective effects of San Qi (Chen and Chen, 2004).

Salvia miltiorrhiza

Salvia miltiorrhiza (Dan Shen) is one of most popular Chinese herbs and has been used for the management of various diseases, especially cardiovascular diseases such as coronary artery disease, ischemic strokes, and cerebral thrombosis. Animal studies demonstrated that Dan Shen extracts produced neuroprotective effects and reversed learning and memory deficits in AD animal models (mice and rats) induced by beta-amyloid peptide treatment (Zhang et al., 2001; Liu and Li, 2007). In VaD models in rats, Dan Shen extracts have also markedly increased the content of ACh and 5-HT and decreased AChE activities in the brain tissue (Yuan et al., 2002).

In addition, it is reported that Dan Shen decreased the number of apoptosis of cranial nerve cells (Huang, 2002). The mechanisms of action underpinning these effects have been suggested to attribute to antioxidant, anti-inflammation, anti-cholinesterase, and CNS-sedatives effects, all of which are relevant to AD (Chen and Chen, 2004). Tanshinone has been suggested to be the key biomarker (He et al., 2009).

Crocus sativus

Crocus sativus (Xi Hong Hua) is commonly used in TCM as an antidepressant, antispasmodic, and anticatarrhal. Data from in vivo and in vitro studies demonstrated that Xi Hong Hua possesses neuroprotective properties. Xi Hong Hua extract has shown to improve learning and memory function in ethanol-induced memory impairment in mice and to ameliorate cerebral ischemia–induced oxidative damage in rat hippocampus (Abe and Saito, 2000; Hosseinzadeh and Sadeghnia, 2005). Crocitin, the principal constituent of Xi Hong Hua, which has a strong antioxidant, is suggested to be largely responsible for saffron's protective effect on the central nervous system (Abe and Saito, 2000).

Camellia sinensis

Data from animal and epidemiological studies suggested that drinking green tea (*Camellia sinensis*) may help to protect the brain against the aging process. There is evidence that suggests a probable inverse correlation between tea consumption and the incidence of AD and other neurodegenerative diseases (e.g., Parkinson's disease) (Sharangi, 2009). The chief bioactive components of tea are polyphenols, caffeine, and amino acids. Polyphenols are responsible for tea's well-known antioxidant properties (Sharangi, 2009). In particular, its main catechin polyphenol constituent, epigallocatechin gallate (EGCG), has been shown to exert neuroprotective/neurorescue activities in a wide array of cellular and animal models of neurological disorders (Mandel et al., 2008). In a human cross-sectional study assessing the effect of green tea on cognitive functions in elderly Japanese participants, it was reported that green tea consumption of two or more cups (100 mL per cup) per day reduced the prevalence of cognitive function impairments in the participant cohort (Kuriyama et al., 2006). More recently, Nurk and his colleagues (2009) examined the relationship between intake of three flavonoid-rich foods (chocolate, wine and tea) in the elderly participants aged between 70 and 74 years. The results showed that the consumption of these foods (especially tea) is associated with enhanced cognitive function in a dose-dependent manner (Nurk et al., 2009).

HERBAL FORMULATION AND SYNERGISTIC EFFECTS

Combination therapy underpins the philosophy of Chinese herbal medicine, where patients are generally treated with *multi-herbal formulations*. There is preliminary evidence that complex chemical mixtures enhance therapeutic efficacy by facilitating synergistic action and/or ameliorating/preventing potential side effects (Kroll and Cordes, 2006; Wagner and Ulrich-Merzenich, 2009). Synergistic effects can occur in many ways, including for example, where constituents from herbal extracts interact with one another to improve their solubility and hence the bioavailability (Wagner and Ulrich-Merzenich, 2009). Furthermore, constituents of complex herbal extracts can affect different targets, which make them ideal therapies for disorders such as dementia, which have multifactorial/multisystem pathophysiological components (Kroll and Cordes, 2006). For example, clinical studies in healthy volunteers have found that the cognitive effects of a ginkgo–ginseng combination are significantly greater than those of either extract delivered alone, suggesting the possibilities of synergistic interactions between the extracts (Kennedy et al., 2001; Scholey and Kennedy, 2002).

Over hundreds of years of TCM clinical practice, numerous complex herbal formulations have been used for managing dementia-like symptoms. For example, Guipi decoction is a multi-herbs herbal formulation (*Atractylodes macrocephala*, *Astragalus henryi*, *Arillus longan*, *Semen zizyphi spinosae*, *Radix ginseng*, *Radix aucklandiae*, *Radix glycyrrhizae*, *Radix Angelicae sinensis*, and *Radix polygalae*), which was first used back in 1253 in China. Data from animal studies suggest that Guipi decoction can significantly improve learning and memory function in AD animal models in mice and rats (Hou and Xu, 2006; Qian et al., 2006; Zou et al., 2006). In the following section, research on a three-herb, standardized herbal formulation, WNK (which the authors have been partially involved) will be discussed to provide an example of the development of new herbal medicine formulations. Using modern chemistry, pharmacognosy, and pharmacology techniques, WNK's chemical and pharmacological profiles were clearly defined. The data from these experiments demonstrate significant improvements in learning and memory functions, pathogenic biochemical parameters in blood and brain tissue, and antioxidant capacity in various experimental dementia models.

In an in vivo study, WNK (11, 22, and 44 mg/kg/day over 15 days) was administered to dysmnesia models in mice induced by scopolamine, reserpine, chlorderazin, sodium nitrite, and alcohol, respectively. Compared with the control, the middle- and high-dose treatment of WNK markedly decreased the error numbers and prolonged the latencies of dysmnesia in all active groups in step-through or step-down tests (Xu et al., 2007). In a chronic cerebral hypoperfusion model induced by bilateral common carotid artery ligation in rats, 8 weeks treatment of WNK (i.g.) significantly shortened the persistent time of finding the platform in Morris Water Maze (Xu et al., 2008). Activity of cholinesterase was also significantly decreased while the ACh level was markedly increased in the brain tissue. In addition, the activity of superoxide dismutase (SOD) was significantly enhanced. The effects of WNK on ACh were also investigated in an amyloid β-protein-induced dementia model in rats (Liu et al., 2004). After 30 days treatment of WNK (15.5 and 31.0 mg/kg/day, i.g.), ACh levels in the brain tissue increased significantly by 18.56% and 19.97%, respectively, when compared with the model group. Similarly, in a PDAPP[v717I] transgenic dementia model in mice, WNK treatment at 31 and 62 mg/kg/day i.g. over 12 weeks significantly increased ACh in both treatment groups, while serotonin (5-HT) levels in the brain tissue decreased significantly in the high-dose WNK group only (Cong et al., 2007). The effects of crocin alone (one of the principal active components of *WNK*) on ischemia/reperfusion (I/R) injury were investigated using a global or bilateral common carotid artery occlusion (BCCAO) model in mice (Zheng et al., 2006). Transient global cerebral ischemia (20 min) followed by 24 h of reperfusion significantly increases the production of nitric oxide (NO) and malondialdehyde (MDA) in cortical microvascular homogenates and decreases the activities of superoxide dismutase (SOD) and glutathione peroxide (GSH-px). Pretreatment with crocin at 20 mg/kg, on the other hand, resulted in a significant decrease in MDA content and a significant elevation in total antioxidant capacity (increased SOD and GSH-px activities).

A pilot randomized, double-blind, placebo-controlled clinical trial of WNK for VaD in humans demonstrated some promising results (Liu et al., 2007). Sixty-two patients (32 in active group, 30 in placebo group) with probable or possible VaD according to the National Institute for Neurologic Disease and Stroke and the Association Internationale pour la Recherche et l'Enseignement en Neurosciences (NINDS-AIREN) criteria were recruited. Patients received 16 weeks treatment of either active or identical placebo after randomization. At completion of treatment, the mean scores of the primary efficacy parameter (ADAS-cog) reduced from 24.5 to 20.3 (mean reduction, 4.18 ± 0.75) in patients receiving WNK and from 18.98 to 17.81 (mean reduction, 1.18 ± 0.58) in patients receiving the placebo. While the minor differences in baseline ADAS-cog scores were not significantly different, the improvement of ADAS-cog scores following WNK treatment was significantly greater than that of the placebo group. WNK also significantly reduced the degree of impairment in quality of life caused by VaD as evidenced by the significant improvement in SF36 scores. Out of eight domains, significant improvements were observed in "role emotion," "mental health," "role physical," and "social functioning" in patients receiving WNK while significant trends toward improvement were also noted in two other domains ("physical functioning" and "bodily pain"). Importantly, the results for ADAS-cog are consistent with the finding from a single photon emission computed tomography (SPECT) sub-study of 18 patients ($n = 7$ WNK; $n = 11$ placebo) within this trial. The SPECT scan results showed that when compared to the placebo, WNK treatment appeared to increase blood flow in the inferior frontal and anterior temporal lobes, an effect which was more marked in the left hemisphere. These regions are known to be associated with cognitive, memory, auditory, and speech functions. No serious adverse events were reported within this trial.

ACUPUNCTURE

Acupuncture is one of the main modalities of TCM and has gained popularity over the past decade in the Western world. In TCM, acupuncture is viewed as complex intervention comprising diagnosis, needling, lifestyle advice, therapeutic alliance, and adjunctive treatments such as moxibustion, cupping, and sometimes even Chinese herbs. Acupuncture is practiced widely by not only TCM practitioners but also medical doctors and other allied health practitioners including physiotherapists, chiropractors, nurses, and osteopaths; however, the latter groups tend to focus on the needling component. Acupuncture needling normally involves inserting fine needles into specific points on the body surface (acupoints) to promote circulation of *qi* and *blood* through the meridian channel systems and to restore the balance of *yin* and *yang* and, in doing so, to treat illness and promote health. Acupuncture has been used to treat various diseases including, pain, musculoskeletal injuries, and depression. There is also some evidence to suggest that acupuncture might provide therapeutic benefits in the management of dementia and associated symptoms (e.g., agitated and aggressive behaviors).

CLINICAL TRIALS OF ACUPUNCTURE

Acupuncture treatment has been found to improve cognitive and memory function in patients with dementia (especially VaD). In a study by Zhao et al. (2009), 90 VaD patients were randomized to receive electro-acupuncture, nimodipine, or electro-acupuncture plus nimodipine for 6 weeks. Significant improvements in MMSE scores were reported in all three groups compared with baseline. A similar finding was reported in a study that evaluated the effects of long-term retention of scalp needle on MMSE, Hastgawa dementia scale (HDS), activity of daily living (ADL), and latency and amplitude of event-related potential P300 of EEG in VaD patients (Chu et al., 2008). Here, 65 patients were randomly allocated to acupuncture ($n = 33$) and almitrine bismesylate (duxil). The total effective rates (compared to baseline) of MMSE, HDS, ADL, and the latency of P300 were significantly greater in the acupuncture group than those in the medication group. The results suggest that the scalp acupuncture treatment can improve cognitive function and activity of daily living in VaD. In another study by Lin et al. (2009), the combination of acupressure (a special acupuncture technique in which pressure is applied to acupoints, rather than needle insertion) and Montessori-based activities was found to be effective in relieving agitated and aggressive behaviors in 133 institutionalized residents with dementia in Taiwan.

In a positron emission tomography (PET) study, Huang et al. (2005) investigated whether acupuncture improved cerebral glucose metabolism in 10 patients with VaD. Half of the patients received acupuncture at LI15, SJ5, LI4, SP10, ST36, SP6, and LR3 (standard hemiplegia needling) whereas the other half received acupuncture at these same points plus DU20, DU26, and HT7 (VaD specific needling). The acupuncture was administered every weekday for four weeks, i.e., 20 sessions. Compared with baseline, standard hemiplegia needling increased glucose metabolism in the lentiform nucleus of the affected hemisphere and the temporal lobe of the non-affected hemisphere. The addition of VaD-specific needling increased glucose metabolism in the frontal lobes and thalami bilaterally as well as in the temporal lobe and the lentiform nucleus of the non-affected hemisphere, which is consistent with the results of a study using the same cohort of participants that showed a significant increase in blood circulation in response to acupuncture to the respective areas of the brain (Huang et al., 2006).

Despite the encouraging results from the aforementioned clinical trials, there are significant methodological limitations to most of these studies. Perhaps most importantly, these studies did not include placebo comparators, making it difficult to determine if the observed effects of acupuncture reflect its efficacy or some other processes, such as the placebo effect. Further, most of the studies had relatively small sample sizes. Weina et al. (2007) recently conducted a Cochrane review of acupuncture for VaD. The authors identified 95 relevant studies of which improvement was reported in up to 71%–90% of the treatment group. However, none of these trials met their inclusion criteria due to inadequate methodology. Of the 95 trials, only 17 were randomized controlled trial and these were limited due to combining interventions or providing pharmaceutical treatment to the control groups. This led to a conclusion that the effectiveness of acupuncture for VaD is uncertain and that high-quality randomized placebo-controlled trials of acupuncture are needed.

ANIMAL STUDIES

There has been substantial research into acupuncture needling on animals, most of which has been published in the Chinese literature. These studies attempt to determine mechanistic actions of acupuncture and are generally consistent with findings from clinical trials in terms of the potential cognition enhancing effects of acupuncture. For example, in VaD rats, animals given acupuncture needling performed better in subsequent tasks in a Morris water maze when compared with those in the control groups (Lai et al., 2000; Wang, 2002; Meng et al., 2009). The following mechanistic pathways have been postulated based on the existing animal studies.

Effects on neurotransmitter: Acupuncture is reported to reduce the activity of cholinesterase and hence slow down the ACh metabolism in the brain of VaD mice (Tian et al., 2002). An increase in expression of ChAT mRNA and ChAT protein–positive cells in the hippocampus of VaD rats was observed after acupuncture treatment, implying enhanced cholinergic neuron function (Ma and Tang, 2009). Acupuncture is also found to be able to reverse the dementia-induced reduction of 5-HT, dopamine, and noradrenalin in cortex, hippocampus, and hypothalamus (Lai et al., 1999). Electro-acupuncture can also increase the level of arginine vasopressin, an important neuropeptide closely related to memory and learning (Mo et al., 2002).

Effects on oxygen free radicals: Electro-acupuncture has been shown to increase the activity of superoxide dismutase (SOD) and decrease malondialdehyde (MDA), suggesting that the intervention promotes free-radical scavenging abilities and hence minimizes the tissue damage caused by the free radicals (Lai et al., 2000; Zhao et al., 2000).

Effects on vasoactive substances: Electro-acupuncture treatment was associated with an increase in nitric oxide (NO) content in hippocampus in VaD rats. This change correlated well with the improvement of learning and memory observed in the same animals (Lai and Yu, 2001). Acupuncture is also shown to increase 6-keto-PGF1 in plasma and this may be, at least partially, responsible for the beneficial antiplatelet aggregation and vasodilation effects of acupuncture in VaD (Tian et al., 2002).

Anti-apoptotic effects: Bcl-2 is an important anti-apoptotic gene. Ear acupuncture has shown to significantly increase expression of the anti-apoptotic Bcl-2 protein in CA1 region of hippocampus. The result suggests that acupuncture can provide protective effects to hippocampal neurons against apoptosis in VaD, and hence contribute to the improvement of learning and memory function (Zhang et al., 2001).

Other mechanisms: It is reported that electro-acupuncture increased the number and density of the neurons in CA1 area of hippocampal gyrus in VaD model in mice (Zhao et al., 2001). A study investigating the effect of electro-acupuncture on the cerebral circulation revealed that the treatment at Du20, Du14, and St36 increased the regional cerebral blood flow in the parietal lobe and hippocampus in VaD rats (He, 2002). Ear acupuncture has also shown to cause increased expression of c-fos mRNA and protein-positive cell in CA1 area of the hippocampus in VaD rat model (Zhang et al., 2001).

While these animal studies are suggestive of potential anti-dementia effects of acupuncture needling, it is unclear if and how these findings might generalize to humans and this needs to be investigated empirically.

OTHER CHINESE MEDICINE INTERVENTIONS FOR DEMENTIA

Other Chinese medicine modalities that have been used in clinical practice to treat dementias and/or alleviate associated symptoms include (but are not limited to): Qigong exercise, Tai Chi, Tuina, nutritional or food therapies. The direct scientific evidence to support the use of these interventions for dementia is currently lacking, although there is some preliminary evidence regarding Qigong and Tai Chi, which is described in the following.

QIGONG EXERCISE

Qigong is an ancient and widely practiced Chinese meditation exercise combining meditation, controlled breathing, and gentle physical movements aimed at directing mental attention to specific area of the body (Posadzki et al., 2010). Qigong has been suggested to be able to control the vital energy (*qi*) of the body and consequently to improve spiritual, physical, and mental health (Jones, 2001). Medical Qigong can be divided into two parts: internal and external. The former is developed by individual practice of Qigong exercise. The latter refers to experienced Qigong practitioners emitting *qi* to a patient for the purpose of healing/curing diseases (Sancier, 1996).

Preliminary evidence suggests that Qigong practice can help to relieve stress and anxiety. For example, a clinical trial involving 24 generally healthy male participants found that Qigong significantly improved anxiety, alertness, depression, fatigue, and tension (Jung et al., 2006). It has also been reported that Qigong could normalize, stabilize, and sustain positive and pleasant emotional states of human mind and as a result improve the well-being and quality of life in humans (Posadzki et al., 2010). In a randomized controlled trial by Lee et al. (2005), heart-rate variability (HRV) was compared in 40 subjects receiving external Qigong or placebo. Here, Qigong significantly reduced heart rate and increased HRV, suggesting that the treatment stabilizes the sympathovagal function when compared with the placebo (Lee et al., 2005). EEG topographic mapping also revealed some unique EEG patterns in response to Qigong practice, which were clearly different from that of resting status with eyes closed (Zhang, 1988). The alpha activity occurred predominantly in the anterior half of brain during Qigong state. There were also significant increased alpha1 component and decreased alpha2 component, suggesting significant slowing of alpha peak frequency.

The mechanisms of action underpinning the observed physiological/psychological effects of Qigong are not clear. However, acute Qigong training has been associated with a significant increase in the levels of human endogenous opioids peptides (e.g., beta-endorphin) and a decrease in stress-related hormones including adrenocorticotrophic hormone (ACTH), cortisol, and dehydroepiandrosterone-sulfate (DHEA-S) (Ryu et al., 1996; Jones, 2001). These indicate that Qigong, when used as a stress coping method, can affect hormonal regulation to maintain neurological homeostasis.

Combined with evidence that Qigong can increase local cerebral blood flow (Cahn and Polich, 2006), this suggests that this type of exercise may have the potential to improve dementia symptoms or delay its onset.

TAI CHI

Tai Chi is form of martial art emphasizing slow movements to enhance body alignment and improve balance. Tai Chi has been shown to be able to reduce the incidence of falls and to reduce fear of falling, improve body balance, and enhance emotion and memory in elderly people (Wahbeh et al., 2008; Yao et al., 2008). Numerous studies have been identified in a recent systematic review that was designed to evaluate the effects of Tai Chi on psychological well-being. Twenty one of 33 randomized and nonrandomized trials reported that Tai Chi significantly reduced stress, anxiety, depression, and enhanced mood in young or elderly healthy as well as in patients with various chronic medical conditions including knee osteoarthritis, HIV infection, depression, breast cancer, adolescents with ADHD, as well as dementia. In a narrative study, the effects of Tai Chi and structured reminiscence were evaluated with nine moderately advanced dementia patients. The study suggested that structured reminiscence with Tai Chi facilitated learning that was focused and insightful beyond the level normally manifested for this group of participants (Gibb et al., 1997).

CONCLUSION

Dementia remains a leading cause of mental and physical disability and the conventional anti-dementia therapies provide only modest relief. Some TCM modalities such as herbal medicine and acupuncture have demonstrated potential health benefits in the treatment of dementia. However, a large proportion of the existing evidence comes from animal and in vitro studies, and epidemiological data. The overall clinical evidence to support these TCM interventions remains weak and often conflicting, with clinical trials frequently producing inconsistent results. Therefore, high-quality clinical research is required to better determine the effectiveness and safety of TCM interventions for dementia, including assessing possible preventive effects. To this end, one potential advantage of Chinese herbal medicine is its multi-component and multitarget approach, given the complex nature of dementia in terms of its pathogenesis.

REFERENCES

Abe, K. and Saito, H. (2000). Effects of saffron extract and its constituent crocin on learning behaviour and long-term potentiation. *Phytotherapy Research*, **14**, 149–152.

Access Economics. (2009). *Keeping Dementia Front of Mind: Incidence and Prevalence*. Canberra, Australian Capital Territory, Australia: Access Economics.

Access Economics for Alzheimer's Australia. (2003). *The Dementia Epidemic: Economic Impact and Positive Solutions for Australia*. Canberra, Australian Capital Territory, Australia: Access Economics.

Aggarwal, N.T. and Decarli, C. (2007). Vascular dementia: Emerging trends. *Seminars in Neurology*, **27**, 66–77.

Barnes, P.M., Bloom, B., and Nahin, R. (2007). CDC National health statistics report # 12. Complementary and alternative medicine use among adults and children: United States. Bethesda, MD: NCAM.

Birks, J. and Evans, J.G. (2009). *Ginkgo biloba* for cognitive impairment and dementia. *Cochrane Database of Systematic Reviews*, (1), CD003120. DOI: 10.1002/14651858. CD003120.pub3.

Cahn, B.R. and Polich, J. (2006). Meditation states and traits: EEG, ERP, and neuroimaging studies. *Psychological Bulletin*, **132**, 180–211.

Chan, P.-C., Xia, Q., and Fu, P.P. (2007). *Ginkgo biloba* leave extract: Biological, medicinal and toxicological effects. *Journal of Environmental Science Health Part C*, **25**, 211–244.

Chen, J.K. and Chen, T.T. (2004). *Chinese Medical Herbology and Pharmacology*. City of Industry, CA: Art of Medicine Press.

Chu, J.M., Bao, Y.H., and Zhou, C. (2008). Effect of long-time retention of scalp needle on the abilities of cognition, daily living activity and P-300 in vascular dementia patients. *Acupuncture Research*, **33**, 334–338.

Cong, W.H., Liu, J.X., and Xu, L. (2007). Effect of extracts of Ginseng and Ginkgo biloba on hippocampal acetylcholine and monoamines in PDAP-PVT1T1 transgenic mice. *Zhongguo Zhang Xi Yi Jie He Za Zhi*, **27**, 810–813.

Drachman, D. (2003). Preventing and treating Alzheimer's disease: Strategies and prospects. *Expert Review of Neurotherapeutics*, **3**, 565–569.

Dubey, N.K., Kumar, R., and Tripathi, P. (2004). Global promotion of herbal medicine: India's opportunity. *Current Science*, **86**, 37–41.

Farlow, M.R., Miller, M.L., and Pejovic, V. (2008). Treatment options in Alzheimer's disease: Maximizing benefit, managing expectations. *Dementia and Geriatric Cognitive Disorders*, **25**, 408–422.

Gibb, H., Morris, C.T., and Gleisberg, J. (1997). A therapeutic programme for people with dementia. *International Journal of Nursing Practice*, **3**, 191–199.

Goel, A., Kunnumakkara, A.B., and Aggarwal, B.B. (2008). Curcumin as "Curecumin": From kitchen to clinic. *Biochemical Pharmacology*, **75**, 787–809.

Guo, C., Wu, J., and Li, R. (2004). The effects of PNS on Alzheimer's disease model of mouse and mechanism of the effects. *China Pharmacy*, **15**, 598–599.

Hao, Z., Liu, M., Liu, Z., and Lv, D. (2009). Huperzine A for vascular dementia. *Cochrane Database of Systematic Reviews*, (2), CD007365.

He, F. (2002). Influence of Electro-acupuncture on learning and memory function cerebral blood flow in vascular dementia rats. *Journal of Anhui Traditional Chinese Medical College*, **21**, 28–30.

He, Z., Pan, Z.-H., and Lu, W.-H. (2009). Neuroprotective effects of tanshinone IIA on vascular dementia in rats. *Lishizhen Medicine and Materia Medica Research*, **20**, 3022–3024.

Heo, J.H., Lee, S.T., Chu, K., Oh, M.J., Park, H.J., Shim, J.Y., and Kim, M. (2008). An open-label trial of Korean red ginseng as an adjuvant treatment for cognitive impairment in patients with Alzheimer's disease. *European Journal of Neurology*, **15**, 865–868.

Hosseinzadeh, H. and Sadeghnia, H.R. (2005). Safranal, a constituent *Crocus sativus* (saffron), attenuated cerebral ischemia induced oxidative damage in rat hippocampus. *Journal of Pharmacy and Pharmaceutical Sciences*, **8**, 394–399.

Hou, Z.F. and Xu, G.C. (2006). Effect on the function of study memory of mouse with taking the pills of Gui-pi. *Journal of Traditional Chinese Medicine*, **25**, 754–755.

Howes, M.-J.R. and Houghton, P.J. (2003). Plants used in Chinese and Indian traditional medicine for improvement of memory and cognitive function. *Pharmacology Biochemistry and Behavior*, **75**, 513–527.

Huang, J. (2002). Mechanism of Fufang Danshen to vascular dementia rats' cranial nerve cell apoptosis. *Journal of Hubei Institute for Nationalities*, **19**, 29–30.

Huang, Y., Chen, J., Lai, X.S.H., Tang, A.W., and Li, D.J. (2005). Effects of needling in Baihui (DU20), Shuigou (DU26) and Shenmen (HT7) on glucose metabolism in the lentiform nucleus in patients with vascular dementia. *Journal of First Military Medical University*, **25**, 1405–1407.

Huang, Y. et al. (2006). Effects of needling in Baihui, Shuigou and Shenmen on regional cerebral blood flow in vascular dementia patients. *China Journal of Traditional Chinese Medicine and Pharmacy*, **21**, 462–464.

Jones, B.M. (2001). Changes in cytokine production in healthy subjects practicing Guolin Qigong: A pilot study. *BMC Complementary and Alternative Medicine*, **1**, 8.

Jung, M.J., Shin, B.C., Kim, Y.S., Shin, Y.I., and Lee, M.S. (2006). Is there any difference in the effects of Qi therapy (external Qigong) with and without touching? A pilot study. *International Journal of Neuroscience*, **116**, 1055–1064.

Kanowski, S., Herrmann, W.M., Stephan, K., Wierich, W., and Horr, R. (1996). Proof of efficacy of the *Ginkgo biloba* special extract EGb 761 in outpatients suffering from mild to moderate primary degenerative dementia of the Alzheimer type or multi-infarct dementia. *Pharmacopsychiatry*, **29**, 47–56.

Kennedy, D.O., Jackson, P.A., Haskell, C.F., and Scholey, A.M. (2007). Modulation of cognitive performance following single doses of 120 mg *Ginkgo biloba* extract administered to healthy young volunteers. *Human Psychopharmacology: Clinical and Experimental*, **22**, 559–566.

Kennedy, D.O., Scholey, A.B., and Wesnes, K.A. (2001). Differential, dose dependent changes in cognitive performance following acute administration of a *Ginkgo biloba/ Panax ginseng* combination to healthy young volunteers. *Nutritional Neuroscience*, **4**, 399–412.

Kennedy, D.O. et al. (2003). Electroencephalograph (EEG) effects of single doses of *Ginkgo biloba* and *Panax ginseng* in healthy young volunteers. *Pharmacology, Biochemistry Behaviour*, **75**, 701–709.

Kroll, U. and Cordes, C. (2006). Pharmaceutical prerequisites for a multi-target therapy. *Phytomedicine*, **13**, 12–19.

Kuriyama, S., Hozawa, A., Ohmori, K., Shimazu, T., Matsui, T., and Tsuji, I. (2006). Green tea consumption and cognitive function: A cross-sectional study from the Tsurugaya Project. *American Journal of Clinical Nutrition*, **83**, 355–361.

Lai, X.S., Mo, F.Z., and Ma, C.X. (1999). Research on effects of the leaning-memory on VD rat by electro-acupuncture. *Acupuncture Research*, **24**, 192–196.

Lai, X.S., Wang, L., and Jiang, X.H. (2000). Research of the effects on learning and memory abilities, SOD and MDA of vascular dementia rats with intervention of electro-acupuncture. *Chinese Acupuncture and Moxibustion*, **20**, 497–500.

Lai, X.S. and Yu, J. (2001). Animal research on vascular dementia with electro-acupuncture intervention. *Chinese Journal of Gerontology*, **21**, 38–40.

Lee, S.T., Chu, K., Sim, J.Y., Heo, J.H., and Kim, M. (2008). *Panax ginseng* enhances cognitive performance in Alzheimer disease. *Alzheimer Disease and Associated Disorders*, **22**, 222–226.

Lee, M.S., Kim, M.K., and Lee, Y.H. (2005). Effects of Qi-therapy (external qigong) on cardiac autonomic tone: A randomized placebo controlled study. *International Journal of Neuroscience*, **115**, 1345–1350.

Li, J., Wu, H.M., Zhou, R.L., Liu, G.J., and Dong, B.R. (2009). Huperzine A for Alzheimer's disease. *Cochrane Database of Systematic Review*, (2), CD005592. DOI: 10.1002/14651858.CD005592.pub2.

Lin, L.C., Yang, M.H., Kao, C.C., Wu, S.C., Tang, S.H., and Lin, J.G. (2009). Using acupressure and Montessori-based activities to decrease agitation for residents with dementia: A cross-over trial. *Journal of the American Geriatrics Society*, **57**, 1022–1029.

Liu, J.G., Chang, D., Chan, D., Liu, J.X., and Bensoussan, A. (2007). A randomized placebo-controlled clinical trial of a Chinese herbal medicine for the treatment of vascular dementia. *2nd International Congress for Complementary Medicine Research*, Munich, Germany, May 2007.

Liu, J.X., Cong, W.H., Xu, L., and Wang, J.N. (2004). Effect of combination of extracts of ginseng and *Ginkgo biloba* (WNK) on acetylcholine in amyloid beta-protein treated rats determined by an improved HPLC. *Acta Pharmacologica Sinica*, **25**, 1118–1123.

Liu, Y-H. and Li, J. (2007). Protective effects of salviol on aphrenia in mice induced by amyloid beta-protein and its mechanism. *Journals of Apoplexy and Nervous Diseases*, **24**, 64–66.

Ma, P. and Tang, Q. (2009). effect of acupuncture combined with rehabilitation on learning and memory in vascular dementia (VD) rats and the expression of CHAT in hippocampus. *Information on Traditional Chinese Medicine*, **26**, 54–57.

Mandel, S.A., Amit, T., Kalfon, L., Reznichenko, L., and Youdim, M.B.H. (2008). Targeting multiple neurodegenerative diseases etiologies with multimodal-acting green tea catechins. *Journal of Nutrition*, **138**, 1578S–1583S.

Meng, P.Y., Sun, G.J., and Mao, J.J. (2009). Study of the effect of acupuncture pretreatment on learning and memory abilities in vascular dementia rats. *Shanghai Journal of Acupuncture and Moxibustion*, **28**, 293–296.

Mo, F.Z.H., Li, J.Q., and Lei, L.P. (2000). Influence of electroacupuncture on learning and memorizing of vascular dementia and its ACTH in the hypothalamus in rats. *Acupuncture Research*, **25**, 170–173.

Ng, T.P., Chiam, P.C., Lee, T., Chua, H.C., Lim, L., and Kua, E.H. (2006). Curry consumption and cognitive function in the elderly. *American Journal of Epidemiology*, **164**, 898–906.

Nurk, E., Refsum, H., Drevon, C.A., Tell, G.S., Nygaard, H.A., Engedal, K., and Smith, A.D. (2009). Intake of flavonoid-rich wine, tea, and chocolate by elderly men and women is associated with better cognitive test performance. *Journal of Nutrition*, **139**, 120–127.

Ogata, J. (1999). Vascular dementia: The role of cerebral infarcts. *Alzheimer Disease and Associated Disorders*, **13**, S38–S48.

Posadzki, P., Parekh, S., and Glass, N. (2010). Yoga and Qigong in the psychological prevention of mental health disorders: A conceptual synthesis. *Chinese Journal of Integrative Medicine*, **16**, 80–86.

Qian, H.N., Chen, L.B., Hu, X.Q., and Liang, Y. (2006). Impairment of learning and memory of the model rats of invigorating spleen and the influence of the soup of spleen. *Chinese Journal of Behavioral Medical Science*, **15**, 2002–2204.

Radad, K., Gille, G., Liu, L.L., and Rausch, W.D. (2006). Use of ginseng in medicine with emphasis on neurodegenerative disorders. *Journal of Pharmacological Science*, **100**, 175–186.

Reay, J.L., Kennedy, D.O., and Scholey, A.B. (2006). Effects of *Panax ginseng*, consumed with and without glucose, on blood glucose levels and cognitive performance during sustained 'mentally demanding' tasks. *Journal of Psychopharmacology*, **20**, 771–781.

Ringman, J.M., Frautschy, S.A., Cole, G.M., Masterman, D.L., and Cummings, J.L. (2005). A potential role of the curry spice in Alzheimer's disease. *Current Alzheimer Research*, **2**, 131–136.

Ryu, H., Lee, H.S., Shin, Y.S., Chung, S.M., Lee, M.S., Kim, H.M., and Chung, H.T. (1996). Acute effect of Qigong training on stress hormonal levels in man. *American Journal of Chinese Medicine*, **24**, 193–198.

Sancier, K.M. (1996). Medical application of Qigong. *Alternative Therapies*, **2**, 40–46.

Schneider, L.S. (2008). *Ginkgo biloba* extract and preventing Alzheimer's disease. *Journal of American Medical Association*, **300**, 2306–2308.

Scholey, A. and Kennedy, D.O. (2002). Acute, dose-dependent cognitive effects of *Ginkgo biloba*, *Panax ginseng* and their combination in healthy young volunteers: Differential interactions with cognitive demand. *Human Psychopharmacology: Clinical and Experimental*, **17**, 35–44.

Sharangi, A.B. (2009). Medicinal and therapeutic potentialities of tea (*Camellia sinensis* L.)—A review. *Food Research International*, **42**, 529–535.

Sun, Y., Lu, C.J., Chien, K.L., Chen, S.T., and Chen, R.C. (2007). Efficacy of multivitamin supplementation containing vitamins B6 and B12 and folic acid as adjunctive treatment with a cholinesterase inhibitor in Alzheimer's disease: A 26-week, randomized, double-blind, placebo-controlled study in Taiwanese patients. *Clinical Therapeutics*, **29**, 2204–2214.

Tian, Y.X., Li, L., and Zhao, J.X. (2002). The effect of the electroacupuncture on the plasma TXB-2,6-Keto-PGF1. *Hebei Journal of Traditional Chinese Medicine*, **24**, 478–480.

Wagner, H. and Ulrich-Merzenich, G. (2009). Synergy research: Approaching a new generation of phytopharmaceuticals. *Phytomedicine*, **16**, 97–110.

Wahbeh, H., Elsas, S-M., and Oken, B.S. (2008). Mind-body interventions: Applications in neurology. *Neurology*, **70**, 2321–2328.

Wang, L. (2002). The improvement of memory in vascular dementia model of rats by electro—Acupuncture. *New Journal of Traditional Chinese Medicine*, **34**, 70–72.

Wang, B.S., Wang, H., Wei, Z.H., Song, Y.Y., Zhang, L., and Chen, H.Z. (2009). Efficacy and safety of natural acetylcholinesterase inhibitor huperzine A in the treatment of Alzheimer's disease: An updated meta-analysis. *Journal of Neural Transmission*, **116**, 457–465.

Weina, P., Zhao, H., Zhishun, L., and Shi, W. (2007). Acupuncture for vascular dementia. *Cochrane Database of Systematic Reviews 2007*, (2), CD004987. DOI: 10.1002/14651858.CD004987.pub2.

Wesnes, K.A., Ward, T., McGinty, A., and Petrini, O. (2000). The memory enhancing effects of a *Ginkgo biloba/Panax ginseng* combination in healthy middle-aged volunteers. *Psychopharmacology*, **152**, 353–361.

Xu, L., Cong, W.H., Wei, C.E., and Liu, J.X. (2007). Effect of Weinaokang (WNK) on dysmnesia in mice model. *Journal of Pharmacological and Clinical Chinese Herbal Medicine*, **23**, 60–61.

Xu, L., Liu, J.X., Cong, W.H., and Wei, C.E. (2008). Effects of Weinaokang (WNK) capsule on intracephalic cholinergic system and capability of scavenging free radicals in chronic cerebral hypoperfusion rats. *China Journal of Chinese Materia Medica*, **33**, 531–534.

Yang, L., Hao, J., and Zhang, J. (2009). Ginsenoside Rg3 promotes beta-amyloid peptide degradation by enhancing gene expression of neprilysin. *Journal of Pharmacy and Pharmacology*, **61**, 375–380.

Yang, F., Lim, G.P., Begum, A.N., Ubeda, O.J., Simmons, M.R., Ambegaokar, S.S., and Cole, G.M. (2005). Curcumin inhibits formation of amyloid beta oligomers and fibrils, binds plaques, and reduces amyloid *in vivo*. *Journal of Biological Chemistry*, **280**, 5892–5901.

Yao, L., Biordani, B., and Alexander, N.B. (2008). Developing a positive emotion-motivated Tai Chi (PEM-TC) exercise program for older adults with dementia. *Research and Theory for Nursing Practice*, **22**, 241–255.

Yuan, D.-P., Huang, Q., and Zhang, M.-L. (2002). Effect of compound Danshen on transmitter in rat brain of vascular dementia. *Shanxi Journal of Traditional Chinese Medicine*, **18**, 47–49.

Zhang, J.-Z. (1988). Statistical brain topographic mapping analysis for EEGs recorded during Qigong state. *Internal Journal of Neuroscience*, **38**, 415–425.

Zhang, S.-Y., Chen, Y., Zuo, P.-P., Yuan, B., Liu, D., Jin, Y.L., and Liu, Y. (2001). Therapeutic effect of 764-3 on rat with Alzheimer's like disease. *Basic Medical Sciences and Clinics*, **21**, 69–72.

Zhao, J.X., Tian, Y.X., and Li, G.M. (2001). The influence of acupuncture (Shenshu, Geshu, Baihui) on cell number of in the area of ca of hippocampal gyrus of mouse with vascular dementia. *Information on Traditional Chinese Medicine*, **18**, 57–58.

Zhao, J.X., Tian, Y.X., and Sun, Y.H. (2000). The effect of Shenshu (BL23), Geshu (BL17), Baihui (DU20) electropuncture on learning and memory of mouse with synthetic VD. *Chinese Journal of Behavioral Medical Science*, **9**, 100–102.

Zhao, L., Zhang, H., Zheng, Z., and Huang, J. (2009). Electroacupuncture on the head points for improving gnosia in patients with vascular dementia. *Journal of Traditional Chinese Medicine*, **29**, 129–134.

Zheng, Y.Q., Liu, J.X., Wang, J.N., and Xu, L. (2006). Effects of crocin on reperfusion-induced oxidative/nitrative injury to cerebral microvessels after global cerebral ischemia. *Brain Research*, **1138**, 86–94.

Zou, Y.P., Huang, F., and Shan, D.H. (2006). Effects of Guipi Decoction on behavior, learning and memory, and neuron number in hippocampus CA3 area in rat model of depression. *Journal Traditional Chinese Medicine*, **21**, 28–29.

11 Epigallocatechin Gallate (EGCG) and Cognitive Function

Melissa Finn, Andrew Scholey,
Andrew Pipingas, and Con Stough

CONTENTS

WHAT IS EGCG?

Epigallocatechin gallate (EGCG) is a phytochemical found in green tea. Green tea, a popular beverage produced from the leaves of the *Cammelia sinesis* plant, is made up of several chemical compounds called polyphenols. Polyphenols are naturally occurring compounds found in most plant-derived foods with tea, red wine, fruits, and vegetables among the richest dietary sources (Kondratyuk and Pezzuto, 2004). Polyphenols can be divided into three classes: tannins, lignins, and flavonoids. The polyphenols in tea are classified as flavonoids. Flavonoids are distinguished by their chemical structure under the categories: anthocyanidins, carotenes, catechins, flavones, flavonols, flavanones, glucosinolates, isoflavones, lavones, and organosulfides. The major flavonoids in tea are catechins (Kaur et al., 2008). In green tea the catechins account for 30%–40%

of the contents in a normal tea bag (Pan et al., 2003). Catechins are present in higher quantities in green tea than in black or oolong tea (Neilson et al., 2006). This is because during processing green tea is exposed to the least amount of oxidation. The catechins in green tea include epicatechin (EC), epigallocatechin (EGC), epicatechin gallate (ECG), and epigallocatechin gallate (EGCG). EGCG is the most abundant catechin found in green tea and accounts for 65% of the total catechin content (González de Mejía, 2003; Zaveri, 2006). A normal cup of green tea contains approximately 200 mg of EGCG (McKay and Blumberg, 2007; Pietta et al., 1998). There is evidence to suggest the catechins in green tea can be used for preventing and treating cancer, cardiovascular diseases, inflammatory diseases, and neurodegenerative diseases (Zaveri, 2006).

Most of the beneficial effects of green tea are attributed to EGCG (Lee et al., 2004). Extracts of green tea, made almost exclusively of EGCG (e.g., Teavigo®), are currently marketed and sold for weight management and improving oral and cardiovascular health. The consumption of EGCG has also been associated with various neurological benefits such as reducing symptoms of Parkinson's disease and Alzheimer's disease (Weinreb et al., 2004) and in improving cognitive function in general (Xie et al., 2008). The beneficial effects of EGCG are often attributed to its potent antioxidant properties (Zaveri, 2006), although recent studies have discovered potential neuroprotective properties of the catechin (Mandel et al., 2005).

Currently there are no clinical trials investigating the relationship between EGCG and cognitive function in humans. The premise that EGCG may improve cognitive function is generated from the results of animal studies, and research on antioxidants, flavonoids, and green tea. The following sections of this chapter outline previous research investigating the relationship between antioxidants, flavonoids, green tea, EGCG, and cognitive function. The biological mechanisms by which EGCG may improve cognitive function and suggestions for future research in the area are also discussed.

ANTIOXIDANTS, FLAVONOIDS, AND COGNITIVE FUNCTION

Several studies have examined the general relationship between antioxidants or flavonoids and cognitive function. Flavonoids are rapidly absorbed by the human body and have been reported to have positive effects on numerous aspects of health, including a reduced risk of coronary heart disease (Hertog and Feskens, 1993), cancer (Luisa Castellani et al., 2007), and neurodegenerative diseases (Rao and Balachandran, 2002). The beneficial effects of flavonoids are often attributed to the antioxidants they provide (Mandel et al., 2005; Zaveri, 2006). Tea consumption has been reported to increase the prevalence of antioxidants in the body (van het Hof et al. 1998; Salah et al., 1995). This is supported by findings demonstrating that tea catechins, particularly EGCG, are more powerful antioxidants than vitamins C and E (Kimura et al., 2002; Kuriyama et al., 2006).

ANTIOXIDANTS AND COGNITIVE FUNCTION

Epidemiological and longitudinal studies investigating the relationship between antioxidants and cognitive function have been conducted. One study found the use of supplemental antioxidants (vitamins C and E) prevented cognitive decline

(Gray et al., 2003). Cognitive function was assessed using the short portable mental status questionnaire (SPMSQ), a brief 10-item cognitive screen. Those who consumed antioxidants regularly were reported to have a 29%–34% lower risk of developing cognitive deficits. Similar results were obtained in a larger study, where the regular consumption of supplemental antioxidants (vitamins A, C, and E) was associated with significant improvements in cognitive function (Grodstein et al., 2003). A telephone version of the mini-mental state examination (MMSE) was used as the main measure to assess cognitive function. The MMSE includes questions pertaining to attention and short-term memory (Kuriyama et al., 2006). However, it is important to note the SPMSQ and MMSE have been criticized for being crude and nonspecific measures of cognitive function (Poole and Higgo, 2006). They were designed as screening instruments for dementia (Cress, 2006), rather than for comprehensive evaluations of cognitive function. Measures of criterion validity have also revealed that the MMSE is not sensitive to mild cognitive impairments or differences in individuals with normal cognitive function (Ng et al., 2009; Wind et al., 1997).

Several studies have found that the dietary intake of antioxidants is associated with a lower risk of Alzheimer's disease (Corrada et al., 2005; Engelhart et al., 2002; Morris et al., 2002). Due to the cognitive nature of Alzheimer's disease, the results of these studies could have important implications for cognitive function. However, some studies have found no association between Alzheimer's disease and antioxidant intake (Laurin et al., 2004; Luchsinger et al., 2003).

Although several studies indicate a relationship between antioxidant consumption and cognitive function, other studies do not. One longitudinal study, using data from a battery of 15 neuropsychological tests measuring cognitive performance, examined whether the use of antioxidant supplements was associated with cognitive function (Mendelsohn et al., 1998). Those who consumed antioxidants performed better than those who did not; however, the results were not significant after participant demographics were controlled. A similar study found no significant association between antioxidant vitamin consumption and cognitive function (Peacock et al., 2000). Cognitive function was assessed using the delayed word recall test, the Wechsler Adult Intelligence Scale-Revised digit symbol subtest, and the controlled oral word association test of the multilingual aphasia examination.

Inconsistent results in the literature may be due to differences in the population samples, such as differences in the distribution of age or socioeconomic status, being studied. Furthermore, unlike several of the studies indicating a positive relationship, Laurin et al. (2004), Lushinger et al. (2003), Mendelsohn et al. (1998), and Peacock et al. (2000) used less comprehensive dietary assessment methods. Therefore, they may have investigated samples consuming smaller doses of antioxidants for shorter periods of time, accounting for nonsignificant results. In addition, a variety of methods of measuring cognitive function have been used throughout the studies, making it difficult to compare findings.

FLAVONOIDS AND COGNITIVE FUNCTION

Epidemiological and longitudinal studies have also been conducted investigating the relationship between flavonoids and cognitive function. A recent study examined flavonoid intake in relation to cognitive function over 10 years (Letenneur et al., 2007). The MMSE, Benton's visual retention test and Isaacs set test were used to assess cognitive function. After adjustments for participant demographics, flavonoid intake was significantly associated with better cognitive function over time. Positive effects on cognitive function in healthy volunteers as well as those with cognitive deficits (Mohsen et al., 2002) have been found using treatment with *Gingko biloba* extract, known to contain flavonoids (Cotman et al., 2002; Mandel et al., 2004b). Burns and Nettelbeck (2003) conducted a 12 week, placebo-controlled study assessing the effects of *G. biloba* on cognitive abilities. Long-term memory assessed by associational learning tasks and tested in immediate and delayed-recall forms showed significant improvement in those consuming the *G. biloba*. A similar 30 day, randomized, double-blind, placebo-controlled clinical trial found significant improvements in speed of information processing working memory and executive processing in participants consuming *G. biloba* (Stough et al., 2001). According to the authors these improvements are likely to be due to the antioxidant and flavonoid characteristics of the extract (Burns and Nettelbeck, 2003; Stough et al., 2001).

METHODOLOGICAL LIMITATIONS OF EPIDEMIOLOGICAL STUDIES INVESTIGATING ANTIOXIDANT AND FLAVONOID CONSUMPTION

Most of the studies investigating the relationship between antioxidant/flavonoid consumption and cognitive function are epidemiological. Unfortunately, epidemiological studies have several methodological limitations. Finding the time and equipment needed for assessing the cognitive function of large groups of people is difficult. Therefore, most studies use short cognitive questionnaires (e.g., MMSE), or small subsets of tests within much larger test batteries, rather than comprehensive cognitive assessments. Furthermore, in most cases no rationale was provided to validate the combination of tests used. Studies using more accurate and specific methods of assessment would provide a clearer understanding of the relationship between cognitive function and antioxidant consumption.

Epidemiological studies also have no control over participants' antioxidant/flavonoid consumption. There are various factors that could influence the effects of supplement consumption. These include "the length of time taking the supplement, the constancy, amount, purity and type of preparation, and the composition of the mixture" (Martin and Mayer, 2003, p. 71). This makes it difficult to examine the effects of individual antioxidants on cognitive function (Gray et al., 2003), given the wide variety of antioxidant preparations publicly consumed in supplements and in everyday diet.

Another limitation of epidemiological study designs is that they do not allow researchers to fully exclude the possibility of confounding by unmeasured factors (Kuriyama et al., 2006). For example, supplement intake may indicate a healthier diet or different social and lifestyle factors in those consuming supplements, which independently affect cognitive function. Furthermore, epidemiological designs do not enable researchers to make causal associations between antioxidant/flavonoid consumption and cognitive function. Even if it is possible to successfully adjust the findings for covariates, it is impossible to ensure there is not any hidden unmeasured covariates that could affect the results (Little and Rubin, 2000).

ANIMAL STUDIES

Several animal studies have investigated the relationship between antioxidants/ flavonoids and cognitive function. Milgram et al. (2004) investigated the long-term effects of antioxidant supplementation on cognitive function, specifically learning, in a longitudinal study of aged dogs. Dogs administered the antioxidant supplement for 1 year presented with significantly improved learning ability on a size discrimination learning task and on a size discrimination reversal learning task. In a similar study, the consumption of an antioxidant-enriched supplement for 6 months resulted in a significant improvement in the ability of aged dogs to perform difficult learning tasks (Cotman et al., 2002). The dogs' cognitive ability improved most on tasks involving oddity discrimination. The authors attributed these results to the cognitive enhancing effects of the flavonoids in the supplements. This is supported by studies on flavonoids. Joseph et al. (1999) found that aged rats fed flavonoid-rich supplemented food (including blueberry, spinach, and strawberry) for 8 weeks significantly improved in spatial learning and memory as assessed by the working memory version of the Morris water maze.

GREEN TEA AND COGNITIVE FUNCTION

Only four studies have investigated the relationship between tea and cognitive function in humans. The first study analyzed data collected from the Tsurugaya Project, an extensive health assessment of elderly Japanese people (Kuriyama et al., 2006). They found a higher consumption of green tea was associated with a lower prevalence of cognitive impairment. There was no association between the consumption of black tea, oolong tea, or coffee and cognitive function. The lower prevalence of cognitive impairment was attributed to the unique effects of EGCG. The second study analyzed the association between tea consumption and cognitive function using a cross-sectional and longitudinal methodological design (Ng et al., 2008). Participants were a cohort of Chinese older adults from the Singapore longitudinal aging study (SLAS). Statistical analysis supported an association between green tea consumption and lowered cognitive impairment. However, this association cannot be separated from the influence of black or oolong tea, given only 10 participants in the study consumed green tea exclusively (Ng et al., 2008). Another limitation is both Kuriyama et al. and Ng et al. used the MMSE as their primary measure of cognitive function. As mentioned earlier, the MMSE was designed as a screening instrument for dementia (Cress, 2006), consequently it does not comprehensively evaluate cognitive function.

There are two more studies that have investigated the relationship between tea and cognitive function in humans. Both of the studies assessed cognitive function using comprehensive batteries of tests. Nurk and associates (2009) found a positive association between flavonoid-enriched foods (including wine, tea, and chocolate) and improved cognitive test performance. Participants who consumed tea performed significantly better on tests assessing attention, perceptual speed, visuospatial skills, and global cognition than those who did not. Several mechanisms by which flavonoid consumption might protect against cognitive impairment were proposed with major emphasis on the antioxidant actions of flavonoids (Nurk et al., 2009). However, the type of tea consumed was not recorded; therefore, it is impossible to determine the unique influence of green tea in comparison to black or oolong tea. The fourth study, a continuation of the study conducted by Ng et al. (2008), analyzed the relationship between tea consumption and cognitive function in a subsample of participants from the SLAS cohort (Feng et al., 2010). Tea consumption was associated with improved memory, executive function, and information processing speed. Coffee consumption and cognitive function were not related, indicating the cognitive effects observed were due to a component specific to tea consumption. Both black/oolong tea consumption and green tea consumption were associated with better cognitive performance. However, of the 716 participants only four participants consumed green tea exclusively, making it difficult to examine the unique effects of green tea.

The studies conducted by Kuriyama et al. (2006), Ng et al. (2008), Nurk et al. (2009), and Feng et al. (2010) are methodologically limited due to the limitations of epidemiological designs. As mentioned earlier, epidemiological studies do not allow researchers to fully exclude the possibility of confounding by unmeasured factors. Tea intake may indicate a healthier diet or more favorable social and lifestyle factors in those consuming tea, which independently improve cognitive function. There is also large intersubject and intrasubject variability in the consumption of tea polyphenols (Chow et al., 2001). As only epidemiological studies have been conducted on this topic, it is impossible to determine a causal relationship between tea consumption and cognitive function.

EGCG AND COGNITIVE FUNCTION

Several animal studies have examined the relationship between EGCG and cognitive function. A recent study found EC consumption improves spatial memory in mice (Praag et al., 2007). The mice were trained in a water maze for 2 weeks with four trials per day for 8 days. Learning was faster and retention longer in EC-treated mice. Praag et al. attributed the memory-enhancing benefits of EC to its ability to increase cortical blood flow. The relationship between catechin consumption and cognitive function has also been investigated in rats. Haque et al. (2006) investigated the effect of long-term oral administration of green tea catechins (60% EGCG) on the spatial learning ability of young rats. Rats administered green tea catechins presented with improved memory-related learning ability as measured by completion of the eight-arm radial maze. The authors attributed this improvement to the antioxidative activity of green tea catechins (Haque et al., 2006). Similarly, another study investigated the effects of green tea extract administration (30% EGCG) on cognitive function, as measured by passive avoidance and an elevated maze task, in young and old rats (Kaur et al., 2008). The extract significantly improved learning and memory in older rats.

EGCG has also been implicated in preventing cognitive decline in cognitively impaired animals. Unno and associates (2006) found the usual decline of memory in senescence-accelerated (SAMP10) mice was significantly slowed when the mice consumed green tea catechins (31.7% EGCG) on a daily basis for 1 year. Memory was assessed using the passive avoidance task. SAMP10 mice are a model of brain senescence with cerebral atrophy and cognitive dysfunction. The administration of green tea polyphenols (60% EGCG) has also been found to reduce cognitive impairments induced by psychological stress in rats (Chen et al., 2009). The animal model of psychological stress was developed by restraint where the rats' movement was limited periodically for 3 weeks. The rats' cognitive function, as measured by performance in an open-field test, step-through test, and water maze, was improved by the consumption of the green tea polyphenols.

EGCG has also been found to enhance long-term potentiation (LTP). In a study conducted by Xie et al. (2008), LTP in Ts65Dn mice was significantly enhanced when hippocampal slices were pre-incubated with EGCG for 1 h prior to the experiment. This is important because LTP is a "well-characterized form of synaptic plasticity that fulfils many of the criteria for a neural correlate of memory" (Cooke and Bliss, 2006, p. 1659), and is widely considered the major cellular mechanism that influences learning and memory (Abraham and Williams, 2003). Furthermore, Xie et al. investigated Ts65Dn mice, a Down syndrome mouse model with deficits in LTP and spatial learning and memory, further implicating the potential of EGCG in influencing organisms with cognitive impairments.

ACUTE COGNITIVE EFFECTS OF EGCG

Of the studies reviewed, the majority have involved the administration of antioxidants, flavonoids, green tea, or EGCG for at least 2 weeks. However, there is evidence that EGCG may also have an acute effect (Pietta et al., 1998; Xie et al., 2008). Several studies have investigated the absorption of EGCG. The time to maximum absorption of catechins has been reported to be approximately 2 h (Duffy et al., 2001; Pietta et al., 1998), with the plasma elimination half-life of approximately 2–3 h (Collie and Morley, 2007; McKay and Blumberg, 2007). In line with these findings, the highest concentrations of catechins in humans have been reported to occur around 1 h after ingestion (Kimura et al., 2002).

Several studies have investigated the effect of EGCG consumption on acute antioxidant activity. In one study a sample of 12 participants were supplemented with single doses of green tea catechins equivalent to 400 mg EGCG, both in free and in phospholipid complex forms (Pietta et al., 1998). Blood samples were collected before and 60, 120, 180, 240, 300, and 360 min after ingestion. A single dose of both forms of EGCG produced an increase in total radical antioxidant parameter, with a peak at 2 h after ingestion (Pietta et al., 1998). Similarly, Kimura and associates (2002) investigated the consumption of a single (164 mg) and double (328 mg) dose of EGCG on ferric-reducing antioxidant power. In contrast to Pietta and associates, there were no significant differences in either dose from baseline 30, 60, or 180 min after ingestion. However, Kimura et al. had a sample of five participants, which may have been insufficient to detect an effect. Furthermore, the studies measured antioxidant activity using different assays that differ in their chemistry and their mechanisms in detecting differences in activity (Pellegrini et al., 2003).

Given its acute physiological effects, two recent studies have examined the acute neurocognitive effects of EGCG administration (Scholey et al., 2012; Wightman et al., 2012). Scholey and colleagues found increased self-rated calmness and reduced stress 2 h following a 300 mg dose of EGCG (Teavigo). The behavioral effects were coupled with treatment-related changes in overall EEG activity—and increases in alpha, beta, and theta waveform activity localized to frontal regions. Wightman et al.'s (2012) results also indicate changes in frontal activity, with decreased blood flow to this region as measured using near infrared spectroscopy following 135 mg of the same extract. The Wightman study found no change in mood or cognitive performance (repeated cycles of cognitively demanding tasks) following 135 or 270 mg of the extract.

The acute effects of flavonoids, more specifically the consumption of *G. biloba* and *Bacopa monniera*, on cognitive function have been investigated. Kennedy et al. (2000) investigated the cognitive effects of an acute administration of *G. biloba*. Their results showed acute *G. biloba* administration enhances cognitive performance in healthy young adults. This cognitive enhancement was most noticeable in tasks assessing attention. The effect was time-dependent, with significant improvements found at 150, 240, and 360 min following ingestion. Contrastingly, a similar study conducted by Nathan et al. (2001) found no relationship between a single dose of *B. monniera*, known to have similar antioxidant activities to EGCG, and cognitive function. The authors suggested a chronic administration of *B. monniera* may be required to significantly improve cognitive function.

BIOLOGICAL MECHANISMS OF EGCG RELATED TO COGNITIVE FUNCTION

ANTIOXIDANT PROPERTIES

Although the mechanisms underlying the cognitive effects of EGCG are not fully understood, its benefits are commonly attributed to its antioxidant properties (Zaveri, 2006). EGCG is a more powerful antioxidant than vitamins C and E (Kimura et al., 2002; Kuriyama et al., 2006). Several studies have investigated the antioxidant power of EGCG. Kimura and associates (2002) investigated the consumption of a single dose of EGCG on ferric-reducing antioxidant power. They found no significant differences from baseline 30, 60, or 180 min after the ingestion of 328 mg of EGCG. Conversely, other studies have found a single dose of green tea causes an increase in ferric-reducing ability of plasma assay (Leenan et al., 2000) and total radical antioxidant parameter (Pietta et al., 1998), measuring antioxidant power. However, in the later studies participants ingested a larger than normal amount of tea catechins (400–900 mg vs. a normal value of 300 mg), suggesting larger amounts of EGCG may need to be consumed to exert its antioxidant power. The regular consumption of green tea catechins, as evidenced in a 30 day study conducted by Pietta and Simonetti (1998), has also been found to provide antioxidant protection.

The antioxidative effect of EGCG may also influence cognitive function by ameliorating the effects of oxidative stress. Oxidative stress refers to the damages in cellular structure and functions that occur due to the increased production of free radicals, reactive species, and oxidant-related reactions (Yu and Chung, 2006). Research has shown the brain is particularly vulnerable to oxidative stress over time because of its high oxygen consumption, 20% of the total body oxygen, and its deficiency in free radical protection (Joseph et al., 1999; Kaur et al., 2008). Antioxidants, in particular EGCG, may improve cellular functioning and minimize oxidative stress in aged organisms (Blokhina et al., 2003; Yu and Chung, 2006). The antioxidant properties of EGCG have also been found to promote the inhibition of xanthine oxidase, which can lower the production of oxygen-free radicals in the brain (Lee et al., 2004; Pietta and Simonetti, 1998). Furthermore, Unno and associates (2006) found oxidative damage in DNA was suppressed in aged mice fed EGCG. The consumption of EGCG has been suggested as a treatment for Alzheimer's disease due to its ability to reduce the effects of oxidative stress (Engelhart et al., 2002).

The metal-chelating properties of green tea catechins are also important contributors to their antioxidative activity (Zaveri, 2006). Although EGCG is a relatively selective chelator of iron (Mandel et al., 2007), it also chelates other metals (Reznichenko et al., 2006). Metal chelators are compounds that bind to metals consequently rendering them motionless and unable to participate in neurodegenerative progression (Chaston and Richardson, 2003). It has been suggested in neurodegeneration metals accumulate in the brain where neuronal death occurs (Mandel and Youdim, 2004). However, the causal direction is unclear. The metal-chelating characteristic of EGCG has been associated with its ability in treating Parkinson's disease and Alzheimer's disease (Mandel et al., 2006).

Researchers have proposed several other explanations for the improvements in cognitive function seen after antioxidant consumption. The consumption of antioxidants may result in reduced inflammatory responses (Joseph et al., 1999) and acetylcholinesterase activity (Kaur et al., 2008), which is part of the central cholinergic system involved in regulating cognitive functions (Kim et al., 004). While some have also proposed an increase in cortical blood flow (Pragg et al., 2007) and in the production of neurons (Spencer, 2009) may explain the effects. This is supported by recent research that has shown flavanol-enriched foods can increase cerebral blood flow velocity (Sorond et al., 2008). Furthermore, the oral administration of EC has been found to enhance angiogenesis, the development of new blood vessels from pre-existing blood vessels (Pragg et al., 2007). Cerebral blood flow has been found to be correlated with cognitive function (Ruitenberg et al., 2005).

NEUROPROTECTIVE EFFECTS

Data from several studies indicate the antioxidant properties of green tea catechins are unlikely to be the sole explanation for their effects (Mandel et al., 2005). Both in vivo and in vitro studies have demonstrated green tea catechins exert a neuroprotective role, and several researchers attribute this to their diverse pharmacological activities (Mandel et al., 2004b, 2005; Mandel and Youdim 2004). Bastianetto et al. (2006) showed that EGCG is the sole catechin to contribute to the neuroprotective effects of green tea. This characteristic of EGCG could be involved in the prevention of cognitive decline. Over the last 5 years research has demonstrated green tea catechins, through their neuroprotective properties, have the ability to effect cell survival or death genes and signal transduction (Kalfon et al., 2006; Lee et al., 2004; Levites et al., 2003; Mandel and Youdim, 2004; Mandel et al., 2004a, 2006; Weinreb et al., 2004; Youdim et al., 2002). These effects have been reported in a variety of models of toxicity (Bastianetto et al., 2006) including that induced by ischemia (Lee et al., 2004), N-Methyl-4-phenyl-1,2,3,6-tetrahydropyridine (Levites et al., 2001), glutamate (Lee et al., 2004), and β-amyloid peptides (Levites et al., 2003).

It is important to note it is still unclear whether these neuroprotective effects are due to antioxidant activities or due to the unique activities of EGCG on a range of molecular targets. Furthermore, most of the mechanisms that have been proposed are based on in vitro studies with amounts of EGCG much higher than those achievable in vivo (Zaveri, 2006). Plasma tea catechin concentrations determined in humans after the oral consumption green tea catechins have been found to be 5–50 times less than the concentrations shown to exert biological activities (Chow et al., 2005). Whether the neuroprotective mechanisms of EGCG can be replicated in vivo is still unknown (Zaveri, 2006).

BIOAVAILABILITY OF EGCG

Understanding the biological effects of tea consumption in humans is made difficult by inadequate information on the bioavailability and biotransformation of tea catechins. There is evidence EGCG can cross the blood–brain barrier and has access to the brain after oral ingestion (Bastianetto et al., 2006; Mohsen et al., 2002). Suganuma et al. (1998) found after 3 h 33% of the total amount of catechin absorbed was found in the mouse brain from a single administration of EGCG. Human studies on the pharmacokinetics of tea catechins have been limited in scope. However, the studies that have been done indicate the bioavailability of EGCG in humans to be limited. Levels in plasma up to a maximum of 7.3 μmol/L (±3.6) have been reported, but more often are in the submicromolar range (Howells et al., 2007; Yang et al., 2008). Higher plasma concentrations have been found in fasting patients compared to those consuming EGCG with food (Naumovski, 2010). Peperine, derived from black pepper, has also been found to enhance the bioavailability of EGCG in mice (Lambert et al., 2004). However, oral consumption of EGCG in humans results in high plasma clearance levels and volume distribution, suggesting the bioavailability of EGCG in the blood may be low (Howells et al., 2007). In addition, a large variability between people in the pharmacokinetics of green tea catechins has been reported (Chow et al., 2005). Although green tea extracts have been marketed as nutritional supplements, it appears large doses may need to be used because of the limited pharmacokinetic mechanisms of the catechin (Zaveri, 2006).

FUTURE RESEARCH

Research investigating the relationship between EGCG and cognitive function is in its infancy. The effect of EGCG consumption on cognitive function in humans has not been adequately investigated. Positive associations between antioxidants, flavonoids, green tea, and cognitive function have been found in several epidemiological studies. However, the inherent limitations of epidemiological designs make it difficult to infer causal relationships from these results. The premise that EGCG improves cognitive function is generated predominantly from the results of several animal studies. Therefore, additional acute and chronic clinical trials investigating the relationship between EGCG and cognitive function in humans are needed.

Further understanding of the bioavailability and pharmacokinetic profiles of EGCG in the human brain is crucial to understanding the influence of EGCG on cognitive function. Previous studies indicate the bioavailability of EGCG to be low. Therefore, it is important to determine regimens which can enhance EGCG bioavailability. The biological mechanisms of EGCG related to brain function also need further research. Several animal studies have validated the use of EGCG in a variety of models of toxicity (Bastianetto et al., 2006). However, no human studies have been conducted.

Nootropics, particularly nutraceuticals, are becoming a popular alternative to conventional medicine (Hill, 2008). However, most nutraceuticals still need support of extensive scientific studies to determine the extent of their effects (Kalra, 2003). As a result, it is becoming increasingly important to establish the efficacy of substances, such as EGCG, which may provide psychological and physiological benefits to everyday cognitive functioning.

REFERENCES

Abraham, W. C. and Williams, J. M. (2003). Properties and mechanisms of LTP maintenance. *Neuroscientist*, 9 (6), 463–474.

Bastianetto, S., Yao, Z., Papadopulos, V., and Quiron, R. (2006). Neuroprotective effects of green and black teas and their catechin gallate esters against B-amyloid induced toxicity. *European Journal of Neuroscience*, 23, 55–64.

Blokhina, O., Virolainen, E., and Fagerstedt, K. V. (2003). Antioxidants, oxidative damage and oxygen deprivation stress: a review. *Annals of Botany*, 91, 179–194.

Burns, J. and Nettelbeck, T. (2003). Effects of ginkgo biloba on cognitive abilities and mood in healthy older adults. *Australian Journal of Psychology*, 55, 74–79.

Chaston, T. B. and Richardson, D. R. (2003). Iron chelators for the treatment of iron overload disease: Relationship between structure, redox activity, and toxicity. *American Journal of Hematology*, 73, 200–210.

Chen, W., Zhao, X., Hou, Y., Li, S., Hong, Y., Wang, D. et al. (2009). Protective effects of green tea polyphenols on cognitive impairments induced by psychological stress in rats. *Behavioural Brain Research*, 202, 71–76.

Chow, S. H., Cai, Y., Alberts, D. S., Hakim, I., Dorr, R., Shahi, F. et al. (2001). Phase I pharmacokinetic study of tea polyphenols following a single-dose administration of epigallocatechin gallate and polyphenon E. *Cancer Epidemiology, Biomarkers and Prevention*, 10, 53–58.

Chow, S., Hakim, I. A., Vining, D. R., Crowell, J. A., Ranger-Moore, J., Chew, W. M. et al. (2005). Effects of dosing condition on the oral bioavailability of green tea catechins after a single-dose administration of polyphenon E in healthy individuals. *Clinical Cancer Research*, 11 (12), 4627–4633.

Collie, A. and Morley, G. (2007). Do polyphenols affect human cognitive function? *Current Topics in Nutraceutical Research*, 5 (4), 145–148.

Cooke, S. F. and Bliss, T. V. (2006). Plasticity in the human central nervous system. *Brain*, 129 (7), 1659–1673.

Corrada, M. M., Kawas, C. H., Hallfrisch, J., Muller, D., and Brookmeyer, R. (2005). Reduced risk of Alzheimer's disease with high folate intake: The Baltimore longitudinal study of aging. *Alzheimer's and Dementia*, 1, 11–18.

Cotman, C. W., Head, E., Muggenburg, B. A., Zicker, S., and Milgram, N. W. (2002). Brain aging in the canine: A diet enriched in antioxidants reduces cognitive dysfunction. *Neurobiology of Aging*, 23, 809–818.

Cress, C. J. (2006). *Handbook of Geriatric Care Management*. Sudbury, Ontario, Canada: Jones & Bartlett Publishers.

Duffy, S. J., Keaney, J. F., Holbrook, M., Gokce, N., Swerdloff, P.L., Frei, B. et al. (2001). Short- and long-term black tea consumption reverses endothelial dysfunction in patients with coronary artery disease. *Circulation*, 104, 151–156.

Engelhart, M. J., Geerlings, M. I., Ruitenberg, A., van Swieten, J. C., Hofman, A., Witteman, J. C. M. et al. (2002). Dietary intake of antioxidants and risk of Alzheimer disease. *Journal of the American Medical Association*, 287 (24), 3223–3229.

Feng, L., Gwee, X., Kua, E. H., and Ng, T. P. (2010). Cognitive function and tea consumption in community dwelling older Chinese in Singapore. *Journal of Nutrition, Health and Aging*, 14 (6), 433–438.

González de Mejía, E. (2003). The chemo-protector effects of tea and its components. *Archivos Latinoamericanos de Nutrición*, 53, 111–118.

Gray, S. L., Hanlon, J. T., Landerman, L. R., Ark, M., Schmader, K. E., and Fillenbaum, G. G. (2003). Is antioxidant use protective of cognitive function in the community-dwelling elderly? *American Journal of Geriatric Pharmacotherapy*, 1 (1), 3–10.

Grodstein, F., Chen, J., and Willett, W. C. (2003). High-dose antioxidant supplements and cognitive function in community-dwelling elderly women. *American Journal of Clinical Nutrition*, 77, 975–984.

Haque, A. M., Hashimoto, M., Katakura, M., Tanabe, Y., Haray, Y., and Shido, O. (2006). Long-term administration of green tea catechins improves spatial cognition learning ability in rats. *Journal of Nutrition*, 136, 1043–1047.

Hertog, M. G. L. and Feskens, E. J. M. (1993). Dietary antioxidant flavonoids and risk of coronary heart disease: The zutphen elderly study. *Lancet*, 342 (8878), 1007–1011.

Hill, B. (2008). Natural health care. *Equities*, 57 (5), 92–93.

van het Hof, K. H., Kivits, G. A. A., Weststrate, J. A., and Tijburg, L. B. M. (1998). Bioavailability of catechins from tea: The effect of milk. *European Journal of Clinical Nutrition*, 52, 356–359.

Howells, L. M., Moiseeva, E. P., Neal, C. P., Foreman, B. E., Andreadi, C. J., Sun, Y., Hudson, A., and Manson, M.M. (2007). Predicting the physiological relevance of in vitro cancer preventative activities of phytochemicals. *Acta Pharmacologica Sinica*, 28 (9), 1274–1304.

Joseph, J. A., Shukitt-Hale, B., Denisova, N. A., Bielinski, D., Martin, A., McEwen, J. J. et al. (1999). Reversals of age-related declines in neuronal signal transduction, cognitive, and motor behavioural deficits with blueberry, spinach, or strawberry dietary supplementation. *Journal of Neuroscience*, 19 (18), 8114–8121.

Kalfon, L., Youdim, M. B. H., and Mandel, S. A. (2006). Green tea polyphenol (−) epigallocatechin-3-gallate promotes the rapid protein kinase C- and proteasome-mediated degradation of bad: Implications for neuroprotection. *Journal of Neurochemistry*, 10, 1111–1122.

Kalra, E. K. (2003). Nutraceutical—Definition and introduction. *American Association of Pharmaceutical Scientists*, 5 (3), 27–28.

Kaur, T., Pathak, C. M., Pandhi, P., and Khanduja, K. L. (2008). Effects of green tea extract on learning, memory, behaviour and acetylcholinesterase activity in young and old male rats. *Brain and Cognition*, 67, 25–30.

Kennedy, D. O., Scholey, A. B., and Wesnes, K. A. (2000). The dose-dependent cognitive effects of acute administration of ginkgo biloba to healthy young volunteers. *Psychopharmacology*, 151, 416–423.

Kim, H. K., Kim, M., Kim, S., Kim, M., and Chung, J. H. (2004). Effects of green tea polyphenol on cognitive and acetylcholinesterase activities. *Bioscience, Biotechnology, and Biochemistry*, 68, 1977–1979.

Kimura, M., Umegaki, K., Kasuya, Y., and Higuchi, M. (2002). The relation between single/double or repeated tea catechin ingestions and plasma antioxidant activity in humans. *European Journal of Clinical Nutrition*, 56, 1186–1193.

Kondratyuk, T. P. and Pezzuto, J. M. (2004). Natural product polyphenols of relevance to human health. *Pharmaceutical Biology*, 42, 46–63.

Kuriyama, S., Hozawa, A., Ohmori, K., Shimazu, T., Matsui, T., Ebihara, S. et al. (2006). Green tea consumption and cognitive function: A cross-sectional study from the Tsurugaya Project 1. *American Journal of Clinical Nutrition*, 83, 355–361.

Lambert, J. D., Hong, J., Kim, D. H., Mishin, V. M., and Yang, C. S. (2004). Piperine enhances the bioavailability of the tea polyphenol (−)-epigallocatechin-3-gallate in mice. *Journal of Nutrition*, 134, 1948–1952.

Laurin, D., Masaki, K. H., Foley, D. J., White L. R., and Launer, L. J. (2004). Midlife dietary intake of antioxidants and risk of late-life incident dementia: The Honolulu-Asia aging study. *American Journal of Epidemiology*, 159 (10), 959–967.

Lee, H., Bae, J. H., and Lee, S. R. (2004). Protective effect of green tea polyphenol EGCG against neuronal damage and brain edema after unilateral cerebral ischemia in gerbils. *Journal of Neuroscience Research*, 77, 892–900.

Letenneur, L., Proust-Lima, C., Le Gouge, A., Dartigues, J. F., and Barberger-Gateau, P. (2007). Flavonoid intake and cognitive decline over a 10-year period. *American Journal of Epidemiology*, 165 (12), 1364–1371.

Levites, Y., Amit, T., Mandel, S., and Youdim, M. B. H. (2003). Neuroprotection and neurorescue against amyloid beta toxicity and PKC-dependent release of non-amyloidogenic soluble precursor protein by green tea polyphenol (−)-epigallocatechin-3-gallate. *Federation of American Societies for Experimental Biology Journal*, 17, 952–954.

Levites, Y., Weinreb, O., Maor, G., Youdim, M. B. H., and Mandel, S. (2001). Green tea polyphenol (−)-epigallocatechin-3-gallate prevents N-methyl-4-phenyl-1,2,3,6-tetrahydropyridine-induced dopaminergic neurodegeneration. *Journal of Neurochemistry*, 78, 1073–1082.

Little, R. J. and Rubin, D. B. (2000). Causal effects in clinical and epidemiological studies via potential outcomes: concepts and analytical approaches. *Annual Review of Public Health*, 21, 121–145.

Luchsinger, A. J., Tang, M., Shea, S., and Mayeux, R. (2003). Antioxidant vitamin intake and risk of Alzheimer disease. *Archives of Neurology*, 60 (2), 203–208.

Luisa Castellani, M., Shaik, Y. B., Narayan Shanmugham, L., Frydas, S., Madhappan, B., Vincenzo, S. et al. (2007). Role of flavonoids and vitamins in cancer. *Biology Forum*, 100 (1), 39–54.

Mandel, S., Amit, T., Bar-Am, O., and Youdim, M. B. H. (2007). Iron dysregulation in Alzheimer's disease: Multimodal brain permeable iron chelating drugs, possessing neuroprotective-neurorescue and amyloid precursor protein-processing regulatory activities as therapeutic agents. *Progress in Neurobiology*, 82, 348–360.

Mandel, S., Amit, T., Reznichenko, L., Weinreb, O., and Youdim, M. B. H. (2006). Green tea catechins as brain-permeable, natural iron chelators-antioxidants for the treatment of neurodegenerative disorders. *Molecular Nutrition and Food Research*, 50, 229–234.

Mandel, S., Avramovich-Tirosh, Y., Reznichenko, L., Zheng, H., Weinreb, O., Amit, T. et al. (2005). Multifunctional activities of green tea catechins in neuroprotection. *Neurosignals*, 14, 46–60.

Mandel, S., Maor, G., and Youdim, M. B. H. (2004a). Iron and a-synuclein in the substantia nigra of MPTP-treated mice. *Journal of Molecular Neuroscience*, 24, 401–416.

Mandel, S., Weinreb, O., Amit, T., and Youdim, M. B. H. (2004b). Cell signalling pathways in the neuroprotective actions of the green tea polyphenol (−)-epigallocatechin-3-gallate: implications for neurodegenerative disease. *Journal of Neurochemistry*, 88, 1555–1569.

Mandel, S. and Youdim, M. B. H. (2004). Catechin polyphenols: Neurodegeneration and neuroprotection in neurodegenerative diseases. *Free Radical Biology and Medicine*, 37 (3), 304–317.

Martin, A. and Mayer, J. (2003). Antioxidant vitamins E and C and risk of Alzheimer's disease. *Nutrition Reviews*, 61 (2), 69–74.

McKay, D. L. and Blumberg, J. B. (2007). Roles for epigallocatechin gallate in cardiovascular disease and obesity: An introduction. *Journal of the American College of Nutrition*, 26 (4), 362S–365S.

Mendelsohn, A. B., Belle, S. H., Stoehr, G. P., and Ganguli, M. (1998). Use of antioxidant supplements and its association with cognitive function in a rural elderly cohort. *American Journal of Epidemiology*, 148 (1), 38–44.

Milgram, N. W., Head, E., Zicker, S. C., Ikeda-Douglas, C., Murphey, H., Muggenberg, B. A. et al. (2004). Long-term treatment with antioxidants and a program of behavioural enrichment reduces age-dependent impairment in discrimination and reversal learning in beagle dogs. *Experimental Gerontology*, 39, 753–765.

Mohsen, A. E. M. M., Kuhnle, G., Rechner, A. R., Schroeter, H., Rose, S., Jenner, P. et al. (2002). Uptake and metabolism of epicatechin and its access to the brain after oral ingestion. *Free Radical Biology and Medicine*, 33, 1693–1702.

Morris, M. C., Evans, D. A., Bienias, J. L., Tangney, C. C., Bennett, D. A., Aggarwal, N. et al. (2002). Dietary intake of antioxidant nutrients and the risk of incident Alzheimer disease in a biracial community study. *Journal of the American Medical Association*, 287 (24), 3230–3237.

Nathan, P. J., Clarke, J., Lloyd, J., Hutchison, C. W., Downey, L., and Stough, C. (2001). The acute effects of an extract of bacopa monniera (brahmi) on cognitive function in healthy normal subjects. *Human Psychopharmacology: Clinical and Experimental*, 16, 345–351.

Naumovski, N. (2010). Effects of ingestion conditions on the oral bioavailability of epigallocatechin gallate (EGCG) after a single-dose administration in healthy humans. *Australasian Medical Journal*, 1 (1), 74–96.

Neilson, A. P., Green, R. J., Wood, K. V., and Ferruzzi, M. G. (2006). High-throughput analysis of catechins and theaflavins by high performance liquid chromatography with diode array detection. *Journal of Chromatography A*, 1132, 132–140.

Ng, T. P., Feng, L., Niti, M., Kua, E. H., and Yap, K. B. (2008). Tea consumption and cognitive impairment and decline in older Chinese adults. *American Journal of Clinical Nutrition*, 88, 224–231.

Nurk, E., Refsum, H., Drevon, C. A., Tell, G. S., Nygaard, H. A., Engedal, K. et al. (2009). Intake of flavonoid-rich wine, tea and chocolate by elderly men and women is associated with better cognitive test performance. *Journal of Nutrition*, 139, 120–127.

Pan, T., Jankovic, J., and Le, W. (2003). Potential therapeutic properties of green tea polyphenols in Parkinson's disease. *Drugs Ageing,* 20 (10), 711–721.

Peacock, J. M., Folsom, A. R., Knopman, D. S., Mosley, T. H., Goff, D. C., and Szklo, M. (2000). Dietary antioxidant intake and cognitive performance in middle-aged adults. *Public Health Nutrition*, 3 (3), 337–343.

Pellegrini, N., Serafini, M., Colombi, B., Del Rio, D., Salvatore, S., Bianchi, M. et al. (2003). Total antioxidant capacity of plant foods, beverages and oils consumed in Italy assessed by three different in vitro assays. *Journal of Nutrition*, 133, 2812–2819.

Pietta, P. G. and Simonetti, P. (1998). Dietary flavonoids and interaction with endogenous antioxidants. *Biochemical Molecular Biology International*, 44 (5), 1069–1074.

Pietta, P., Simonetti, P., Gardana, C., Brusamolino, A., Morazzoni, P., and Bombardelli, E. (1998). Relationship between rate and extent of catechin absorption and plasma antioxidant status. *Biochemistry and Molecular Biology International*, 46 (5), 895–903.

Poole, R. and Higgo, R. (2006). *Psychiatric Interviewing and Assessment*. Cambridge, U.K.: Cambridge University Press.

Praag, H. V., Lucero, M. J., Yeo, G. W., Stecker, K., Heivand, N., Zhao, C. et al. (2007). Plant-derived flavanol (–)epicatechin enhances angiogenesis and retention of spatial memory in mice. *Journal of Neuroscience*, 27 (22), 5869–5878.

Rao, A. V. and Balachandran, B. (2002). Role of oxidative stress and antioxidants in neurodegenerative diseases. *Nutritional Neuroscience*, 5 (5), 291–309.

Reznichenko, L., Amit, T., Zheng, H., Avsarmovich-Tirosh, Y., Youdim, M. B. H., Weinreb, O. et al. (2006). Reduction of iron-regulated amyloid precursor protein and B-amyloid peptide by (–)-epigallocatechin-3-gallate in cell cultures: Implicated for iron chelation in Alzheimer's disease. *Journal of Neurochemistry*, 10, 1–10.

Ruitenberg, A., den Heijer, T., Bakker, S. L., van Swieten, J. C., Koudstaal, P. J., Hofman, A. et al. (2005). Cerebral hypoperfusion and clinical onset of dementia: the Rotterdam study. *Annals of Neurology*, 56 (6), 789–794.

Salah, N., Miller, N. J., Paganga, G., Tijburg, L., Bolwell, P. G., and Rice-Evans, C. (1995). Polyphenolic flavanols as scavengers of aqueous phase radicals and as chain-breaking antioxidants. *Archives of Biochemistry and Biophysics*, 322 (2), 339–346.

Scholey, A., Downey, L., Ciorciari, J., Pipingas, A., Nolidin, K., Finn, M., Wines, M., Catchlove, S., Terrens, A., Barlow, E., Gordon, L., and Stough, C. (2012) Acute neurocognitive effects of epigallocatechin gallate (EGCG). *Appetite,* 58, 767–770.

Sorond, F. A., Lipsitz, L. A., Hollenberg, N. K., and Fisher, N. D. I. (2008). Cerebral blood flow response to flavanol-rich cocoa in healthy elderly humans. *Neuropsychiatric Disease and Treatment*, 4 (2) 433–440.

Spencer, J. P. E. (2009). The impact of flavonoids on memory: Physiological and molecular considerations. *Chemical Society Reviews*, 38, 1152–1161.

Stough, C., Clarke, J., Llloyd, J., and Pradeep, N. J. (2001). Neuropsychological changes after 30-day Gingko biloba administration in healthy participants. *International Journal of Neuropsychopharmacology*, 4, 131–134.

Suganuma, M., Okabe, S., Oniyama, M., Tada, Y., Ito, H., and Fujiki, H. (1998). Wide distribution of [3H](–)-epigallocatechin gallate, a cancer preventive tea polyphenol, in mouse tissue. *Carcinogenesis*, 19, 1771–1776.

Unno, K., Takabayash, F., Yoshida, H., Choba, D., Fukutomi, R., Kikunaga, N. et al. (2006). Daily consumption of green tea catechin delays memory regression in aged mice. *Biogerentology*, 10 (8), 36–43.

Weinreb, O., Mandel, S., Amit, T., and Youdim, M. B. H. (2004). Neurological mechanisms of green tea polyphenols in Alzheimer's and Parkinson's diseases. *Journal of Nutritional Biochemistry*, 15, 506–516.

Wightman, E. L., Haskell, C. F., Forster J. S., Veasey R. C., and Kennedy, D. O. (2012). Epigallocatechin gallate (EGCG), cerebral blood flow parameters, cognitive performance and mood in healthy humans: A double-blind, placebo-controlled, crossover investigation. *Human Psychopharmacology—Clinical and Experimental,* 27, 177–186.

Wind, A. W., Schellevis, F. G., Van Staveren, G, Scholten, R. P, Jonker, C., and Van Eijk, J. T. (1997). Limitations of the mini-mental state examination in diagnosing dementia in general practice. *International Journal of Geriatric Psychiatry*, 12, 101–108.

Xie, W., Ramakrishna, N., Wieraszko, A., and Hwang, Y. (2008). Promotion of neuronal plasticity by (–)-epigallocatechin-3-gallate. *Neurochemical Research*, 33, 776–783.

Yang, C. S., Sang, S., Lambert, J. D., and Lee, M. (2008). Bioavailability issues in study-
    ing the health effects of plant polyphenolic compounds. *Molecular Nutrition and Food
    Research*, 52, S139–S151.
Youdim, K. A., Spencer, J. P., Schroeter, H., and Rice-Evans, C. (2002). Dietary flavonoids as
    potential neuroprotectants. *Journal of Biological Chemistry*, 383, 503–519.
Yu, B. P. and Chung, H. Y. (2006). Adaptive mechanisms to oxidative stress during aging.
    Mechanisms of Ageing and Development, 127, 436–443.
Zaveri, N. T. (2006). Green tea and its polyphenolic catechins: Medicinal uses in cancer and
    noncancer applications. *Life Sciences*, 78, 2073–2080.

12 Assessing the Utility of *Bacopa monnieri* to Treat the Neurobiological and Cognitive Processes Underpinning Cognitive Aging

Con Stough, Vanessa Cropley, Matthew Pase, Andrew Scholey, and James Kean

CONTENTS

With increasing life expectancies and the maturation of the "baby boom" generation, adapting to the challenges posed by the aging population has been identified as one of the major issues facing contemporary Australian society (Australian Productivity Commission, 2005). For Australia, like many Western nations, *human aging has significant societal, economic, health, and, importantly, personal costs.* In purely economic terms, the costs of aging reflect *decreased productivity* as well as *increased levels of reliance on public services to health and social support* but this also has obvious ramifications for older citizens' *ability to lead fulfilling lives.* Increasing age is associated with a cluster of illnesses involving oxidative stress, cardiovascular and respiratory disease, and, importantly, neurological conditions such as Parkinson's disease and Alzheimer's disease. The New Zealand Treasury has estimated that the *cost to the public health system* alone of individuals over 65 years of age is *five times that of people under 65* (Bryant and Sonerson, 2006). The same report concludes that *33% of these increased costs could be offset by measures aimed at maintaining improved health, which of course also involves brain and cognitive processes.*

A time-honored and much empirically supported method of promoting optimal health throughout the life span has been through the adoption and maintenance of an appropriate, healthy diet. Recent research suggests that this principle not only applies to protection from "physical ailments" such as cardiovascular problems, but may also extend to ameliorating the effects of cognitive decline associated with increased age. The maintenance of brain health underpinning intact cognition is a key factor to maintaining a positive, engaged, and productive lifestyle. In the light of this, the role of diet including supplementation with nutritional and even pharmacological interventions capable of ameliorating the neurocognitive changes that occur with age constitutes vital areas of research.

WHAT IS COGNITIVE AGING?

Individual age-related changes in cognition vary greatly. However, research in cognitive aspects of aging (typically in 60–90 year-olds) has identified consistent deficits in reasoning and decision making, spatial abilities, perceptual-motor and cognitive speed, and, most robustly, memory (e.g., Christensen and Kumar, 2003). Longitudinal studies of aged populations illuminate the time course of cognitive deterioration. Using 5–10 year retest intervals, significant decrements across most cognitive capacities become evident. A recent review of longitudinal aging studies concludes that crystallized intelligence (e.g., factual knowledge) remains intact until late aging whereas measures of speed, information processing, and aspects of memory (e.g., working memory) are more sensitive to decline from age 60 (Christensen and Kumar, 2003).

BRAIN AGING AND OXIDATIVE STRESS

Neuroimaging studies reveal that increasing age is reliably associated with ventricular enlargement, reduction in gross brain volume, reductions in frontal and temporoparietal brain volume, higher levels of cortical atrophy, and increased white-matter hyperintensities (Looi and Sachdev, 2003). Ultimately, shrinkage of cortical volume reduces cognitive capacity (MacLullich et al., 2002), and age-related increases in neuropathological events such as β-amyloid protein deposition and formation of neurofibrillary tangles represent significant risk factors. Neuropathological events such as β-amyloid deposition are not exclusive to neurodegenerative disorders such as AD, in fact occurring in a large proportion of cognitively intact individuals. For example, in one study, the proportion of nonclinical subjects with β-amyloid deposits ranged from 3% in a 36–40 age group to 75% in an 85+ age group (Braak and Braak, 1997). Alongside age-associated cortical degeneration (MacLullich et al., 2002), there exist numerous microscopic insults related to oxidative stress. Free radicals formed in the brain produce significant cellular damage and mediate processes that result in neural cell death on large scales (Packer, 1992). Between 95% and 98% of free radicals and reactive oxygen species (ROS), O_2^-, HO, and H_2O_2 are formed by mitochondria as by-products of cellular respiration. Studies of mitochondria isolated from the brain show that 2%–5% of total oxygen consumed yields ROS, and these highly reactive molecules make a significant contribution to the peroxidation of principal cell structures (e.g., membrane lipids) (Papa and Skulachev, 1997). Brain tissue is particularly susceptible due to its disproportionately high metabolic rate and levels of oxygen, the cytotoxic actions of glutamate, and its high concentrations of peroxidizable unsaturated fatty acids (Packer, 1992). Aging decreases the brain's ability to combat the actions of free radicals. Aging is associated with increased levels of pro-oxidant mediators and decreases in antioxidants (Artur et al., 1992). The relationship between cognition and oxidative stress is evident in the extensive damage caused by free radicals in age-related neurological conditions (Coyle and Putfarcken, 1993; Smith et al., 1996) and animal models of age-related oxidative injury with central cognitive and behavioral impairments (Forster et al., 1996). Concurrent with the normal age-related cognitive changes are increases in the formation of brain ROS resulting in significant damage to DNA, proteins, and in particular membrane lipids (Smith et al., 1991). Although multiple factors precipitate oxidative stress throughout the body, the brain is particularly vulnerable, and its cumulative effects may account for the delayed onset and progressive nature of Alzheimer's and Parkinson's dementias as well as normal age-related mental deterioration (Coyle and Putfarcken, 1993).

ANTIOXIDANTS AND COGNITION

The central role of oxidative stress in age related cognitive decline and neurode generative diseases has driven numerous studies examining the potential benefits of antioxidants in altering, reversing, or forestalling neuronal and behavioral changes (e.g., Sano et al., 1997). Antioxidant supplementation results in improved cognition and behavior in aged animals and concurrent decreases in oxidative insult to neural structures (Socci et al., 1995). Human research in this area is largely limited to epidemiological studies. These have identified positive associations in aged individuals between biological levels of dietary antioxidants (vitamins E and C) and working memory measures including the Wechsler Memory test (Goodwin et al., 1983). Less reliable than biological measures, large-scale studies (3000+ participants) have also identified positive relationships between dietary intake of vitamins C and E and standardized memory measures (Masaki et al., 2000). While these nonclinical trials do not demonstrate causality, the consensus that memory is the main cognitive variable affected by antioxidant status is consistent with patterns of age-related cognitive decline and the in vivo neuroanatomy of lipid peroxidation (Sram et al., 1993). Three controlled studies of active antioxidant supplementation in aged individuals over periods of 1 year or longer reported improved performance on tests of short-term memory, verbal learning, and nonverbal memory (Sram et al., 1993; La Rue et al., 1997; Chandra, 2001). However, these studies did not incorporate indicators of oxidative stress, making it impossible to determine the role of antioxidants in the cognitive changes. Despite the great promise that antioxidant supplementation holds for understanding age-related mental deterioration, studies published in the area have been methodologically inadequate. In particular, human studies have thus far been severely limited by inappropriate cognitive measures, lack of biochemical indicators, uncontrolled subject populations, and unspecific antioxidant supplementations. One particular herbal medicine that may have some utility in treating pathological changes in the brain associated with age-related cognitive decline and that has been used in our laboratory is *Bacopa monnieri* (BM).

BACOPA MONNIERI

Bacopa monnieri (BM) is a botanical medicine from India that has been used for over 3000 years as a traditional ayurvedic treatment for asthma, insomnia, epilepsy, and as a "memory tonic" (Russo and Borrelli, 2005). BM has been used in traditional ayurvedic medicine for various indications including memory decline, inflammation, pain, pyrexia, epilepsy, and as a sedative (Russo and Borrelli, 2005). BM contains Bacoside A and bacoside B that are steroidal saponins believed to be essential for the clinical efficacy of the product. While BM has been reported to have many actions, its memory enhancing effects have attracted most attention and are supported by the psychopharmacology literature. Behavioral studies in animals have shown that BM improves motor learning, acquisition, retention, and delay extinction of newly acquired behavior (Singh and Dharwan, 1997). Although the exact mechanisms of action remain uncertain, evidence suggests that BM may modulate the cholinergic system and/or have antioxidant and metal-chelating effects (Agrawal, 1993; Bhattacharya et al., 1999). BM may also have antiinflammatory (Jain, 1994), anxiolytic and antidepressant actions (Bhattacharya and Ghosal, 1998), relaxant properties in blood vessels (Dar and Channa, 1999), and adaptogenic activity (Rai et al., 2003). Chronic administration of BM inhibits lipid peroxidation in the prefrontal cortex, striatum, and hippocampus via a similar mechanism to vitamin E (Bhattacharya et al., 2000). In an animal model of AD, there was a dose-related reversal by BM of cognitive deficits produced by the neurotoxins colchicine and ibotenic acid (Bhattacharya et al., 1999). In rodents, BM inhibited the damage induced by high concentrations of nitric oxide in astrocytes (Russo et al., 2003). Memory deficits following cholinergic blockade by scopolamine were reversed by BM treatment. In animal studies, BM reduced lipid peroxidation induced by $FeSO_4$ and cumene hydroperoxide, indicating that, similarly to the chelating properties of EDTA, it acts at the initiation level by chelating Fe^{++} (Tripathi et al., 1996). More recently, in transgenic mice, BM supplementation reduced specific amyloid peptides by up to 60% while also improving memory performance (Holcomb, 2006). Thus, BM appears to have multiple modes of action in the brain all of which may be useful in ameliorating cognitive decline in the elderly. These include (1) direct procholinergic action, (2) antioxidant (flavonoid) capacity, (3) metal chelation, (4) antiinflammatory effects, (5) increased blood circulation, (6) adaptogenic activity, and (7) removal of β-amyloid deposits.

EXTRACTS OF BM

Extracts of BM contain significant levels of saponic bacosides A, B, and C; and bacosapoinins D, E, and F, in addition to other chemical constituents including alkaloids, flavonoids, and phytosterols (Pengelly, 1997; Heinrich et al., 2004). The main

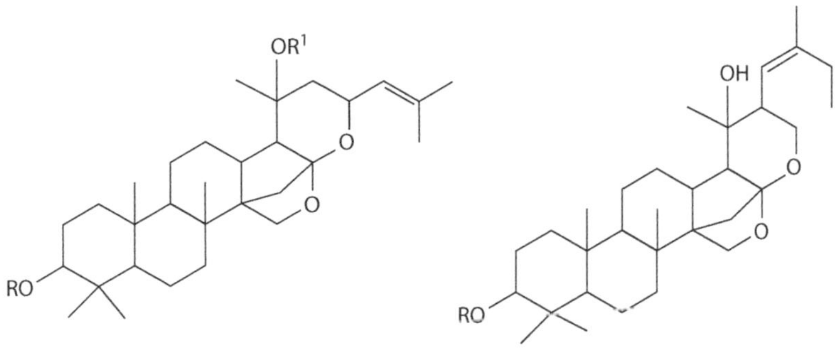

FIGURE 12.1 Chemical compound schematic of bacoside A and bacoside B.

chemical constituent of BM is bacoside A (see Figure 12.1), which has been postulated to be responsible for the memory facilitating action of the plant (Russo and Borrelli, 2005). Bacoside A usually cooccurs with bacoside B that differs only in terms of its optical rotation. BM has been available in a standardized form since 1996 for clinical research (Singh and Dhawan, 1997).

Importantly, for both research and the community, the bacoside content does vary between manufacturers, as does the quality of the extract. Thus, clinical evidence from standardized high quality extracts cannot be extrapolated to other extracts. Higher level clinical studies typically use BM with bacoside content standardized to 50%–55%. Currently, most clinical evidence for a cognitive related effect from BM stems from one to two extracts including the extract CDRI08, which has been developed and studied extensively preclinically and in animals by the Indian Government (particularly the CDRI). Progressively, human trials on both CDRI08 and BM are now appearing with a particular emphasis on improving cognitive performance including memory.

HUMAN CLINICAL TRIALS

A systematic review of the literature (Pase et al., in press) found eight human randomized controlled trials that met entry requirements (i.e., double-blinded, high quality studies). Of these studies, seven used chronic administration of BM while one was an acute study using a single 300 mg dose. No acute studies have to date shown a positive cognitive enhancement although there are current trials underway, and new data may shed some light on this possibility. However, based on the animal and in vitro studies on BM, it seems more likely that the mechanisms to improve cognition will exert influence chronically rather than acutely. Usually, acute cognitive enhancers or nootropic substances exert cognitive benefits via blood flow or direct neurotransmitter release. BM extracts are more likely to exert cognitive enhancement through inflammatory, antioxidant, or even removal of beta amyloid. The exact mechanism(s) are yet to be confirmed although large-scale mechanistic studies are now underway such as the Australian Research Council Longevity Intervention (Stough et al., 2012).

Chronic studies typically utilized a daily dose of 300 mg (standardized for bacosides) for the duration of the study—nearly all studies use 3 months administration or similar (Stough et al., 2001; Roodenrys et al., 2002; Calabrese et al., 2008; Stough et al., 2008; Morgan and Stevens, 2010). In Roodenrys' (2002) study, an increased dose of 400 mg/day was given to participants weighing over 90 kg. However, lower (250 mg/day) (Raghav et al., 2006) and higher doses of 450 mg/day (Barbhaiya et al., 2008) have been used chronically.

The human studies reviewed by Pase et al. (2012) provide evidence of highly promising results in areas of cognition, memory, and speed of processing tasks. Some studies used a healthy young adult population (Stough et al., 2001; Nathan et al., 2004; Stough et al., 2008), and others used "middle aged" (Roodenrys et al., 2002) or "elderly" populations (Calabrese et al., 2008). Progressively, aging popula- tions will be targeted for BM supplementation, given current evidence for mode of action on the brain.

Regardless of the age of the population, BM consistently improved selected cog- nitive functions. For example, BM was shown to improve working memory (Stough et al., 2008), learning rate and memory consolidation, and other components of the Rey Auditory Verbal Learning Test (Stough et al., 2001; Calabrese et al., 2008; Morgan and Stevens, 2010) as well as improvements in memory measured by the Wechsler Memory Scale found in the Raghav et al. (2006) study. Furthermore, exec- utive functioning tasks, such as the stroop (Calabrese et al., 2008) and inspection time (Stough et al., 2001), have also shown to be improved by intervention with BM. The effect sizes on many domains were moderate to strong. BM on various out- come measures consistently improved cognition, memory, and speed of processing in older adults with an overall moderate effect on all measure of memory (mean $d = 0.58$). BM was also found to improve learning and memory (mean $d = 0.61$), working memory and executive function (mean $d = 0.54$), and visual processing and attention (mean $d = 0.28$).

BM has also been studied in the context of age-associated memory impairment (AAMI). Raghav et al. (2006) tested a sample of middle aged to elderly participants with AAMI. Participants receiving a standardized BM extract (250 mg/day) showed statistically significant improvement across subtests of the Wechsler Memory Scale from week 4 onward. Tasks of mental control, logical memory, and paired associated learning showed the greatest improvement. Clearly, these data are preliminary and need to be replicated. However, they support and reinforce the memory and cognitive enhancing effects of BM as well as an appropriate age-related target for intervention.

To date, only one study has been reported to assess the efficacy of BM in a sample of participants with dementia. A small 6 month pilot study has been carried out by Morgan and colleagues (see Morgan and Stevens, 2010) on participants diagnosed with mild–moderate dementia using BM (300 mg daily) as an adjunct intervention to their standard treatment. Participants ($n = 5$) were tested using the mini mental state examination (MMSE) and Alzheimer's disease assessment scale (ADAS-cog; cogni- tive subscale) at baseline and 6 months. Both scales are valid measures of AD decline (Folstein et al., 1975; Kolibas et al., 2000). The BM intervention improved scores on both scales with 4/5 patients improving on the MMSE and 3/5 improving on the ADAS-cog (although there was a dissociation between the patient who improved on the two scales). These results are an indication that BM may have potential as an adjunct treatment for AD.

CONCLUSIONS

High quality extracts of bacopa such as CDRI08 appear to be highly promising for the chronic treatment of age-associated cognitive decline. Like many herbal extracts, BM may have multiple mechanisms to act on the central nervous system including anti-inflammation, removal of beta amyloid, and increased antioxidant properties. More recent animal and in vitro research has suggested other potential mechanisms (e.g., improvement in cardiovascular functioning and increased dendritic connectivity). The interesting animal and in vitro research and the largely positive results from several cognitive RCTs in human suggest the utility of BM extracts and the treatment of cognitive decline. Future research should plan longer interventions using BM extracts like CDRI08 that assess the role of different biological mechanisms on cognitive change in the elderly.

REFERENCES

Agrawal, A. (1993). A comparative study of psychotropic drugs and bio-feedback therapy in the prevention and management of psychosomatic disorder. Thesis. Banaras Hindu University, Varanasi, India.

Artur, Y. et al. (1992). Age-related variations of enzymatic defenses against free radicals and peroxides. *Experientia Supplementa*, **62**, 359–367.

Barbhaiya, H.C., Desai, R.P., Saxena, V.S., Pravina, K., Wasim, P., Geetharani, P., Allan, J.J., Venkateshwarlu, K., and Amit, A. (2008). Efficacy and tolerability of BacoMind® on memory improvement in elderly participants—A double blind placebo controlled study. *Journal of Pharmacology and Toxicology*, **3(6)**, 425–434.

Bhattacharya, S.K., Bhattacharya, A., Kumar, A., and Ghosal, S. (2000). Antioxidant activity of *Bacopa monniera* in rat frontal cortex, striatum and hippocampus. *Phytotherapy Research*, **14(3)**, 174–179.

Bhattacharya, S. and Ghosal, S. (1998). Anxiolytic activity of a standardized extract of *Bacopa monniera*: An experimental study. *Phytomedicine*, **5(2)**, 77–82.

Bhattacharya, S.K., Kumar, A., and Ghosal, S. (1999). Effect of *Bacopa monniera* on animal models of Alzheimer's disease and perturbed central cholinergic markers of cognition in rats. *Research Communications in Pharmacology and Toxicology*, **4(3–4)**, 1–12.

Braak, H. and Braak, E. (1997). Frequency of stages of Alzheimer-related lesions in different age categories. *Neurobiology of Aging*, **18(4)**, 351–357.

Bryant, J. and Sonerson, A. (2006). Gauging the Cost of Aging, *Finance and Development*, International Monetary Fund, **43(3)**.

Calabrese, C., Gregory, W.L., Leo, M., Kraemer, D., Bone, K., and Oken, B. (2008). Effects of a standardized *Bacopa monnieri* extract on cognitive performance, anxiety, and depression in the elderly: A randomized, double-blind, placebo-controlled trial. *Journal of Alternative and Complementary Medicine*, **14(6)**, 707–713.

Chandra, R.K. (2001). Effect of vitamin and trace-element supplementation on cognitive function in elderly subjects. *Nutrition*, **17**, 709–712.

Christensen, H. and Kumar, R. (2003). Cognitive changes and the ageing brain. In: *The Ageing Brain*. Ed. Sachdev, P.S. Lisse, the Netherlands: Swets, pp. 75–96.

Coyle, J.T. and Putfarcken, P. (1993). Oxidative stress, glutamate, and neurodegenerative disorders. *Science*, **262(29)**, 689–695.

Dar, A. and Channa, S. (1999). Calcium antagonistic activity of *Bacopa monniera* on vascular and intestinal smooth muscles of rabbit and guinea-pig. *Journal of Ethnopharmacology*, **66(2)**, 167–174.

Folstein, M.F. et al. (1975). 'Mini mental state'. A practical method for grading the cognitive state of patients for the clinician. *Journal of Psychiatric Research*, **12**(3), 189–198.

Forster, M.J. et al. (1996). Age-related losses of cognitive function and motor skills in mice are associated with oxidative protein damage in the brain. *Neurobiology*, **93**, 4765–4769.

Goodwin, J.S. et al. (1983). Association between nutritional status and cognitive functioning in a healthy elderly population. *Journal of the American Medical Association*, **249**, 2917–2921.

Heinrich, M. et al., Eds. (2004). *Fundamentals of Pharmacognosy and Phytotherapy*. London, U.K.: Churchill and Livingstone.

Holcomb, L.A., Dhanasekaran, M., Hitt, A.R., Young, K.A., Riggs, M., and Manyam, B.V. (2006). *Bacopa monniera* extract reduces amyloid levels in PSAPP mice. *Journal of Alzheimer's Disease*, **9**(3), 243–251.

Jain, S.K. (1994). *Ciba Found Symposium*, **185**, 153–164; Discussion 64–68.

Kolibas, E. et al. (2000). ADAS-cog (Alzheimer's Disease Assessment Scale-cognitive subscale)—Validation of the Slovak version. *Bratislavske Lekarske Listy*, **101**(11), 598–602.

La Rue, A. et al. (1997). Nutritional status and cognitive functioning in a normally aging sample: a 6-y reassessment. *American Journal of Clinical Nutrition*, **65**, 20–29.

Looi, J. and Sachdev, P.S. (2003). Structural neuroimaging of the ageing brain. In: *The Ageing Brain*. Ed. Sachdev, P.S. Lisse, the Netherlands: Swets, pp. 49–62.

MacLullich, A.M.J., Ferguson, K.J., Deary, I.J., Seckl, J.R., Starr, J.M., and Wardlaw, J.M. (2002). Intracranial capacity and brain volumes are associated with cognition in healthy elderly men. *Neurology*, **59**(2), 169–174.

Masaki, K.H., Losonczy, K.G., Izmirlian, G., Foley, D.J., Ross, G.W., Petrovitch, H., Havlik, R., and White, L.R. (2000). Association of vitamin E and C supplement use with cognitive function and dementia in elderly men. *Neurology*, **54**(6), 1265–1272.

Morgan, A. and Stevens, J. (2010). Does *Bacopa monnieri* improve memory performance in older persons? Results of a randomized, placebo-controlled, double-blind trial. *Journal of Alternative and Complementary Medicine*, **16**(7), 753–759.

Nathan, P.J., Tanner, S., Lloyd, J., Harrison, B., Curran, L., Oliver, C., and Stough, C. (2004). Effects of a combined extract of *Ginkgo biloba* and *Bacopa monniera* on cognitive function in healthy humans. *Human Psychopharmacology*, **19**(2), 91–96.

Packer, L. (1992). Free radical scavengers and antioxidants in prophylaxy and treatment of brain diseases. In: *Free Radicals in the Brain*. Eds. Packer, L., Prilipko, L., and Christen, Y. New York. Springer-Verlag.

Papa, S. and Skulachev, V.P. (1997). Reactive oxygen species, mitochondria, apoptosis and aging. *Molecular and Cellular Biochemistry*, **174**(1–2), 305–319.

Pase, M.P., Kean, J., Sarris, J., Neale, C., Scholey, A.B., and Stough, C. (2012). The cognitive enhancing effects of Bacopa monnieri: A systematic review of randomized, controlled human clinical trials. *Journal of Alternative and Complementary Medicine*, **18**(7), 647–652.

Pengelly, A. (1997). *The Constituents of Medicinal Plants*. Glebe, New South Wales: Fast Books.

Productivity Commission. (2005). Economic implications of an ageing Australia. Productivity Commission Research Report. Canberra, Australian Capital Territory, Australia.

Raghav, S., Singh, H., Dalal, P.K., Srivastava, J.S., and Asthana, O.P. (2006). Randomized controlled trial of standardized *Bacopa monniera* extract in age-associated memory impairment. *Indian Journal of Psychiatry*, **48**(4), 238–242.

Rai, D., Bhatia, G., Palit, G., Pal, R., Singh, S., and Singh, H.K. (2003). Adaptogenic effect of *Bacopa monniera* (Brahmi). *Pharmacology Biochemistry and Behavior*, **75**(4), 823–830.

Roodenrys, S., Booth, D., Bulzomi, S., Phipps, A., Micallef, C., and Smoker, J. (2002). Chronic effects of Brahmi (*Bacopa monnieri*) on human memory. *Neuropsychopharmacology*, **27**(2), 279–281.

Russo, A. and Borrelli, F. (2005). *Bacopa monniera*, a reputed nootropic plant: An overview. *Phytomedicine*, **12(4)**, 305–317.

Russo, A. et al. (2003). *Life Science*, **73(12)**, 1517–1526.

Sano, M. et al. (1997). A controlled trial of selegiline, alpha-tocopherol, or both as treatment for Alzheimer's disease. *New England Journal of Medicine*, **336(17)**, 1216–1222.

Singh, H.K. and Dharwan, B.N. (1997). Neuropsychopharmacological effects of the Ayurvedic nootropic *Bacopa monniera* linn (Brahmi). *Indian Journal of Pharmacology*, **29**, S359–S365.

Smith, C.D. et al. (1991). *Proceedings of the National Academy of Sciences*, **88**, 10540–10543.

Smith, M.A. et al. (1996). Oxidative damage in Alzheimer's. *Nature*, **382**, 120–121.

Socci, D.J., Crandall, B.M., and Arendash, G.W. (1995). Chronic antioxidant treatment improves the cognitive performance of aged rats. *Brain Research*, **693(1–2)**, 88–94.

Sram, R.J. et al. (1993). Effect of antioxidant supplementation in an elderly population. Basic life sciences 1993, 61: 459–477. *Basic Life Sciences*, **61**, 459–477.

Stough, C., Downey, L.A., Lloyd, J., Silber, B., Redman, S., Hutchison, C., Wesnes, K., and Nathan, P.J. (2008). Examining the nootropic effects of a special extract of *Bacopa monniera* on human cognitive functioning: 90 Day double-blind placebo-controlled randomized trial. *Phytotherapy Research*, **22(12)**, 1629–1634.

Stough, C., Lloyd, J., Clarke, J., Downey, L.A., Hutchison, C.W., Rodgers, T., and Nathan, P.J. (2001). The chronic effects of an extract of *Bacopa monniera* (Brahmi) on cognitive function in healthy human subjects. *Psychopharmacology*, **156(4)**, 481–484.

Stough, C. et al. (2012). A randomized controlled trial investigating the effect of Pycnogenol and Bacopa CDRI08 herbal medicines on cognitive, cardiovascular, and biochemical functioning in cognitively healthy elderly people: The Australian Research Council Longevity Intervention (ARCLI) study protocol (ANZCTR12611000487910). *Nutrition Journal*, **11**, 1–9.

Tripathi, Y.B. et al. (1996). *Indian Journal of Experimental Biology*, **34(6)**, 523–526.

13 Memory-Enhancing and Associated Effects of a Bacosides-Enriched Standardized Extract of *Bacopa monniera* (BESEB—CDRI-08)

Hemant K. Singh

CONTENTS

Learning, the formation of memory, and the consequent storage and retrieval of individual experiences thus acquired have been the most important aspects of evolution. Learning and memory are considered to be the distinguishing factors that determine the phylogenetic development of the adaptation and have thus continuously contributed to the process of the survival of species.

At the individual level, learning and memory are fundamental prerequisites for cognitive functions, emotional reactions, consciousness, speaking, thinking, and the ability to construct internal models of the surrounding milieu within the framework of time and space and thus able to take crucial decisions at critical moments. This helps the individuals to shape, broaden, and govern their living space and also to predict the future from the past (Matthies, 1989). Memory is the most important aspect of maintaining the identity and individuality of a personality. Thus, every great civilization had recognized the close correlation between memory and individual personality. In the great book of wisdom *Srimad Bhāgwat-Gitā*, Lord Krishna has said *smriti bhranshādbudhināso budhināsatprsyati*, meaning memory impairment leads to loss of identification and destruction of personality (Goenka, 1967). In Greek mythology, the mother of Muses, the nine Goddesses presiding over arts and science, was Mnemosyne, which literally means remembrance and memory.

Hence, the capacity of the individual organism to learn can be considered as the most important mechanism of adaptation. However, the various theoretical concepts of learning and memory have not been clear on the question of the nature of the performance of memory. This state of affairs seems to stem mainly from the fact that the various definitions of learning have large variation. Let us take the concept of "a relatively permanent change," which is included in various definitions of learning. Should this requirement imply that all memories are permanent or that some are transient? Various theories of learning are usually silent on this vital question. There does not seem to be any logic to assume that all consequences of experience are permanent. In fact, there are several overriding reasons to suggest that this cannot be the case. Many of the stimuli surrounding an individual and evoking a specific functional reaction have little or no adaptive significance. It makes very little sense in terms of economy or adaptation that all stimuli should produce enduring consequences in the brain. On the other hand, it is of adaptative value to be able to store long-term representation of experiences that are either particularly meaningful (in terms of consequences) or are frequently repeated.

This plausible theory, supported in the daily experience of life, has been succinctly explained in *A Study in Scarlet* by Sir Arthur Conan Doyle. Dr. James Watson was astonished that Sherlock Holmes was supremely unaware of the Copernican theory and the composition of the solar system.

"You appear to be astonished" he (Sherlock Holmes) said, smiling at my (Watson's) expression of surprise. "Now that I do know it I shall do my best to forget it."

"To forget it!" (Watson)

"You see" he (Holmes) explained, "I consider that a man's brain originally is like a little empty attic, and you have to stock it with such furniture as you choose. A fool takes in all the lumber of every sort that he comes across, so that the knowledge which might be useful to him gets crowded out, or at best is jumbled up with a lot of other things, so that he has a difficulty in laying his hands upon it. Now the skilful workman is very careful indeed as to what he takes into his brain-attic. *He will have nothing but the tools which may help him in doing his work, but of these he has a large assortment, and all in the most perfect order.* It is a mistake to think that little room has elastic walls and can distend to any extent. Depend upon it there comes a time when for every addition of knowledge you forget something that you knew before. It is of the highest importance, therefore, not to have useless facts elbowing out the useful ones."

"But the Solar System!" I (Watson) protested.

"What the deuce is to me?" he (Holmes) interrupted impatiently: "you say that we go round the sun. *If we went round the moon it would not make a pennyworth of difference to me or to my work.*" (emphasis added) (Doyle, 1984)

This prescient description by Doyle bears testimony when we consider the currently accepted concept of executive function. In understanding the executive function, the import of the "working memory" (the ability to hold a goal or relevant information, not allowing any distraction and interference from other sources) has to be considered. The efficiency of executive function is closely related to the converse ability to keep interfering information out of working memory and also out of the set response options. Implicit in this postulation is the ability to withhold or suppress such responses that would become incompatible with the goal when the context changes dynamically. Such ability is also termed as "response suppression" and is emphasized, for instance, in various theories of attention deficit/hyperactivity disorder (Barkley, 1997; Schachar et al., 1993).

The individual memorizes only those information that are useful in life and that help him to survive. Thus, the ability to suppress unwanted memories and also to avoid distractive memory through executive control, as postulated by several researchers (Anderson and Grew, 2001; Anderson et al., 2004), assumes a very significant importance in efficient functioning of day-to-day life. This is one vital aspect that one has to methodologically evolve to study either the drugs influencing or the various mechanisms underlying the processes of learning and memory in experimental animals or human beings.

BACOPA: HISTORICAL BACKGROUND

Bacopa monniera (Linn) Wettst [Synonyms: *Herpestis monniera* (Linn) HB&K; *Grahola monniera* Linn; *Monniera cunerfolia* Michx (Vernacular: *Brahmi*)]; family: Scorphulariacae is a perennial herb found throughout India in wet and marshy regions. It is frequently mentioned in the religious, social, and medical treatises of India since the time of Vedic civilization. Its antiquity can be traced to the time of the *Athar-Veda* (science of well-being) written in 800 BCE. The various treatises of ancient and medieval India have laid emphasis on the ability of bacopa in curing various diseases, particularly its unique ability to improve memory, sharpen intellect, facilitate acquisition of newer information, and promote learning. Because of its unique properties many eminent Indologists (Roy, Chandra, personal communication) and modern researchers (Russo and Borrelli, 2005) believe that the word *brāhmi* originates from Lord Brahmā, the creator of the Universe and the originator of *Āyurveda* in Hindu mythology.

Bacopa is an extremely important plant of the Indian system of medicine, viz., *Āyurveda*. In the classical Āyurvedic text of Çaraka (Sharma, 2009), bacopa has been classified as *medhya-rasāyan* (*medhya*: memory enhancing and *rasāyan*: rejuvenating). Çaraka has described the most unique features of the efficacy of bacopa in alleviating old age and age-related diseases, promoting memory and intellect, enhancing life span, providing nourishment, excellence, clarity of voice, complexion and luster (Sharma, 2009). Çaraka has also described a disease that he has termed as *atathvabhēnēveshām* and has also diagnosed it to be the most dreadful mental illness. The symptoms of the disease resemble a combination of hallucination, schizophrenia, obsessive compulsive disorder, and severe psychosis (Sharma, 2009) and can be cured only by taking expressed juice of bacopa. The bacopa juice has also been prescribed as cure for epilepsy (Sharma, 2009). In another treatise, viz., *Suśruta-Samhita*, it has been mentioned that a 3 week course of bacopa juice will produce photographic memory and a person can retain hundred words uttered only twice daily (Singhal et al., 2009). Bacopa with clarified butter has been prescribed to be highly beneficial for leprosy, intermittent fever, epilepsy, and insanity (Singhal et al., 2009). A bacopa-based drink has been prescribed for uroliathiasis (Singhal et al., 2009) and a bacopa-based liquor to cure skin disorder (Singhal et al., 2009).

EARLIER INVESTIGATIONS

Because of the traditional importance and unique therapeutic claims of bacopa, it attracted the early attention of chemists in India. Some rudimentary investigations were done earlier by the chemists of Banaras Hindu University (BHU), Varanasi

(Bose and Bose, 1931; Sastri et al., 1959). The initial neuropharmacological investigations were also done in BHU. Malhotra and Das (1959) reported a sedative effect of glycosides named hersaponins by them. Aithal and Sirsi (1961) found that the alcoholic extract and to a lesser extent the aqueous extract of the whole plant and chlorpromazine improved the performance of rats in motor learning. Sinha (1971) had reported that a single dose of the glycoside hersaponin is better than pentobarbitone in facilitating acquisition and retention of brightness discrimination reactive.

It is difficult, however, to interpret these results in the context of the known traditional claims of improving learning and memory.

INVESTIGATIONS DONE AT CENTRAL DRUG RESEARCH INSTITUTE, LUCKNOW

CHEMICAL INVESTIGATIONS

Rastogi and Dhar (1960) at the Central Drug Research Institute (CDRI) were the first to undertake a systematic chemical examination of the plant. The following constituents were isolated (Chatterjee et al., 1963):

Name of the Constituent	m.p. (0°C)	Yield (On the Weight of Dry Plant)	%
Bacoside A	250–1	1.54	64.28
Bacoside B	203	0.65	27.11
Betulic acid	315	0.11	4.58
D-Mannitol	166	0.02	0.83
Stigmasterol	170	0.013	0.54
β-Sitosterol	137	0.014	0.58
Stigmastanol	141	0.05	2.08

The activity of the ethanolic extract was traced to a mixture of triterpenoid saponins, designated as bacosides A and B. Both the bacosides showed single spots on TLC on silica gel, while bacoside A was obtained as colorless needles (Chatterjee et al., 1965) and bacoside B was colorless powder (Basu et al., 1967). Bacoside A is levo-rotatory and bacoside B is dextro-rotatory.

Bacoside A was the major component of the plant and comprised two sets of saponins. One set was derived from pseudojujubogenin as the genuine aglycone. The pseudojujubogenin on acid hydrolysis furnished four triterpenoid transformation products, viz., bacogenins A_1, A_2, A_3, and A_5 (Kulshreshtha and Rastogi, 1973). The jujubogenin on acid hydrolysis yielded two triterpenoids with triene side chains as transformation products. These triterpenoids were designated as bacogenins A_4 (trans) and (cis).

Seasonal Variations

To monitor the seasonal variations of bacosides, fresh plant materials were collected every month, extracted with ethanol and fractionated. It was carried over a period of

14 months commencing from March. Thus, it covered all the five seasons of India, viz., spring, summer, monsoon, autumn, and winter, taking care of a wide range of temperature and humidity. From this study it was concluded that bacosides A and B are available in May. In the rest of the months, other compounds start appearing and disappearing (Rastogi et al., 1996).

Bacosides-Enriched Standardized Extract of Bacopa

It is essential for plant extracts in order to be therapeutically effective to leave all the constituents intact because the therapeutic effect is generally the result of concerted activity of several active constituents as well as the most of the accompanying substances. Although these inert accompanying substances do not directly affect the therapeutic mechanism, it is reasonable to use the complex mixtures of components provided by a medicinal plant because these inert components might influence bioavailability and have an optimum effect on the pharmacodynamics and pharmacokinetics of the active components. Further, inert plant components would augment the stability as well as minimize the possible side effects. If there are different active compounds present in a therapeutically active plant product, they might have additive and potentiating effect (Williamson, 2001). The memory-enhancing effect of bacopa is much more than can be explained by its bacosides contents. Similar observations have been made with several other plants as well (Dhanukar, 1998; Dhawan 1997; Singh et al., 1996). This approach of keeping all the chemical constituents intact has certain inherent advantages, and now, in traditional herbal medicines, such multiple components are purposely combined to provide physiological balance and to harmonize drug effect (Handa, 1998).

Therefore, a standardized extract of bacopa containing a minimum of 55% ± 5% of bacosides with an optimum concentration of bacogenins (especially bacogenin A_3) vis-a-vis memory-enhancing effect was developed. This was termed as BESEB CDRI-08.

Further, it becomes imperative to mention the fact that damarene-type-triterpenoid saponins are major constituents of several reputed herbal drugs including ginseng, used for centuries. Jujubogenin glycosides have also been isolated from a number of reputed medicinal plants of rhamnaceae and scrophulariaceae families. However, what is unique about bacopa is that in addition to jujubogenin glycosides, pseudojujubogenin glycosides are also present and these have been reported together so far only from this medicinal plant (Garai et al., 1996; Russo and Borrelli, 2005).

MEMORY-ENHANCING EFFECT

If Rastogi and coworkers at CDRI were the first to undertake a systematic chemical examination of bacopa, Singh and Shankar (1996) also of the CDRI, were the first to elucidate a targeted memory-enhancing effect of the bacosides-enriched standardized extract of *B. monniera* (BESEB CDRI-08).

Certain Methodological Considerations

The preparation, discovery, research, and development of drugs with psychological effects, in general, and memory effects, in particular, have occupied the interest and energy of humans since the inception of civilization. The experiments of Lashley (1917) demonstrating the facilitation of maze learning by low dose of strychnine sulfate is perhaps the first recorded scientific investigation on a drug affecting learning and memory. Ever since then, there has been a tremendous upsurge in the study of drugs influencing learning and memory. With a greater widening of horizon, technological advancement of laboratory research equipments, easy availability of target pharmacological tools, the refinement of conceptual framework has also significantly improved. All these have resulted in the study of drugs facilitating, impairing, and neuromodulating learning and memory in greater depth and detail.

However, it will not be wise to overlook the fact that the evaluation of a new drug affecting learning and memory is still beset with several methodological problems. Even though these have been dealt in detail elsewhere (Singh and Dhawan, 1992), it would still be pertinent to summarize some of the overriding issues.

At the outset, one important issue is to distinguish the drug effect on learning vis-a-vis its effect on performance. A performance in the absence of a reward or motivation would only reflect latent learning. A true measure of the actual learning can only be witnessed when the experimental animals are motivated or when the reward is introduced (Blodget, 1929).

It is also important to dissociate the effect of a drug under the conditions when the learning by the experimental animals takes place in the drugged state. Such learning performance is not reflected during the normal state when the influence of the drug wears off. Generally, any drug having a depressant effect, for example, chlordiazepoxide, chlorpromazine, and pentobarbital, will produce such dissociation effects.

Similarly, it is of great theoretical significance to consider that the effect of a given drug on learning and memory will also depend upon the fact whether the primary action of the drugs is on the receptor or effector systems. Any drug having only peripheral effects will still indirectly alter the CNS function by modifying the inputs to the CNS. Conversely, any centrally acting drug will also alter the functions of peripheral nervous system.

The type of apparatus chosen to study the effect of a drug on learning and memory is also a serious consideration. Small differences in the design of a maze can produce discrepant results in the studies of the effects of drugs on learning. For example, the addition of retracing doors has been shown to increase the reliability of mazes (Silverman, 1978).

Certain Conceptual Considerations

Apart from these methodological considerations, our endeavor to formulate a strategy to elucidate the memory-enhancing effect of BESEB was also substantially influenced by different conceptual and experimental postulates available in the 1970s. If the classical idea of Ivan Petrovich Pavlov about the formation of temporary connection provided an excellent starting point, the hypothesis of Hebb (1949) of convergence of pathways and the coactivity of neurons resulting in modifiable synapses provided the additional fillip. An extremely important consideration was the result obtained by McGaugh (1966), who taking a cue from the consolidation hypothesis evolved from the studies of electroconvulsive shock (ECS) induced amnesia (Dunccan, 1949; Glickman, 1961), suggested that the memory formation is a time-dependent process.

The experimental demonstration that ECS intervention was effective only when applied immediately after the completion of a learned task and was found to be ineffective if the time lapse between completion of the learned task and application of ECS exceeded even 15 min was considered to be of great significance in determining the time of drug application to influence consolidation. Generally, a drug should be applied immediately after the training, but the fact that pretraining applications are also known to selectively influence consolidation cannot be overlooked. Pretraining application will thus have a biphasic effect both on the acquisition and the consolidation processes. However, there have been certain notable exceptions. For instance, it has been reported that naloxone enhanced the consolidation of Morris Water Maze learning with pretraining but not with post-training application (Decker and McGaugh, 1989). The nootropic CGS5649B when administered immediately after training failed to improve the task performance in maze learning, but it significantly enhanced task performance when administered 8 or 24 h after training (Mondadori, 1990).

Another important consideration is that short term memory (STM) does not consolidate into long term memory (LTM) in a linear curve manner. This has been suggested by the classical studies of the German Scientist Herman Ebbinghaus at the end of the eighteenth century and has been supported by the investigations of Müller and Pilzecker (1900).

These researches have demonstrated that during the process of consolidation when information was stored in STM, the consolidation of long-term trace entailed certain time-dependent processes. As stated earlier, they also found that during STM and consolidation of LTM, the memory trace was very sensitive to interventions, but when the memory was consolidated, it became stable and hence could not be influenced by any intervention. Later, the time course of both STM and the consolidation of LTM were more precisely elucidated by several investigations (Agranoff et al., 1967; Baronodes and Cohen, 1967; Bennett et al., 1977; Duncan, 1949; Flexner et al., 1963; Flood et al., 1986; McGaugh, 1966). However, the investigations of Kamin (1957, 1963) have indicated that there are peak and trough in the retention curve during the course of LTM. These observations have led to the hypothesis that memory formation takes place in three or four stages, which has been substantiated by corresponding experimental data (Gibbs and Ng, 1976; Frieder and Allweis, 1982; Matthies, 1974; Ott and Matthies, 1978).

Learning Models

However, the differences in the results obtained in the time course of consolidation, on different memory stages as well as the duration of amnesic effects, can only be accounted for by considering the following two factors: first, the stimuli applied in different experimental tasks were of different intensities; second, the signals thus converge at different levels of functional hierarchy (Figure 13.1).

This is of great methodological significance and has influenced our choice of a suitable learning model to study the effect of BESEB or learning and memory. We also, in the first instance, rejected learning at a relatively low level of complexity and neuronal organization, for example, an one trial passive avoidance task, even though this offers the advantage of rapid acquisition. In order to obtain a confirmatory memory-enhancing effect, our choice was a model that should induce well-distributed and

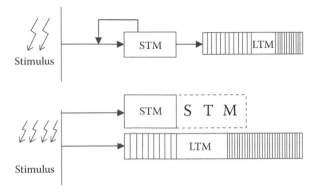

FIGURE 13.1   Signals coverage at different levels of functioning learning hierarchy depending upon the intensity of the stimulus. STM can precede LTM if the stimulus is weak and STM and LTM can be parallel when the stimulus is strong. A very strong stimulus can lead to disruption of learning; hence, it is desirable that the intensity of stimulus is optimal.

extensive cellular changes during the learning discrimination task and that would also involve a reversal of innate behavior. Our first choice was a foot-shock-motivated brightness discrimination task on rats in a semiautomatic Y-maze (Figure 13.2) as the standard procedure in most of the studies done on the elucidation of the memory-enhancing effect of BESEB CDRI-08.

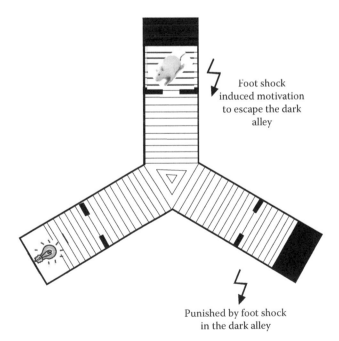

FIGURE 13.2   Foot-shock motivated brightness discrimination task on rats in a semi-automatic Y-maze.

This method has been described in detail by Ott et al. (1972), Singh et al. (1975), and Ott (1977). Briefly, rats were trained to escape a foot shock by running into one of the other two arms. The entry into the dark alley was punished by a foot shock (1 mA). For the next trial, the former goal box was used as a start box, which also helped in avoiding handling during the training session. The intertrial interval was approximately 60 s. The direction of alley illumination was changed after every three trials to avoid position discriminative. The training consisted of 40 trials (45 min). The retention was tested 24 h after training using the same paradigm as in the learning session. The number of negative trials, i.e., the number of foot shocks received during training (T_s) and relearning (R_s), were used to calculate percent saving, which was the difference between the two, expressed as percent of the former:

$$\% \text{ savings}: T_s - R_s \times \frac{100}{TS}$$

The results obtained from the relearning test was also analyzed on the basis of increase in positive response during relearning (δR), i.e., the difference between R_R (positive response during relearning) and T_R (positive response during training). A response was considered to be positive when the rat ran immediately into the lighted alley in the last run prior to and the first run after the change in the direction of alley illumination.

In this training procedure, both the percent saving and δR appear to be less dependent on influences of performance than is true of latency measures. However, it is also important to use a battery of diversified tests whenever attempting to measure the effect of any drug on level of motivation or emotionality (Miller and Teuber, 1968). Therefore, we decided to use different test models for the evaluation of the memory-enhancing effects of BESEB CDRI-08. The battery of tests used consisted of positive (reward) as well negative reinforcement (punishment), labile phase of memory (when the memory is in formative stage), and stable phase of memory (when the memory formation has taken place). Moreover, the training should be such that it allows a demarcation in three phases of memory, i.e., acquisition, consolidation, and retention. The model should also clearly distinguish between the successive stages of short-term, intermediate, and long-term memories (T. Ott and H. Matthies, personal discussion, 1975). The foot-shock-motivated brightness discrimination reaction fulfilled most of these criteria. However, it was a training with negative reinforcement (foot shock), the training was completed in one session, and the memory formation was labile. The active avoidance, Sidman's conditioned avoidance, and conditioned taste aversion responses were initially used to confirm the memory enhancing effect of CDRI-08. Later, one-trial passive avoidance response in few experiments were used.

Other Learning Models

In the active conditioned avoidance response, which was a modified method of Cook and Weidley (1957), the rats had to jump on a wooden pole to avoid a foot shock. The method was based upon the hypothesis expounded by many neurobehavioral scientists that learning is essentially of two types, viz., procedural learning and declarative learning.

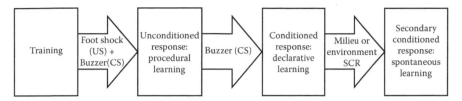

FIGURE 13.3   Succession of events leading to the formation of SCR. The series of actions ultimately compute into a subroutine.

In the procedural learning, the animals have to learn that presentation of an unconditioned stimulus (US) is preceded by conditioned stimulus (CS). In other words, they learn that a particular response or an event (presentation of US in the form of a foot shock of 1 mA intensified for a duration of 15 s) is reinforced when a discriminative signal or CS (a buzzer presented for 15 s) is given. Thus, the active conditioned avoidance response essentially consists of a series of actions conducted in a certain order and manner that ultimately computes into a subroutine. The succession of events leading to these three types of learning is shown in Figure 13.3.

A well-trained animal not only shows procedural learning but also exhibits declarative learning by responding to CS. In few cases, animals attain a very high level of training by exhibiting spontaneous learning and respond to the milieu or environment (secondary conditioned stimulus [SCS]) without waiting for the CS or the US.

The third learning schedule was the continuous avoidance response as evolved by Sidman (1953, 1956). It is a relatively complicated task for rats, as no contingencies between avoidance behavior and exteroceptive stimulus (e.g., a tone as conditioned stimulus) are involved. The training comprised a daily session of 1 h with lever pressing selected as the avoidance response. The programming is done in such a manner that each lever press delayed the shock for 25 s. A minimum interval of 15 s was assured between two successive shocks, i.e., when no avoidance behavior occurred. However, when the avoidance behavior occurred, i.e., when the lever was pressed, the shock was further delayed by 10 s. In other words the next shock was delivered after 25 s. Thus, the shock–shock interval was 15 s and shock–response interval was 25 s (SS15RS25). A trained animal responded repeatedly by pressing the lever at a stationary rate to avoid shocks (Kuribara et al., 1975, 1976).

The three learning schedules employed in these experiments consisted of a negative reinforcement, i.e., a foot shock to motivate the animals. Hence, a fourth confirmatory test involving positive reinforcement, i.e., reward in the form of conditioned taste aversion (CTA) response was performed.

One of the most important tasks of neuroethology is to develop laboratory versions of ethological experiments that would make it possible to investigate the underlying brain mechanisms. Particularly important in this respect are the neural processes controlling food intake and licking or aversion to some items of food. In the CTA response, single male rats were kept in individual cages containing openings for two 30 mL Richter tubes with 0.5 mL gradations. The rats had access to standard food pellets but were deprived of water overnight. Each rat was allowed to drink tap water from the Richter tube for 1 h only every morning. After a few exposures, all rats started drinking after presentation.

These rats were then tested in gustatory discrimination apparatus by exposing them simultaneously to two drink spouts: one delivering sucrose, the other containing an aversive fluid Lithium Chloride (LiCl). The CTA test is based on the capability of rats to avoid a novel food or liquid, the ingestion of which leads to undesirable side effects (moderate gastrointestinal disorder) (Barker et al., 1977; Garcia et al., 1955; Milfram et al., 1977; Rozin and Kalat 1971). A trained rat rapidly finds the sucrose-containing spout and starts drinking there from. After the stabilization of licking sucrose is suddenly replaced by the aversive LiCl fluid. The position of the spouts is changed through motarized rotation by 180°. An important prerequisite of stationary retrieval condition is that prolonged testing does not cause CTA extinction. High motivation of the animals to avoid the aversive concentrations eliciting moderate gastrointestinal disorder after the ingestion of a few milliliters of isotonic solution of LiCl (0.15 M) can serve as a combined CS (salty taste) and US (symptom of gastrointestinal disorders).

Animals

The choice of experimental animals were either inbred albino Charles Foster or Sprague Dawley rats, which because of their similarity with human neuromorphological, neurophysiological, and neurochemical parameters are considered to be excellent infrahumans.

Dose Schedule

The dose schedules of BESEB CDRI-08 were chosen after careful considerations (Pandey, personal discussion, 1976). The best course to determine the dosage schedule was to perform a series of preliminary experiments with daily oral administration of different doses for varying number of days. It was established that the administration of 40 mg/kg p.o. for 3 days was optimal in producing a significant effect.

However, the doses, frequency of oral administration, and training–relearning intervals varied according to the experimental design, schedule, and level of motivation. These are schematically shown in Figure 13.4a through c.

The dose regimen for CTA is summarized in Table 13.1.

Summary of Results

The results of these investigations confirm that BESEB CDRI-08 has facilitatory effect on memory and learning in a wide variety of responses. The effect of BESEB CDRI-08 is manifest both in negative reinforcement (shock-motivated brightness discrimination reaction, conditioned and continuous avoidance responses) as well as positive reinforcement (conditioned taste aversion). This is significant as BESEB CDRI-08 facilitated the responses that are susceptible to the effects of punishment as well as reward. BESEB CDRI-08 is thus markedly effective in increasing the memory in a wide variety of responses. (Singh and Dhawan, 1978, 1982; 1974a,b; Singh et al., 1990a,b, 1996a,b, 1988).

The results are shown in Figures 13.5 through 13.8.

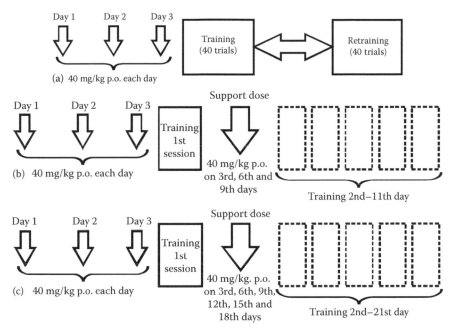

FIGURE 13.4 (a) Dose and training regimens for foot-shock-motivated brightness discrimination reaction. The training was completed in one session of approximately 1 h duration. (b) Dose and training regimens for active conditioned avoidance response. This was an interval training consisting of multiple sessions. Each session was of approximately 1 min duration and continued for 10 days. (c) Dose and training regimens for Sidman continuous avoidance response. This was also interval training and daily session of 1 h each continued for 18 days.

TABLE 13.1
Summary of Dosage Regimen for CTA

Group	Treatment	Dose (mg/kg po)
I	Saline	Saline 0.5 mL
II	BESEB-CDRI-08	2.5
III	BESEB-CDRI-08	5.0
IV	BESEB-CDRI-08	7.5
V	BESEB-CDRI-08	5.0[a]

[a] This group was not intubated with Li Cl on day 1.

EFFECT ON EARLY PHASES OF MEMORY CONSOLIDATION

A feature of memory formation across the animal kingdom is its progression from short-lived labile form to a long-lasting stable form. But as already described earlier, memory formation is not a direct flow of neuronal activity from short-term to long-term storage. Congruent lines of evidences have pointed instead to an intricate, multiphasic pathway of consolidation. Based on their experimental investigations, Ott and Matthies (1978)

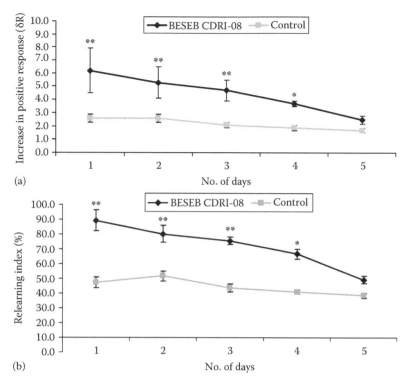

FIGURE 13.5 BESEB CDRI-08 improves the consolidation of memory significantly. *p < 0.05, **p < 0.01 as compared to controls as evidenced by increase in both positive responses (a) and relearning index (b).

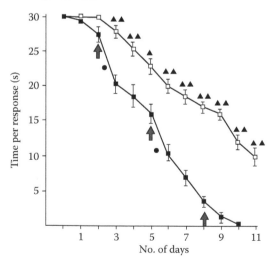

FIGURE 13.6 The effect of BESEB CDRI-08 on reaction time of rats during training for conditioned avoidance response. (■) denotes experimental group; (□) denotes control group., ▲▲p < 0.01, ▲p < 0.05, ●p < 0.05 between the day of administration of a support dose of BESEB CDRI-08 and the following day. The administration of support dose is indicated by arrows.

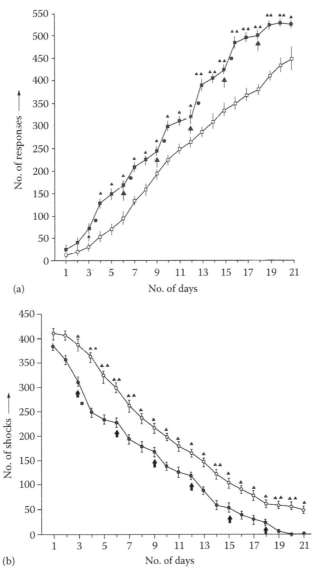

FIGURE 13.7 (a) BESEB CDRI-08 produces the significant increase in lever pressing response as compared to controls during Sidman continuous avoidance response. (b) BESEB CDRI-08 produces the significant reduction in the number of shocks received as compared to controls during Sidman continuous avoidance response. (■) denotes experimental group; (□) denotes control group., ▲▲p < 0.01, ▲p < 0.05, ●p < 0.05 between the day of administration of a support dose of BESEB CDRI-08 and the following day. The administration of support dose is indicated by arrows.

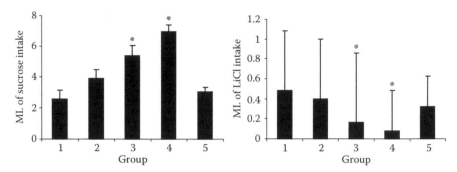

FIGURE 13.8   Dose-dependent facilitatory effect of BESEB CDRI-08 is found in the significantly improved intake of sucrose and significantly less intake of LiCl during a sucrose-versus-LiCl discrimination in a conditioned taste aversion response. The group treatment is explained in Table 13.1. *p < 0.05, **p < 0.01.

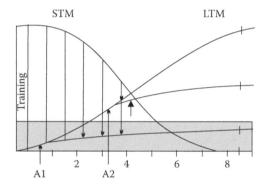

FIGURE 13.9   Deficit in learning curve (↑) occurs at a point when STM overlaps with LTM.

have postulated that memory essentially exists in two forms, viz., short term and long term. Both these forms start simultaneously and a deficit in retention curve is obtained at a point where these two forms overlap, i.e., it is V-shaped (Figure 13.9).

The hypothesis was tested in the foot-shock-motivated brightness discrimination reaction in the Y-maze and it was found that the retention curve instead of being V-shaped was W-shaped, i.e., there were deficits occurring at two points, viz., 1.5 h and 4.0 training–relearning intervals. Therefore, it was presumed that, at least in this experimental model, there are perhaps three forms of memory, i.e., short term (few seconds to minutes) and long term (few hours to days) and in between there exists an intermediate form of memory (few minutes to hours). The two deficits apparently occur at points where one memory overlaps with the other memory as shown in Figure 13.10.

The BESEB CDRI-08 in a dose of 20 mg/kg p.o. when given 30 min prior to training in a shock-motivated brightness discrimination reaction abolished these deficits when relearning was done at various time intervals after training. A support dose of 10 mg/kg p.o. given 30 min prior to the 24 h relearning test also produced a significant enhancement in the relearning index. These results suggest that the facilitatory effect of BESEB CDRI-08 is apparently due to their ability to consolidate the retention at the earliest form, i.e., short-term memory. The facilitatory effect of BESEB

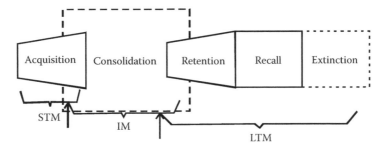

FIGURE 13.10 Functional relationship between acquisition, consolidation, and retention. Deficit in learning curve (↑) retention is postulated to occur where one form of memory overlaps with the other form.

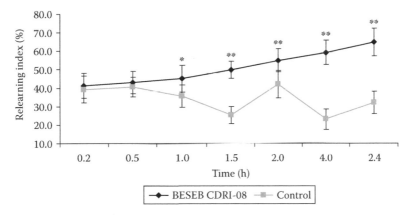

FIGURE 13.11 BESEB CDRI-08 abolishes the learning deficits by facilitating short-term, intermediate, and long-term memories. *p < 0.05, **p < 0.01 as compared to respective controls.

CDRI-08 persists even when the other two forms, i.e., intermediate and long-term memories occur (Singh and Dhawan, 1992, Singh et al., 1996a,b) (Figure 13.11).

ANTIAMNESIC EFFECT

During the period of consolidation, memory can be disrupted through administration of a wide variety of amnesic agents. ECS, hypothermia, and hypoxia are some of the noninvasive procedures that induce retrograde amnesia. While most of the amnesic agents used induce retrograde amnesia, there are others that cause temporary amnesia. For instance, diethylthiocarbamate, an inhibitor of synthesis of noradrenaline, disrupts intermediate memory when administered prior to training. Similarly, hypoxia disrupts intermediate memory specifically when administered immediately after training. Therefore, it appears that memory consolidation involves both serial and parallel processing of information.

In order to have comparable information, the experimental protocols were kept uniform, and retrograde amnesia was produced in rats by three interventions, viz., immobilization stress administered for 17 h, ECS (0.5 mA, 50 Hz, 0.5 s) administration and scopolamine. All these interventions were administered immediately after the completion of training. The training schedule used was a brightness discrimination reaction. BESEB CDRI-08 used was in a dose of 20 mg/kg p.o. × 3 day. The amnesic treatments led to a significant disruption of consolidation in control groups, as measured by percent savings in the relearning test performed 24 h after the completion of training. The BESEB CDRI-08 pretreatment significantly attenuated the amnesic effects in all these treatments (Singh and Dhawan 1997, 1998).

A qualitative similar effect was reported by Sethy et al. from Upjohn Company, Kalamazoo, MI. BESEB CDRI-08 attenuated the scopolamine-induced amnesia (Sethi et al., 1994) (Figure 13.12).

Another investigation was done by Manjrekar (1996) on the effect of CDRI-08 on scopolamine-induced amnesia in rats and mice employing elevated plus maze test, step-down passive avoidance test, and Cook and Weidley pole-jumping avoidance test. It was found that a single dose of 10 mg/kg BESEB CDRI-08 to be either equipotent to or more effective than similarly administered piracetam (150 mg/kg), cyclandelate (200 mg/kg) or *Gingkgo biloba* and *Withania somnifera* extracts (100 mg/kg p.o. for 4 days) in attenuating the amnesic effect of scopolamine. This finding was consistent for all the three test models used.

Das et al., (2002) have also found BESEB CDRI-08 to be equipotent to *Ginkgo biloba* extract in counteracting the scopolamine-induced transient amnesia on transfer latency time (Figure 13.13).

As can be evidenced, *Ginkgo biloba* and BESEB CDRI-08 have qualitatively comparative effect on memory. The insignificance difference can be attributed to suppression of fear induced by BESEB CDRI-08 because of its anxiolytic effect (Shanker and Singh 2000). The Ginkgo biloba extract has not been reported to possess any anxiolytic property.

Prabhakar et al. (2007) and Saraf (2009) also evaluated the effect of BESEB CDRI-08 on diazepam-induced amnesia in mice in Morns Water Maze. Benzodiazepines are known to produce amnesia through the involvement of GABA ergic system and hence the reversal of memory deficit induced by diazepam was investigaged.

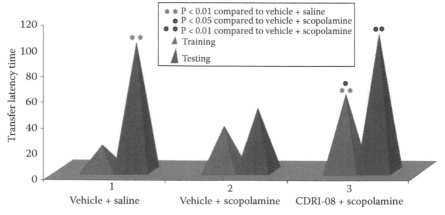

FIGURE 13.12 Antagonism of scopolamine-induced amnesia in mice by BESEB CDRI-08.

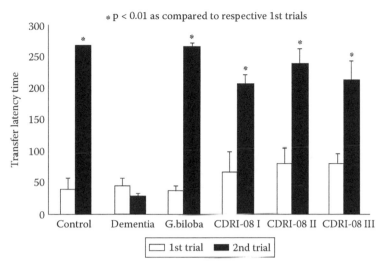

FIGURE 13.13 Equipotency of BESEB CDRI-08 and *Ginkgo biloba* in attenuating scopolamine-induced amnesia.

The results revealed antiamnesic effect of CDRI-08 (120 mg/kg p.o. administered 60 min prior to training) on diazepam (1.75 mg/kg p.o. administered 30 min prior to training)-induced amnesia. The degree of reversal by BESEB CDRI-08 was significant as it progressively reduced escape latency time when mice treated with diazepam were subjected to acquisition trials 7 days after training. The results suggest that the antiamnesic effects of BESEB CDRI-08 follow a gamma-aminobulyric acid—benzodiazepine pathway, possibly affecting long-term potentiation.

ANXIOLYTIC AND OTHER EFFECTS

Singh et al. (1979a,b) and Singh and Singh (1980) were the first to suggest an antianxiety effect of a bacopa extract based on its effect on gross behavior and prolongation of barbiturate hypnosis in rats. Shanker and Singh (2000) have also reported anxiolytic activity of BESEB CDRI-08 by observing the effect on prolongation of barbiturate-induced sleeping time, aggregation of amphetamine hyperactivity, inhibition of clonidine-induced fighting and biting behavior, and increased survival under hypoxia. The BESEB CDRI-08 also has a mild antidepressant activity as evidenced by the reversal of reserpine-induced syndromes and the enhancement of immobility time in swim despair test in mice (Singh et al., 1996b). It has also been found that BESEB CDRI-08 protects metrazol-induced convulsions both in multiple (ED_{50}: 216 mg/kg p.o. × 5 day) and single (ED_{50}: 647 mg/kg p.o.) doses in mice (Singh et al., 1996a,b). Although there is no clinical evidence available, on the strength of other supportive data it can be postulated that BESEB CDRI-08 may alleviate epilepsy by its ability to stabilize seizures (electrical activities) in the (cell membrane of) brain.

ADAPTOGENIC EFFECT

In recent times, adaptation has emerged as a unique concept as it helps living beings to combat day-to-day stress and is therefore considered to be essential for survival.

Experimental animals were exposed to two types of stress, i.e., acute and etronic. In acute stress, rats were confined for 150 min inside an acrylic hemicylindrical plastics tube for a day, whereas in chronic stress the confinement was for the same duration but for 7 consecutive days. The BESEB CDRI-08 (80 mg/kg p.o. × 3 day) exhibited pronounced adaptogenic effect by normalizing the changes induced by stress, e.g., prevented the formation of ulcer, reversed adrenal and thymus hypertrophy, spleen hypotrophy, normalized blood sugar, and aspartate aminotransferase (AST), alanine aminotransferase (ALT), and creatine kinase (CK) activities. A comparative study was made with *Panax quinquefolium* (PQ), one of the recognized varieties of ginseng (Rey, 2009). PQ produced similar effect except that unlike BESEB CDRI-08, PQ failed to reverse the formation of ulcer, adrenal hypertrophy, and hyperglycemia even at a higher dose of 100 mg/kg p.o. × 3 day (Rai et al., 2003).

These findings were further confirmed by Sheikh et al. (2007) by investigating the effect of CDRI-08 on stress-induced changes in brain corticosterone and brain monoamines in rats. The acute stress was the same as in the study of Rai et al. (2003), but instead of chronic stress, a chronic unpredictable stress (CUS) was administered. The CUS regimen involved exposing two different stressors of variable intensity every day in an unpredictable manner for seven consecutive days. These stresses include immobilization (150 min), forced swimming (20 min), overnight soiled cage bedding, foot shock (2 mA for 20 min), day–night reversal, and fasting (12 h). The salient findings of this study are that BESEB CDRI-08 normalized the stress-induced alterations in plasma corticosterones and levels of monamines noradrenaline, serotonin, and dopamine both in the cortex and hippocampus regions of the brain, both of which are more vulnerable to stressful conditions.

The importance of this finding lies in the fact that changes in monaminergic activity result in behavioral changes as well as a cascade of hormonal release from hypothalamus–pituitary–adrenal (HPA) axis. These abnormal alterations in monoamines under prolonged stress has been associated with a wide range of central and peripheral disorders like depression, anxiety, drug abuse, obsessive compulsive disorder, eating and sleeping disorders, hyperglycemia, and decreased immune response. The present-day lifestyle accompanied by the advent of various stress-related disorders entails unduly heavy physiological and psychological demands. It is widely felt by health planners that there is an urgent need to develop agents to overcome these abnormalities. The BESEB CDRI-08 has thus the required potential to fulfill this demand.

ANTISTRESS EFFECT

The antistress effect of CDRI-08 was studied at doses of 20 and 40 mg/kg p.o. administered for seven consecutive days in rats (Kar Chowdhuri et al., 2002). For administering the stress, the modified method of Ramchandran et al. (1990) was used. Briefly, rats were fasted overnight and subjected to a multiple stress of restraint, cold (5°C), and low-oxygen tension (428 mm Hg pressure, equivalent to an altitude of 15,000 ft) in a vertical decompression chamber. Under such stressful conditions, the core temperature of the experimental animals starts falling. The treatment is terminated when a core temperature of $23° ± 1°C$ is attained. The modulatory effects of BESEB CDRI-08 on stress-induced changes in expression of HSP70 and activities of superoxide dismutase (SOD) and cytochrome P450-dependent 7-pentoxyresorufin-D-alkylase (PROD) and 7-ethoxy-resorufin-O-deethylase (EROD) were estimated. The data clearly demonstrated the potential of BESEB CDRI-08 in reducing stress by modulating the expression of Hsp70 and the activities of P450s and SOD, the enzymes known to be involved in the production and scavenging of reactive oxygen species in different brain regions of the brain.

NEUROPROTECTIVE EFFECT

Amar Jyoti et al. (2006, 2007) investigated the neuroprotective role of the BESEB CDRI-08 against aluminium-induced oxidative stress in the hippocampus and cerebral cortex of rat brain and compared with L-deprenyl (seleginine). L-Deprenyl is an MAO-B inhibitor and neuroprotectant offering protection against the effects of neurotoxins and excitatory amino acids (Ebadi et al., 2002). It is also used as a therapeutic agent for the neurodegenerative disorder Parkinson's disease involving oxidative stress. L-Deprenyl prevents the apoptosis of dopaminergic neurons associated with Parkinson's disease by altering the expression of a number of genes such as SOD, Bcl-2, Bcl-XL, NOS, CJUN (Ebadi and Sharma, 2003). L-Deprenyl is also a very well-known antioxidant (Xu et al., 1993) and is postulated to have antiaging effects as it has shown to properly extend life expectancy (Bickford et al., 1997).

The data generated clearly indicated that coadministration of BESEB CDRI-08 together with aluminium treatment prevented the latter's oxidative stress effects. Aluminium reduced superoxide dismutase (SOD) activity significantly. CDRI-08 attenuated this reduction and restored the SOD activity to near normal. Similarly, CDRI-08 prevented the aluminium-induced increase in thio-barbituric acid-reactive substance (TBA-RS) and carbonyls. This protective role is further supported by the microscopic observations, which showed that BESEB CDRI-08 prevented the aluminium-induced lipofuscin accumulation and ultrastructural changes in the hippocampal CA1 field. The hippocampal CA1 field was selected for microscopic studies as its pyramidal neurons are potentially more vulnerable to aluminium-induced neurotoxicity (Sreekumaran et al., 2003) and hypoxia (Kawasaki et al., 1990) than CA2 and CA3 hippocampal fields.

The antioxidative effect of the BESEB CDRI-08 was found to be similar to L-deprenyl. This is an important finding as oxidative stress (lipid peroxidation, lipofuscin accumulation, etc.) is postulated to be an extremely important contributory component of the ageing process. Because of its antilipidperoxidative and antilipofuscinogenitric effects, the BESEB CDRI-08 can thus be considered as a potential antiaging substance. This postulation is further strengthened by the similarity between the effects of the BESEB CDRI-08 with those of L-deprenyl, which itself is considered to be a candidate antiaging drug. This is a further indication of the antiaging potential of the BESEB CDRI-08. The importance of this finding can be further gauged by the fact that neuroprotective effects are observed both in hippocampus and cortex. The severity of aluminium induced oxidative stress is more pronounced in the hippocampus and neocrotex regions than any other area of the central nervous system. Such oxidative damage during aging and oxidative stress is postulated to significantly contribute to the impairment of cognitive functions like learning and memory (Fukui et al., 2000). On the strength of these findings the authors have concluded that BESEB CDRI-08, because of its antilipidperoxidative and antilipofuscinogenistic effects, could be considered as a potential antiaging substance (Kaur et al., 2003).

MECHANISM OF ACTION

In order to study the mechanism of action, investigations were designed to (1) establish the biochemical correlates of the memory-enhancing effects of BESEB CDRI-08 and (2) to elucidate its cellular mechanism of action.

Biochemical investigations showed that BESEB CDRI-08 enhanced the protein kinase activity in hippocampus. It also induced an increase in protein and serotonin and lowered the epinephrine levels in hippocampus. Similar changes were also observed in hypothalamus and cerebral cortex (Singh and Dhawan, 1997). The enhancements of protein and serotonin, and the depletion of norepinephrine levels are indicative of the facilitatory effect of BESEB CDRI-08 on long-term and intermediate forms of memory (Angers et al., 1998; Byrne and Kandel, 1996; Cohen et al., 2003; Crow et al., 2001; Kandel, 1987; Kandel et al., 1986; Menaces, 1995, 2003).

In brain tissue, nitric oxide causes activation of the soluble guanyl cyclase enzyme, resulting in an increase in the cellular levels of cyclic GMP (cGMP). A common stimulus for nitric oxide is the activation of a particular class of receptors for the excitatory neuro transmitter glutamate, i.e., the NMDA receptors.

Investigations were done at the Wolfson Institute of Biomedical Research University College, London, to study (1) an effect of CDRI-08 itself to indicate whether it has the ability per se to release nitric oxide and (2) an effect of CDRI-08 in the presence of a submaximally effective concentration of NMDA to elucidate a potentiating or inhibiting effect or whether it simulates nitric oxide formation. The results obtained showed that although CDRI-08 did not possess an intrinsic effect on nitric oxide release, it potentiated NMDA-mediated nitric oxide and cGMP production in the cerebellum at a concentration of 100 µg/mL.

The hippocampus is a brain area associated with memory formation. At the cellular level, memory is considered to be stored in the form of long-lasting changes in the strength of the synaptic connections between neurons. Such changes in hippocampus are induced through tetanization, i.e., brief stimulation of synaptic connectivity at high frequency.

Using electrophysiological recording techniques, the effect of BESEB CDRI-08 on synaptic transmission in rat hippocampal slices were also studied at the Wolfson Institute of Biomedical Research, University College, London (1997).

The investigations showed that BESEB CDRI-08 possessed clear intrinsic biological effects on synaptic transmission in the hippocampus by producing a synaptic depression during application. This depression reverted to an enduring potentiation

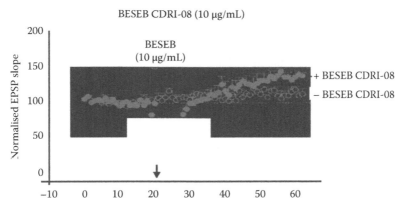

FIGURE 13.14 **(See color insert.)** Intrinsic biological effect of BESEB CDRI-08 on synaptic transmission in the hippocampus. The initial depression reverted to an enduring potentiation when the BESEB CDRI-08 was washed out.

when the BESEB CDRI-08 was washout. The changes thus observed are similar to those produced by nitric oxide under identical conditions. Such a change is identified as a natural neuroprotective device (Figure 13.14).

The result obtained clearly indicated that BESEB CDRI-08 in a concentration of 10 µg/mL was able to change the properties of the hippocampal synapses in such a way that a weak stimulus (which normally has little or no lasting effect) was able to elicit an enduring potentiation. This result has a direct relevance to the memory-enhancing effect of BESEB CDRI-08 (Garthwaite, 1997).

ACUTE AND REGULATORY TOXICITY

The LD_{50} of the BESEB CDRI-08 was determined in rats and mice by both oral and i.p. routes. It was >3 g/kg body weight by oral route in both the species, but by i.p. route it was much lower and was found to be 205 mg/kg [with the confidence limit (CL) ranging from 230 to 182] in rats and 224 mg/kg (CL 260–193) in mice.

GROSS PHARMACOLOGY

In the gross observations, the BESEB CDRI-08 (500 mg/kg p.o.) did not produce any significant alterations in the total, ambulatory, and stereotype behaviors in India. The product produced no effect on respiration and undesirable effects like writh-ing, ataxia, tremor, convulsions, etc. were absent. It did not produce any blockade in posture and tones and the eye movements were normal. The pineal, corneal, and righting reflexes were normal, and there was no effect on other reflexes. BESEB CDRI-08 did not produce any neurological deficit and all the innate and motivated behavioral responses were intact. The BESEB CDRI-08 at a dose of 50 mg/kg p.o. did not show anti-inflammatory activity nor produced any change in blood pres-sure, heart rate, respiration, or nictitating membrane contraction in rats. Standard vasopressor response of adrenaline (2–4 µg) and vasopressor responses of acetyl-choline (1–2 µg) were unaltered. The BESEB CDRI-08 at a dose of 100 mg/kg p.o.

did not produce any significant diuretic or antidiuretic effect and no fall from a rota rod in a forced locomotor activity test. Graded concentrations of BESEB CDRI-08 (10 and 50 µg/mL) produced an inhibition of serotonin-induced contraction in isolated guinea pig ileum preparation (Singh and Shanker, 1996).

SUBACUTE TOXICITY IN RODENT AND NONHUMAN PRIMATE BY ORAL ROUTE

These investigations were done in the Division of Toxicology, CDRI.

RODENT

BESEB CDRI-08 was given once daily orally for 90 days at the doses of 50, 100, and 200 times the effective dose of 40 mg/kg/p.o./day to rats of Charles Foster strain divided into three groups, each consisting of 15 male and 15 female animals. A fourth group of 15 male and 15 female animals were fed with corresponding volumes of vehicle and served as controls. Weekly monitoring of body weights of all the animals was done. Hemograms and urine analysis parameters of the animals were recorded initially and then at monthly intervals of the study. All the animals were sacrificed at the end of 90 days of the study, and terminal blood biochemistry and histopathology of all the important organs and tissues were studied.

Animals continued to remain active and healthy throughout the period of experimentation. The treated and control animals gained body weight well in comparison to one another. The laboratory investigations showed no indication of treatment-induced damage in the various hematological and biochemical, parameters, organ weights (both absolute and relative of important organs), and histopathological examinations.

BESEB CDRI-08 was thus found to be safe in rats in the 90-day subacute toxicity by oral route at the aforementioned dose levels.

NONHUMAN PRIMATE

BESEB CDRI-08 suspended in 1% aqueous gum acacia was given daily by oral route at the doses of 25, 50, and 100 times the effective dose of 10 mg/kg/p.o./day body weight in a single bolus to rhesus monkeys divided into three groups, each consisting of three male and three female animals. Another group of three male and three female animals were given corresponding volumes of 1% aqueous gum acacia alone in a similar manner and served as control. The treatment continued for 3 months. Average 24 h food consumption of the animals was recorded initially and then at weekly intervals throughout the period of the study. Body weights of the animals were recorded initially and then at monthly intervals. Hemograms, urinalysis, and serum biochemistry were done on weeks 0, 5, 9, and 13 of the study. All the animals were sacrificed at the end of the study and bone marrow was examined. The histopathology of all the important organs and tissues was studied.

The animals continued to remain active and healthy throughout the duration of treatment. Animals of both drug-treated and control groups showed irregular trends of gain in body weight. The laboratory investigations showed no indication of drug-induced damage in urinalysis and hematological and serum biochemical parameters. Elevations in SGPT (ALT) levels of few animals of the treated group were noted. But there was no functional abnormality in the liver as evidenced by constantly normal levels (treated vs. controls) of serum bilirubin, serum albumin, total serum protein, and prothrombin time estimations. Moreover, there was no evidence of intrahepatic biliary stasis as the levels of serum alkaline phosphatase remained within the ranges of normalcy in the treated groups of animals throughout the study. The autopsy studies (including absolute and relative weights of important organs) and gross and histopathological examinations did not reveal any sign of target organ toxicity.

BESEB CDRI-08 was thus found to be safe in rhesus monkeys in a 3 month sub-acute toxicity study by oral route at the aforementioned dose levels (Srivastava and Sudhir, 1996).

TERATOGENICITY STUDY

The purpose of the study was to delineate the harmful effects of BESEB CDRI-08 on pregnancy and unborn offsprings with the intention to provide information on the potential hazards to the developing fetus that may arise due to administration of the compound to the pregnant rats (Charles Foster) and rabbits (New Zealand white) by oral route during their major period of organogenesis, i.e., from day 6 to 15 and day 6 to 18 of the pregnancy period, respectively.

Pathogen-free male rats (175–250 g) and rabbits (1.75–2.25 kg) of proven fertility and sexually mature, randomly chosen nulliparous female rats (150–225 g) and rabbits (1.75–2.00 kg) were obtained from the National Laboratory Animal Center (NLAC) of the CDRI. A total of 10 female and 5 male rats and 2 male and 2 female rabbits were used for the study.

Each rat received a daily oral low dose of 50 mg/kg and high oral dose of 100 mg/kg and each rabbit received a daily oral low dose of 26.66 mg/kg and high oral dose of 53.32 mg/kg

PROCEDURE

Female test animals of timed pregnancy were treated with the test substance daily throughout the appropriate treatment period, i.e., from day 6 to 15 of gestation in rats and day 6 to 18 of gestation in rabbits. Signs of toxicity were recorded as and when they were observed, indicating the time of onset, the degree, and the duration. Females showing signs of abortion or premature delivery were sacrificed and subjected to thorough macroscopic examination. The posttreatment observation period continued till 1 day prior to term. During the treatment and observation period cage side observations included changes in skin and eye and mucus membrane, as well as condition of orifices, behavior pattern, etc. Food and water consumption were also observed every week if they were satisfactory or not (in the case of rabbits, only food consumption was measured). Animals were also weighed.

Observations Made

Standard parameters like general signs, mortalities/morbidity, food and water consumption, and body weight were observed for parent animal and gross, visceral, and skeletal anomalies (minor differences from normal that were detected either in gross and visceral or in skeletal examinations were observed. The percent postimplantation loss was calculated as the percentage of difference between total number of implantation and total number of fetuses.

Results

The results of parent animals revealed that all animals remained healthy during the period of study and there was no obvious change or any sign of reaction to drug in any of the treated rats or rabbits. No mortality was observed and food and water consumption were normal. A steady gain in weight was recorded among all the rats and rabbits of all the groups. The number of Corpora lutea and implantations were comparable and no intrauterine death (still birth) occurred. The litter sizes of all the groups were comparable and well within the range of normal limit.

The fetal abnormalities observed in treated and control groups were all incidental and have been commonly reported upto 10% even among normal rats and rabbits under normal conditions. The incidences observed were not dose dependent, thus ruling out any correlation with BESEB CDRI-08.

In the gross fetal abnormalities, only 2 out of 128 fetuses (1.56%) of control group, 2 out of 130 fetuses (1.53%), and 3 out of 141 fetuses (2.12%) revealed external anomalies. In rabbits, the total incidences among the examined fetuses were 1/56 (1.53%) in control group, 0/29 (−) and 0/25 (−) of low and high doses of BESEB CDRI-08, respectively.

The visceral and skeletal examinations of the fetuses showed no significant anomalies in any of the groups in all the animals.

Therefore, it is concluded that oral administration of BESEB CDRI-08 at the dose levels of 50 and 100 mg/kg in rats and 26.66 and 53.22 mg/kg in rabbits, during their major organogenesis, has not revealed any teratogenic effects (Srivastava and Sudhir, 1996).

GENOTOXICITY STUDIES

Chromosome Aberrations and Sister-Chromated Exchanges after In Vivo Exposure of BESEB CDRI-08 in Bone Marrow Cells of Mice

Sister-chromated exchange (SCE) and chromosome aberrations (CAs) are the important cytogenetic endpoints used extensively to study the genotoxic effects of drugs. In the present study, SCE and CA were carried out after in vivo exposure of BESEB CDRI-08 in bone marrow cells of Swiss albino mice of both sexes 10–12 weeks old and weighing 30 g.

In the study, 5-bromo deoxyuridine (BrdU) tablets (50 mg), Boehrringer Mannheim Biochemicals, Colchicines and Cyclophosphami de, Sigma Chemical Company, St. Louis, Mitomycin C, Andrew, Milwaukee, besides BESEB CDRI-08 were used.

CHROMOSOME ABERRATIONS ASSAY

For CA analysis, three doses (20, 40, and 80 mg/kg) of BESEB
  CDRI-08 were suspended in distilled water and injected i.p. in a volume of
100 µL per mouse. Five animals (three males and two females) were used for
each group and also for the control. Negative control mice were injected with an
equal volume of distilled water (100 µL per mouse) while positive control animals
received 25 mg/kg cyclophosphamide. After 22 h, the animals were injected (i.p.)
with colchicine (2 mg/kg) and 2 h later they were sacrificed by cervical dislocation.
Bone marrow was expelled from the femur bone with 0.075 M KCl. After hypotonic
treatment (0.075 M KCl at 37°C) for 20 min, cells were fixed three times with methanol: acetic acid (3:1). Slides were prepared for bone marrow chromosomes and
the slides were stained with giemsa. All the slides were coded and 100 well-spread
metaphase cells per animal were scored for CA, i.e., a total of 500 metaphase calls
were scored per dose tested. Mitotic index (MI) was scored from 1000 cells/animal
and was expressed as percentage. CA was scored following the standard WHO
guidelines. The aberration frequencies per cell for chromatid and chromosome types
were calculated. Gaps were recorded and not included either as a percentage or as
the frequency of aberrations per cell.

IN VIVO SISTER CHROMATED EXCHANGE ASSAY

Paraffin-coated (approximately 80% of the surface) BrdU tablets (50 mg each)
were implanted subcutaneously in the flank of mice under diethyl ether anesthe-
sia for in vivo SCE study and cell replication kinetic analysis. The test chemical
was administered as a single i.p. injection 1 h after the tablet implantation. Three
doses (20, 40, and 80 mg/kg) of BESEB CDRI-08 were injected i.p, in distilled
water (100 µL/mouse) to different groups of five animals (three males and two
females) each. Negative control mice were injected with 100 µL of distilled water
while mitomycin C was used as a positive control at a dose of 1.5 mg/kg of body
weight. For SCE analysis colchicine (4 mg/kg) was injected (i.p.) 22 h after BrdU
tablet implantation. Two hours later, bone marrow was expelled with 0.075 M
KCl. After hypotonic treatment (0.075 M KCl at 37°C) for 20 min cells were
fixed three times with methanol: acetic acid (3:1). The slides were prepared and
chromosomes were differentially stained with fluorescence-plus-Giemsa technique. All slides were coded and 30 s division metaphase cells (40 ± 2 chromosomes) per animal were scored for SCE frequencies, i.e., a total of 150 cells were
scored per dose tested. Randomly selected metaphase cells (100 cells per animal)
were scored for cell replication kinetic analysis by their staining patterns as first
(M_1), second (M_2), and third (M_3) division metaphases. The replicative indices
(RI) were calculated as follows:

$$RI = \frac{1M_1 + 2M_2 + 3M_3}{100}$$

RESULTS AND DISCUSSION

No significant increase in the SCE and CA was observed when compared with respective negative controls. No significant changes were observed either in replicative index (RI) in SCE study or in the MI in CA study when compared with respective controls. Trend tests for the evidence of dose response effects wave also negative for both SCE and CA studies. The present study of CA and SCE indicates that bacosides A and B were not genotoxic in bone marrow cells of mice.

The overall results of CA and SCE indicate that the new BESEBB CDRI-08 was not genotoxic in bone marrow cells in vivo in this present experimental condition described earlier.

MUTAGENICITY ASSAY OF BESEB CDRI-08 IN *SALMONELLA TYPHIMURIUM*

Ames *Salmonella* mutagenicity assay has been extensively used to determine the mutagenicity of different drugs. The *Salmonella* test was first validated in a study of 300 chemicals, most of which are known carcinogens.

In the present study, mutagenicity tests were carried out from BESEB CDRI-08 using tester strains TA97a and TA100 both with and without metabolic activation systems. Inbred Charles River male rats, weighing between 150 and 175 g were used for the preparation of rat liver homogenate (S9) for mutagenicity assays.

Biotin, histidine, dimethyl sulfoxide (DMSO) as surfactant, nicotine adenine dinuccleotide phosphate (NADP), glucose-6-phosphate, crystal violet, ampicillin trihydrate, sodium azide, 4-nitro-o-phenylenediamine (NPD), 2-aminofluorene (2-AF), and sodium ammonium phosphate of Sigma Chemical Company (St. Louis, MO) and Agar and beef extract HI Media besides BESEB CDRI-08 were used for the study.

BACTERIAL STRAINS

Salmonella tester strains TA97a (showing frame shift mutagenesis) and TA100 (showing base pair mutagenesis) were used for Ames bacterial mutagenecity assay. Before the start of the experiment the genotypes of both the tester strains were confirmed.

The overall results of Ames mutagenicity assay of BESEB CDRI-08 indicate that it was not mutagenic in the aforementioned two strains under the present experimental conditions (Giri and Khan, 1990).

HUMAN CLINICAL TRIALS

The spurt in preclinical investigation has led to several clinical trials on the various extracts of *B. monniera*. Here again the investigators from BHU were the first to undertake the clinical trial of the plant. Thus, two single-blind open clinical studies have found memory- and learning-enhancing effects of chronic treatment with alcoholic extract and expressed juice (sweetened with fruit syrup) of *B. monniera* both in patients with anxiety neurosis (Singh and Singh, 1980) and in children (Sharma et al., 1987).

However, the lead to undertake systematic double-blind, randomized placebo-controlled studies to clinically evaluate the effect of BESEB CDRI-08 on behavioral and cognitive functions in children (6–12 years) suffering from attention deficit hyperactivity disorder (ADHD) and adults (55–70 years) suffering from age associated memory impairment (AAMI) was taken by CDRI.

Clinical Trials in Children Suffering from ADHD

In the clinical trials on ADHD, only those children diagnosed as cases of ADHD as per the diagnostic criteria of *Diagnostic and Statistical Manual of Mental Disorder*, 4th edition (DSM-IV), of American Psychiatric Association's Committee were included. Initially, 88 ADHD children were screened, out of which 40 subjects fulfilled the diagnostic criteria and were included in the study. Twenty children received two 50 mg capsules of BESEB CDRI-08 and 20 children were given two capsules of placebo, identical in shape, size, and color. There were two dropouts from the BESEB CDRI-08 group and seven from the placebo group. The BESEB CDRI-08 was administered for 12 weeks daily and from 13th week to 16th week all the children were given placebo only. They were evaluated initially on day 0, and thereafter at 4, 8, 12 weeks of drug administration. The last evaluation was done after 4 weeks of stopping the medication when all children were given placebo only. The tests applied were personal information, mental control, sentence repetition, logical memory, word recall (both meaningful and non-meaningful words), digit span, picture recall, delayed response, and paired associate learning. Significant to highly significant results were obtained in all the parameters after 4–8 weeks. Some of the results are shown in Figures 13.15 and 13.16.

Clinical Trials in Adults Suffering from AAMI

These clinical trials were carried out at two centers, viz., BRD Medical College, DDU Gorakhpur University, Gorakhpur in 2000 (Sharma, 2000) and KG Medical College, University of Lucknow in 2001 (Raghav et al., 2006; Singh and Sangeetha, 2001).

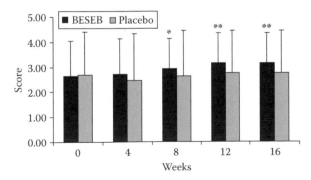

FIGURE 13.15 Distribution of effect of BESEB CDRI-08 and placebo on delayed response learning test score. *p < 0.05, **p < 0.01 as compared to week 0.

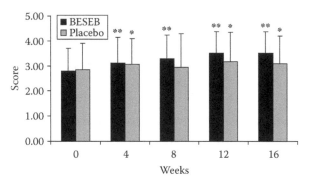

FIGURE 13.16 Distribution of effect of BESEB CDRI-08 and placebo on picture recall test score. *p < 0.05, **p < 0.01 as compared to week 0.

Subjects with complaints of memory loss in everyday life, e.g., difficulty in remembering names, misplacing objects, difficulty in remembering works which should be done, etc., were recruited by announcement in radio programs, advertisements in local news papers, and distribution of hand bills, etc. The inclusion criteria were the following: subjects should be 55 years and older, should have logical memory subtest of Wechsler's memory scale with a cut off score of ≤6, a score of 24 or higher on Mini Mental Scale Examination (MMSE) to exclude dementia. Evidence of delirium, confusion, neurological disorder, psychiatric disorder, malabsorption disorder, history of alcoholism, drug abuse, medication known to influence CNS, history of serious disorders of heart, kidney, or bone marrow, etc., were excluded from the study.

Thus, 20 cases in BESEB CDRI-08 and also in placebo groups were each taken at both the centers. The test tools administered were Wechsler Memory Scale—Revised (Wechsler, 1987), MMSE (Folstein et al., 1975), and dosage record treatment emergent symptom scales (DOTES, 1985). In the study, the Hindu version of some parts of Wechsler Memory Scale were also used (Chandra, 1980). The following parameters were used in the Wechsler Memory Scale: personal and current, orientation, mental control, passages, digits forward, digits backward, visual reproduction, associated learning, total raw score, correct score, and mental quotient (Figure 13.17).

CLINICAL TRIALS TO STUDY REJUVENATING EFFECT

In another randomized control double-blind, crossover design study, in which participants (n = 94) were randomly allocated one of the two treatments, viz., one capsule of 300 mg BESEB CDRI-08 (n = 41) or one capsule of placebo identical in shape, color, and size (Kumar et al., 2010). After 6 months, the volunteers were switched to alternate treatment (crossover) BESEB CDRI-08 significantly improved treatment and period effect in hemoglobin, oxygen intake capacity, hand grip, etc. A significant improvement in stress and anxiety levels was also observed, confirming the *medhya rasayana* (memory enhancing and rejuvenating) effects of BESEB CDRI-08. Besides, BESEB CDRI-08 provides significant improvement in range of

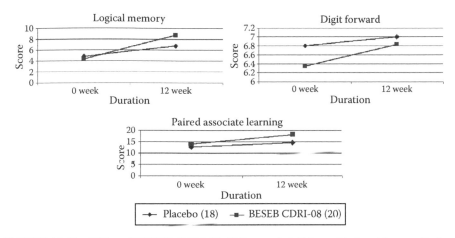

FIGURE 13.17 Effect of BESEB CDRI-08 (150 mg × 2 daily) on elderly subjects suffering from age associated memory impairment.

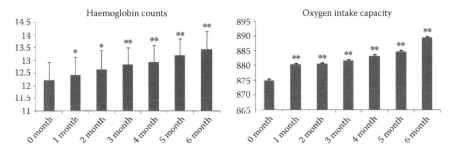

FIGURE 13.18 *Rasayana* (rejuvenating) effect of BESEB CDRI-08 (300 mg × 1 daily) on elderly patients (55–70 years).

motion, joint pain, etc. in patients suffering from sciatica (Amardeep et al., 2010) and significantly improves both quantity and quality of sleep in postmenopausal women (Kala et al., 2010) (Figure 13.18).

REFERENCES

Agranoff B W, Davis R E, Casola L, and Lim R (1967) Actinomycin blocks formation of memory of shock-avoidance in gold fish. *Science* **158**: 1600–1601.

Aithal H N and Sirsi M (1961) Pharmacological investigations on *Herpestis monniera. Indian J. Pharm.* **23**: 2–5.

Amardeep K T, Kashif M, and Singh H K (2010) Randomised control double blind study to clinically assess the effect of standardised extract *(Bacopa monniera)* BESEB CDRI-08 family (Scrophulariacae) in sciatica pain. *International Symposium on Brain Aging and Dementia: Basic and Translational Aspects,* Brain Research Centre, Department of Zoology, Banaras Hindu University, Varanasi, India. Abstract: P. 32.

Anderson M C and Green C (2001) Suppressing unwanted memories by executive centre. *Nature* **410**: 366–369.

Anderson M C, Ochsner K N, Kuhl B et al. (2004) Neural systems underlying the suppression of unwanted memories. *Science* **303**: 232–235.

Angers A, Storozhuk M V, Duchaine T et al. (1998) Cloning and functional expression of an *Aplysia* 5-HT receptor negatively coupled to adenylate cyclase. *J. Neurosci.* **18**: 5586–5593.

Barker L M, Best M R, and Dorijan M (1977) *Learning Mechanism of Food Selection*, Baylor University Press, Waco, TX, p. 632.

Barkley R A (1997) Behavioral inhibition, sustained attention, and executive function constructing a unified theory of ADHD. *Psychol. Bull.* **121**: 65–94.

Barondes S H and Cohen H D (1967) Comparative effects of cyclohexime de and puromycin on cerebral protein synthesis and consolidation of memory in mice. *Brain Res.* **4**: 44–51.

Basu N, Rastogi R P, and Dhar M L (1967) Chemical Examination of *Bacopa monniera* Wettst: Part III—Bacoside B. *Indian J. Chem.* **5**: 84–86.

Bennett E L, Rosenzweij M R, and Flood J F (1977) Protein synthesis and memory studied with anisomycin. In Roberts et al. (Ed.), *Mechanisms, Regulation and Special Function of Protein Synthesis in the Brain*, Elsevier/North Holland Biochemical Press, Amsterdam, the Netherlands, pp. 319–330.

Bickford P C, Adams C E, Boyson S J, Curella P, Gerhardt G A, Heron C et al. (1997) Long-term treatment of male 344 rats with deprenyel: Assessment of effects on longevity, behavior and brain function. *Neurobiol. Aging* **118**: 309–318.

Blodgett H C (1929) Drugs affecting learning and memory. *Univ. Calif. Pub. Psychol.* **4**: 113–114.

Bose K. C. and Bose N. K. (1931) Observations on the actions and uses of *Herpestis monniera*. *J. Indian Med. Assoc.* **1**: 60.

Byrne J H and Kandel E R (1996) Presynaptic facilitation revisited: State and time dependence. *J. Neurosci.* **16**: 425–435.

Central Drug Research Institute (CDRI) (2001) Clinical evaluation of *Bacopa monniera* extract on behavioral and cognitive functions in children suffering from attention deficit hyperactivity disorder.

Chandra S. (1980) Memory changes in primary depression, MD thesis, K G Medical College, University of Lucknow, Lucknow, India.

Chandra S, Personal Communication.

Chatterjee N, Rastogi R P, and Dhar M L (1963) Chemical examination of *Bacopa monniera* Wettst: Part I—Isolation of chemical constituents. *Indian J. Chem.* **1**(5): 212–215.

Chatterjee N, Rastogi R P, and Dhar M L (1965) Chemical examination of *Bacopa monniera* Wettst: Part II—The constituents of bacoside A. *Indian J. Chem.* **3**: 24–29.

Cohen J E, Onyike C U, McElory V L et al. (2003) Pharmacological characterization of an adenylyl cyclase serotonin receptor in *Aplysia*: Comparison with mammalian 5-HT receptors. *J. Neurophysiol.* **89**: 1440–1455.

Cook, W J and Weidley E (1957) Behavioral effects of some psychopharmacological agents. *Ann. N. Y. Acad. Sci.* **66**: 740–752.

Crow T, Xue Bian J J, Siddiqui V et al. (2001) Serotonin activation of the ERK pathway in *Hermisseuda*: Contribution of calcium—Dependent protein Kinase C. *J. Neurochem.* **78**: 358–364.

Dahanukar Sharadomi A, Rage Nirmala N, and Thatte Urmila N. Adaptogens. In KN Johry (Ed.), pp 143–146.

Dahanukar S A, Rage N N, Thatte U M (1997) Adaptogens. In Johry K N (Ed.) *Medicinal Plants: Their Bioactivity, Screening and Evaluation: Proceedings of the International Workshop*, Lucknow, India, December 2–5, 1997, pp. 143–164.

Das A, Shanker G, Nath C, Pal R, Singh S, and Singh H K (2002) A comparative study in rodents of standardized extracts of *Bacopa monniera* and *Ginkgo biloba*: Anticcholinesterase and cognitive enhancing activities. *Pharmacol. Biochem. Behav.* **73**: 893–9000.

Decker, M N and McGough J L (1989) Drugs affecting learning and memory. *Brain Res* **477**: 29–37.

Dhawan, B N (1997) R&D Issues and concerns in the development of herbal drugs. In Wozniak DA, Yuem A, Gasett M, Shuman TM (Eds) *Proceedings of International Symposium on Herbal Medicine: A Holistic Approach*, San Diego State University, San Diego, CA, pp. 12–156.

Dhawan B N and Singh H K (1996) Pharmacology of Ayurvedic nootropic *Bacopa monniera*. In *World Psychiatric Association: Section Meeting: International Convention of Biological Psychiatry Association: Section Meeting: International Convention of Biological Psychiatry: Theme: Future Mental Health Care & Biological Psychology*, Mumbai, Abstract NR-59.

Dhawan B N and Singh H K (1998) Anti-amnesic activity of bacosides, active constituents of *Bacopa monniera* Linn, Abstract: XXI CINP Congress, Glasgow, U.K.

DOTES (1985) Dosage record treatment emergent symptom scale (DOTES). *Psychopharmacol. Bull.* **21(3)**.

Doyle, Sir Arthur Conan (1984) A Study in Scarlet. In *The Penguin Complete Sherlock Holmes*, with a preface by Christopher Morley, Penguin Books, Harmondsworth, Greater London, U.K., p. 21.

Duncan C P (1949a) The retroactive effect of electroshock on learning. *J. Comp. Physiol. Psychol.* **42**: 32–44.

Duncan C P (1949b) The retroactive effect of electroshock on learning. *J. Comp. Physiol. Psychol.* **42**: 441–442.

Ebadi M and Sharma S K (2003) Preoxinitite and mitochondrial dysfunction in the pathogenesis of Parkinson's disease. *Antioxid. Redox. Signal* **5**: 319–335.

Ebadi M, Sharma S, Shavali S, and El-Refacy H (2002) Neuroprotective action of selegiline. *J. Neurosci. Res.* **67**: 285–289.

Flexner I B, Flexner L B, and Steller E (1963) Memory in mice as affected by intracerebral puromycin. *Science* **141**: 51–59.

Flood J F, Smith G E, Bennett E L, Alberti M H, Grme. A E, and Jarrik M E (1986) Neurochemical and behavioral effects of catecholamine and protein synthesis inhibition in mice. *Pharmacol. Biochem. Behav.* **24**: 631–645.

Folstein M F, Folstein S E, and McHugh P R (1975) "Mini-Mental State": A practical method for grading the cognitive state of patients for the clinicians. *J. Psychiatry Res.* **12**: 189–198.

Frieder B and Allweis C (1982) Memory consolidation: Further evidence for the four phase model from the time course of diethyldithiocarbamate and ethacrinic acid amnesia. *Physiol. Behav.* **29**: 1017–1075.

Fukui K, Onodera K, Shinkai T, Suzuki S, and Uranos S (2000) Impairment of learning and memory in rats caused by oxidative stress and aging, and changes in antioxidative effects of L-deprenyl in aged rat brain regions. *Biogerontology* **4**: 105–111.

Garai S, Mahato B, Ohtant K, and Yamasaki K (1996) Bacopasaponin D: A pseudojujubogenin glycoside from *Bacopa monniera. Phytochemistry* **43(2)**: 447–449.

Garcia J, Kimelford D J, and Kvelling RA (1955) A conditioned aversion towards sacchasin resulting from exposure to gamma radiation. *Science* **122**: 157.

Garthwaite, J (1997).

Gibbs M E and Ng KT (1976) Memory Formation: A new three-phase model. *Neurosci. Lett.* **2**: 165–169.

Giri A K and Khan K A (1990) The internal report of the Central Drug Research Institute, Lucknow, India.

Glickman S E (1961) Preservative neural processes and consolidation of memory trace. *Psychol. Bull.* **58**: 218–233.

Goenka, Hari Krishna Das (Translator) (1967) *Srimad Bhāgwat-Gitā: Sankarbhasyā*, Gita Press, Gorakhpur, Uttar Pradesh, India, Chapter 2: Verse 63, p. 69.

Handa S S (1997) Quality control: Medicinal plants and plant products. In Johr K N (Ed.) *Proceedings of International Workshop on Medicinal Plants: Their Bioactivity, Screening and Evaluation*, Lucknow, India. Centre for Science and Technology of the Non-aligned and Other Developing Countries with the support of United Nations Industrial Development Organisation and Central Drug Research Institute (CSIR). December 2–5, 1997, pp. 281–296.

Handa S S Quality control. In Johry K N (Ed.) *Medicinal Plants: Their Bioactivity, Screening and Evaluation: Proceedings of the International Workshop*, Lucknow, India, December 2–5, 1997, pp. 281–296.

Hebb D O (1949) *The Organisation of Behavior*. Wiley: New York.

Jyoti A, Pallavi S, and Sharma D (2007) *Bacopa monniera* prevents from aluminium neurotoxicity in the cerebral cortex of rat brain. *J. Ethnopharmacol.* **111**: 56–62.

Jyoti A and Sharma D (2006) Neuroprotective role of *Bacopa monniera* extract against aluminium-induced oxidative stress in the hippocampus of rat brain. *Neuro. Toxicol.* **27**: 451–457.

Kala M, Kumar T, Kaur V, and Singh H K Randomised control double blind study of clinically assess the effect of standardised *Bacopa monniera* extract (BESEB CDRI-08) on sleep of postmenopausal women. In *International Symposium on Brain Aging and Dementia: Basic and Translational Aspects*, Brain Research Centre, Department of Zoology, Banaras Hindu University, Varanasi, India, Abstract, p. 34.

Kamin L J (1957) The retention of an incompletely learned avoidance response. *J. Comp. Physiol. Psychol.* **50**: 457–460.

Kamin L J (1963) Retention of an incompletely learned avoidance response: Some further analysis. *J. Comp. Physiol. Psychol.* **56**: 713–718.

Kandel E R (1987) The long and short of memory in *Aplysia*: A molecular perspective. In E. Costa (Ed.), *Fidia Research Foundation Neuroscience Award Lectures 1986*, Liviana Press, Padova, Italy, pp. 7–47.

Kandel E R, Klein M, Castellucci V F, Schacher S, and Godet P (1986) Some principles emerging from the study of short-and long-term memory. *Neurosci. Res.* **3**: 498–520.

Kar Chowdhury D, Parmar D, Kakkar P, Seth P K, and Srimal R C (2002) Antistress effects of bacosides of *Bacopa monniera*: Modulation of Hsp 70 expression, superoxide dismutase and cytochrome P450 activity in rat brain. *Phytother. Res.* **16**: 639–645.

Kaur J, Singh S, Sharma D, and Singh R (2003) Aluminium induced enhancement of ageing-related biochemical and electrophysiological parameters in rat brain regions. *Indian J. Biochem. Biophys.* **40**: 330–339.

Kaur J, Singh S, Sharma D, and Singh R (2003a) Neurostimulatory and antioxidative effects of L-deprenyl in aged rat brain regions. *Biogerontology* **4**: 105–111.

Kaur J, Singh S, Sharma D, and Singh R (2003b) Aluminium treated enhancement of ageing-related biochemical and electrophysiological parameters in rat brain regions. *Indian J. Biochem. Biophys.* **40**: 330–339.

Kawasaki K, Traynelis S F, and Dingledine R (1990) Difference responses of (A_1 and CA_3 regions to hypoxia in rat hippocampal slice. *J. Neurophysiol.* **63**: 385–394.

Kulshreshtha D K and Rastogi R P (1973) Identification of eblelin lactone from Bacoside A and the nature of genuine saponin. *Phytochemistry* **12**: 2074–2076.

Kumar T, Wahi A K, Singh R, Srivastava M, and Singh, H K (2010) Randomized control, double blind cross-over study to clinically assess the *rasayana* effect of a standardised extract of *brahmi (Bacopa monniera)* in adult human volunteers. In *International Symposium on Brain Aging and Dementia: Basic and Translational Aspects*, Brain Research Centre, Department of Zoology, Banaras Hindu University, Varanasi, India. Abstract, p. 33.

Kuribara H, Ghashi K, and Tadokoro S (1976) Rat strain differences in the acquisition of conditioned avoidance responses in the effects of diazepam. *Jpn. J. Pharmacol.* **26**: 725–735.

Kuribara H, Okuizumi K, and Tardokoris S (1975) Analytical study of acquisition of pre-operant avoidance for evaluation of psychotropic drugs in rats. *Jpn. J. Pharmacol.* **25**: 541–548.

Lashley K.S. (1917) Drugs affecting learning and memory. *Psychobiology* **1**: 141–1770.

Malhotra C K and Das P K (1959) Pharmacological studies of *Herpestis monniera* Linn (Brahmi). *Indian J. Med. Res.* **47**: 294–305.

Manjrekar N A (1996) Experimental and clinical evaluation of putative cognitive enhancers. PhD thesis, University of Bombay, Mumbai, Maharashtra, India.

Matthies H (1974) The biochemical basis of learning and memory. *Life Sci.* **15**: 2017–2031.

Matthies H (1989) In search of cellular mechanism of memory. *Prog. Neurobiol.* **32**: 27–349.

McGough, J L (1966) Time dependent processes in memory storage. *Science* **153**: 1109–1113.

Meneses A (1995) 5 HT—System and cognition. *Neurosci. Biobehav. Rev.* **23**: 1111–1125.

Meneses A (2003) A pharmacological analysis of an associative task: 5HT$_1$ to 5HT$_7$ receptor subtypes on a Pavlovian/institutional autoshaped memory. *Learn Mem.* **10**: 363–372.

Milfram N W, Krames L, and Allowaj T (1977) *Food Aversion Learning*, Plenum Press, New York, p. 263.

Miller B and Teuber H L (1968) In L. Weiskrantz (Ed) *Analysis of Behavioral Change*. New York: Harper & Row, pp. 242–257.

Mondadori C (1990) New "nootropic" compounds can facilitate memory even if administered upto 24 hours after the learning experience. In *IV Conference on the Neurobiology of Learning and Memory*, Center for Neurobiology of Learning and Memory, University of California, Irvine, CA, October 17–20, p. 112.

Müller G E and Pilzecker A (1900) Experimentelle Beitragezur Lehre vom Gedächtnis. *Z. Psychol. Physiol. Sinnesorgane Ergänz. Band* **1**: 1–300.

Negi K S, Singh, Y D, Kushwaha K P, Rastogi C K, Rathi A K, Srivastava J S, Asthana O P, and Gupta R C (2000) Clinical evaluation of memory enhancing properties of standardised extract of *Bacopa monniera* in children with attention deficit hyperactivity disorder. *Indian J. Psychiatry* **42(2)**: 47 Supplement.

Ott T, Doske A, Thiemann W, and Matthies H (1972) Einetcilautomatisicrtc Lernanlage für optische Discriminierungsversuche mit Ratten. *Acta Biol. Med. German.* **29**: 103–108.

Ott T (1977) *Mechanismen der Gedächtuisbildung: Brain and Behavior Research Monograph Series*, Vol. 7, VEB Gustav Fischer Verlag, Jena, Germany.

Ott T, Dosske A, Thiemann W, and Matthies H (1977) Eine feilautonaatisicste Lermanlage für optische Diskriminierungs Versuche mit Ratten, *Ada Biol. Med. Germ.* **29**: 1103–1108.

Ott T and Matthies H (1978) Lernen und Gedäcchtnis. In Stamm RR and Zerer H. (Eds), *Die Psychologie des 20, Jahrhunderts. Band IV: Lorenz und die Folgen*, Kindler Verlag, Zürich, pp. 988–1018.

Prabhakar S, Saraf M K, Pandhi P, and Anand A (2007) *Bacopa monniera* exerts antiamnesic effect on diazepam-induced anterograde amnesia in mice. *Psychopharmacology*. DOI 10, 1007/S 00213-007-1049-8.

Prakash J C and Sirsi M (1962) Comparative study of the effects of Brahmi (*Bacopa monniera*) on learning in rats. *J. Sci. Ind. Res.* **21**: 93–96.

Raghav S, Singh H, Dalal P K, Srivastava J S, and Asthana O P (2006) Randomised controlled trial of standardised *Bacopa monniera* extract in age-associated memory impairment. *Indian J. Psychiatry* **48**: 238–242.

Rai D, Bhatia G, Palit G, Pal R, Singh S, and Singh H K (2003) Adaptogenic effect of *Bacopa monniera (Brahmi)*. *Pharmacol. Biochem. Behav.* **75**: 823–830.

Ramchandran U, Divekar H M, Grover S K, and Srivastava K K (1990) New experimental model for the evaluation of adaptogenic products. *J. Ethnopharmacol* **29**; 275–281.

Rastogi R P and Dhar M L (1960) Chemical examination of *Bacopa monniera* Wettst. *J. Sci. Ind. Res.* **19B**: 455–456.

Rastogi S, Mehrotra B N, and Klshrestha D K (1996) In Jain S K (Chief Editor) *Proceedings of IV International Congress of Ethnobiology*, National Botanical Research Institute, Lucknow, India, November 17–21, 1994, Deep Publications, New Delhi, p. 93.

Rey J (2009) *Panax ginseng* (G115®) improves aspects of working memory performance and mood in healthy young adults. In *Fapronatura 2009: The 2nd International Symposium on Pharmacology of Natural Products*, Varadero, Cuba, June 3–7, 2009.

Roy, Personal communication.

Rozin P and Kalat J W (1971) Specific hungers and poison avoidance as adaptive specialisation of learning. *Psychol. Rev.* **78**: 459.

Russo, A and Borrelli F (2005) *Bacopa monniera*, a reputed nootropic plant: An overview. *Phytomedicine* **12**: 305–317.

Saraf M (2009) Memory—Mechanisms, tools and aids: Comprehensive review. *Ann. Neurosci.* **166**: 3, 119–222.

Sastri M S, Dhalla N S, and Malhotra C L (1959) Chemical Investigation of *Herpestis monniera* Linn (Brahmi). *Indian J. Pharmacol.* **21**: 303–304.

Schachar R, Tannock R, and Lojan G (1993) Inhibiting control impulsiveness, and attention deficit hyperactivity disorder. *J. Abnorm. Child Psychol.* **23**: 411–437.

Sethy V H, Im W B, Smith M W, and Tang A H (1994) Neuropharmacology of Bacosides, Technical Report submitted by Dr. Vimla H. Sethy, Senior Scientist, Upjohn Company, Kalamazoo, MI, to the Council of Scientific and Industrial Research (CSIR), New Delhi, April 5, 1994.

Shanker G and Singh H K (2000) Anxiolytic profile of standardised *Brahmi* extract. *Indian J. Pharmacol.* 32: 152.

Sharma D, (2000) Double blind placebo controlled trial of standardised *Bacopa monniera* extract in age associated memory impairment. MD thesis, BRD Medical College, DDU University, Gorakhpur, Uttar Pradesh, India.

Sharma P (2008) (Ed. Trans.) Çaraka-Samhita (Agniveśa's treatise refined and annotated by Çaraka and redacted by Drdhábālā), Chaukhamba Orientalia, Varanasi, India.

Sharma P (Ed. Trans.) (2009a) *Charak Samhita: Agnivesa's Treatise Refined and Annotated by Charak and Redacted by Dridhbala*, Text with English Translation, vols. 7–8, Chaukhamba Publications, Varanasi, India, pp. 1–2.

Sharma P (Ed. Trans.) (2009b) *Charak Samhita: Agnivesa's Treatise Refined and Annotated by Charak and Redacted by Dridhbala*, Text with English Translation, vol. 29, Chaukhamba Publications, Varanasi, India, p. 22.

Sharma P (Ed. Trans.) (2009c) *Charak Samhita: Agnivesa's Treatise Refined and Annotated by Charak and Redacted by Dridhbala*, Text with English Translation, vol. 69, Chaukhamba Publications, Varanasi, India, p. 139.

Sharma R, Chaturvedi C, and Tewari P V (1987) Efficacy of *Bacopa monniera* in revitalizing intellectual functions in children. *J. Res. Educ. Indian Med.* **1**: 12.

Sheikh N, Ahmad A, Siripurapu K B, Kuchibhotla V K, Singh S, and Palit G (2007) Effect of *Bacopa monniera* on stress induced changes in plasma artisone and brain monoamines in rats. *J. Ethnopharmacol* **111**: 671–676.

Sidman M (1953) Avoidance conditioning with brief shock and no exteroceptive warning signal. *Science* **118**: 157–158.

Sidman M (1956) Drug-behavior interaction. *Ann. N. Y. Acad. Sci.* **65**: 282–302.

Silverman P (1978) *Animal Behavior in the Laboratory*. London, U.K.: Chapman & Hall, pp. 110–119.

Singh, H K and Dhawan B N (1978) The effect of *Bacopa monniera* on the learning ability of rats. *Indian J. Pharmacol.* **10**: 72.

Singh H K and Dhawan B N (1982) Effect of *Bacopa monniera* Linn (Brahmi) on avoidance responses in rats. *J. Ettnopharmacol.* **5**: 205–214.

Singh, H.K. and Dhawan B N (1992) Drugs affecting learning and memory. In Tandon P N, Bijlani V, and Wadhwa S (Eds.), *Lectures in Neurobiology*, New Delhi, India: Wiley Eastern Limited, pp. 189–207.

Singh H K and Dhawan B N (1994a) Pre-clinical neuro-psychopharmacological investigations on Bacosides: A nootropic memory enhancer. *Update Ayurveda* **94**, Mumbai.

Singh H K and Dhawan B N (1994b) Improvement of learning and memory by saponins of *Bacopa monniera. Can. J. Physiol. Pharmacol.* (Abstract) **7251**: 407.

Singh H K and Dhawan B N (1997) Neuro-psychopharmacological effects of the Ayurvedic nootropic *Bacopa monniera* Linn (*Brahmi*). *Indian J. Pharmacol.* **295**: 359–365.

Singh H K and Dhawan B N (1974a) Pre-clinical neuro-psychopharmacological investigation on Bacoside: A nootropic memory enhancer. Update Ayurveda 94, Mumbai, Abstract No. T3, p. 3.

Singh H K and Dhawan B N (1974b) Improvement of learning and memory by saponins of *Bacopa monnieri. Can. J. Physiol. Pharmacol.* **725I**: 407.

Singh H K, Ott T, and Matthies H (1975) Effect of intrahippocampal injection of atropine on different phases of a learning experiment. *Psychopharmacology (Berl)* **38**: 247–258.

Singh H K, Rastogi R P, Srimal R C, and Dhawan B N (1988) Effect of bacosides A and B on avoidance responses in rats. *Phytother. Res.* **2**: 70–75.

Singh H and Sangeeta R (2001) Clinical evaluation of standardized *Bacopa monniera* extract in elderly subjects with age associated memory impairment, MD thesis, K G Medical College, University of Lucknow, Lucknow, India.

Singh H K and Shanker G (1996) The internal report of the Central Drug Research Institute, Lucknow, India.

Singh H K, Shanker G, and Patnaik G K (1996a) Facilitation of memory by bacosides: Naturally occurring saponins. Abstract: III European Pharmacological Colloquium, Genova (Italy) p. 35.

Singh H K, Shanker G, and Patnaik G K (1996b) Neuro-pharmacological and anti-stress effects of bacosides: A natural memory enhancer. *Indian J. Pharmacol.* **28**: 47.

Singh R H and Singh L (1980) Studies on the anti-anxiety effect of the medhya rasayana drug Brahmi (*Bacopa monniera* Wettst) Part II. *J. Res. Ayurveda Sidha* **1**: 133–148.

Singh R H, Singh B N, Sarkar F H, and Udupa K N (1979a) Comparative biochemical studies on the effects of four medhya rasayana drugs described by Charak on some central neurotransmitters in normal and stressed rats. *J. Res. Indian Med. Yoga Homeo* **14**: 7–14.

Singh R H, Singh L, and Sen P (1979b) Studies on the anti-anxiety effect of the medhya rasayana drug Brahmi (*Bacopa monniera* Linn) Part I (Experimental studies). *J. Res. Indian Med. Yoga Homeo* **14**: 1–6.

Singh A, Singh B, Shanker G, and Singh H K (1996) Integration of traditional medicine into modern medicine in India. In *Abstract: 1st International Conference of Anthropology and History of Health and Disease*, Geneva, Italy.

Singh H K, Srimal R C, and Dhawan B N (1990a) Neuro-psychopharmacological investigations on bacosides from *Bacopa monniera*. In *Ethnopharmacologie: Methods, Objectives: Acta 1st Collogus European d'Ethnopharmacologie*, Metz, France, pp. 319–322.

Singh H K, Srimal R C, Srivastava A K, Garg N K, and Dhawan B N (1990b) Neuro-psychopharmacological effects of bacosides A and B. In *Abstract: IV International Conference of Neurobiology of Learning and Memory*, Centre for Neurobiology of Learning and Memory, University of California, Irvine, CA, p. 144.

Singh H K and Dhawan B N (1997) Neuropsychopharmacological effects of the Ayurvedic nootropic *Bacopa monnieri* Linn (Brahmi). *Indian J. Pharmacol.* **29**(5): S359–S365.

Singhal, GD (Chief Editor) (2009) *Suśruta—Samhita of Suśruta*. New Delhi, India: Chaukhamba Sanskrit Pratishthan.

Sinha M M (1971) Some empirical behavioral data of concomitant biochemical reactions. In *Proceedings of Indian Science Congress, Part II*, Bangalore, India, pp. 1–26.

Sreekumaran R, Ramakrishna T, Madhav T R, Anandh D, Prabhu B M, Sulekha S et al. (2003) Loss of dendritic connectivity in CA_1, CA_2 and CA_3 neurons in hippocampus in rat under aluminium toxicity: Antidotal effect of pyridoxine. *Brain Res. Bull.* **59**: 421–427.

Srivastava S (1996) The internal report of the Central Drug Research Institute, Lucknow, India.

Wechsler D A (1987) *Wechsler Memory Scale: Revised Manual*. San Antonio, TX: The Psychological Corporation/Harcourt, Barce Jovanonich.

Williamson E M (2001) Synergy and other interactions in phytomedicines. *Phytomedicine* **8**: 401–409.

Xu N, Majidi V, Markesbery W R, and Ehmann W D (1992) Brain aluminium in Alzheimer's disease using an improved GFAAS method. *Neurotoxicology* **13**: 735–744.

14 Botanical Anxiolytics, Antidepressants, and Hypnotics

Jerome Sarris

CONTENTS

INTRODUCTION

The treatment of mood and anxiety disorders with botanical medicine goes back to antiquity, with various cultures applying energetic models such as humoral medicine or traditional Chinese medicine models to prescribe a range of plant-based medicines.[1,2] Until recently, understanding of depression and anxiety was rudimentary, with such conditions classified broadly as "melancholia" or "hysteria."[3] In present times, use of herbal medicine and complementary and alternative medicine (CAM) is prevalent among sufferers of mood and anxiety disorders. Data from a nationally representative sample of 2055 people interviewed during 1997–1998 revealed that 57% of those with anxiety attacks and 54% of those with severe depression reported using CAM therapies during the previous 12 months to treat their disorder.[4] Twenty percent of the sample with anxiety and 19% of those with severe depression visited a CAM practitioner for treatment during the year. Interviews of 82 psychiatric North American inpatients hospitalized for acute care for various psychiatric disorders revealed that 63% had used one or more CAM modalities within the previous 12 months.[5] The most frequently used CAM intervention was herbal medicine, with 44% using the therapy during the previous 12 months. Most did not discuss this use with their medical practitioner. A study involving 52 patients from an Australian psychiatric teaching hospital revealed that 52% used CAM treatments over the preceding 18 months.[6] Eighty-five CAM treatments were used by the sample, and 37% did not inform their medical practitioner of this use. Research on CAM or herbal medicine in psychiatry is still in its infancy, although there has been a 50% increase

289

in the literature over the last quinquennium in the combined area of herbal medicine and psychiatry.[7] A Medline search in late 2009 of controlled and uncontrolled clinical trials using the terms "Complementary Medicine" AND "Depression" OR "Anxiety" OR "Psychiatry" revealed 1663 hits. This compares with 10828 hits when the search term "Antidepressants" OR "Cognitive Behavioral Therapy" was substituted for "Complementary Medicine."

Herbal medicine products contribute to a significant part of the modern "CAM industry," and are a vital component of CAM practice. Phytotherapy (the practice of herbal medicine) today is as much science as it is art, with our understanding of herbal psychopharmacology advancing over the past two centuries after the isolation of active constituents such as morphine from opium poppies.[8] Research into psychoactive plants that may affect the CNS has since flourished with an abundance of preclinical in vitro and in vivo studies validating many phytotherapies as having profound biopsychological effects.[9] Aside from notable psychoactive plants (usually containing alkaloids) such as cocaine from *Erythroxylon coca* (coca), morphine from *Papaver somniferum* (opium poppy), or arecoline from *Areca catechu* (betel nut), other less potent plants are developing over the last several decades, which is rich evidence of beneficial therapeutic activity.[10]

Mechanisms of action for herbal medicines used for the treatment of psychiatric disorders primarily involve modulation of neuroreceptor binding and alteration of neurotransmitter formation and activity.[11] Other actions may involve stimulating or sedating CNS activity, and regulating or supporting the healthy function of the hypothalamic pituitary adrenal axis (HPA-axis).[9,11] Herbal medicines have a range of psychotherapeutic actions that include antidepressant, anxiolytic, nootropic (cognitive enhancing), sedative, hypnotic, and analgesic effects (see Table 14.1). Other traditional actions that may not follow orthodox pharmacy include adaptogens and tonics, which provide increased adaptation to exogenous stressors and enhance vitality of

---

TABLE 14.1
Botanical Psychopharmacology: Actions, Mechanisms, and Applications

Traditional Action	Proposed Mechanisms	Applications
Nervines (tonics, stimulants)	HPA-axis-modulation, beta-adrenergic activity	Depression, fatigue, convalescence
Adaptogens, thymoleptics, antidepressant, tonics	Monoamine interactions, HPA-axis modulation	Depression, fatigue, convalescence
Anxiolytics, hypnotics, sedatives	GABA or adenosine-receptor binding or modulation	Anxiety disorders, insomnia
Antispasmodics, analgesics	Calcium/sodium channel modulation	Muscular tension (dysmenorrhea, irritable bowel syndrome,
	Substance P or enkephalin effects	headaches), visceral spasm, pain
Cognitive enhancers	Cholinergic activity, acetylcholine esterase inhibition	Cognitive decline, dementia

Source: Sarris J., in *Clinical Naturopathy: An Evidence Guide to Practice*, eds. Sarris, J. and Wardle, L., 2010, Elsevier, Sydney, New South Wales, Australia.

body/mind via complex effects on neurochemistry and the HPA-axis.[12] These actions may be applied clinically in a range of psychiatric disorders, including generalized anxiety, depression, and insomnia. These conditions are the focus of this chapter, as they are prevalent psychiatric disorders that often comorbidly occur, having a marked socioeconomic effect.[13]

Clinical depression is diagnosed by Diagnostic and Statistical Manual of Mental Disorders-IV (DSM-IV) criteria as a condition uncomplicated by grief, comorbid medical conditions, or substance misuse, presenting with 2 weeks or more of low mood (dysphoria) and/or loss of pleasure (anhedonia), in combination with various somatic and psychological effects (e.g., fatigue, sleep disturbance, digestive changes, excessive guilt, and suicidality).[14] Some antidepressant herbal medicines offer promise for the treatment of this disorder via psychoactive actions such as re-uptake of monoamines such as serotonin, dopamine, and noradrenaline, enhanced binding and sensitization of serotonin receptors, monoamine oxidase inhibition, and HPA-axis modulation.[9,11] Other pathways of activity may include GABAergic effects, cytokine modulation (especially in depressive disorders with a comorbid inflammatory condition), and opioid and cannabinoid system effects.[10] Anxiety disorders such as generalized anxiety disorder (GAD), social phobia, and post-traumatic stress disorder present with a marked element of psychological anxiety and distress, and are accompanied by a range of somatic symptoms such as palpitations, hyperthermia, shortness of breath, dizziness, and digestive disturbance.[14] For GAD to be diagnosed according to DSM-IV criteria, in addition to uncontrollable worrying, there must also be at least three of six somatic symptoms (restlessness, fatigue, concentration problems, irritability, tension, or sleep disturbance) occurring for a period of at least 6 months.[14] Significant distress or impaired functioning from the condition must also be present. Phytotherapies that may benefit anxiety disorders are classed as "anxiolytics," and usually have effects on the γ-aminobutyric acid (GABA) system, either via inducing ionic channel transmission by voltage-gated blockage or alteration of membrane structures, or less commonly via binding with benzodiazepine receptor sites (e.g., GABA-α).[9,11] Primary chronic insomnia, as opposed to transient insomnia caused by acute stress or environmental change, is diagnosed in the DSM-IV as 1 month or more of sleep disturbance (long latency, poor maintenance, and unrestorative sleep), which causes a marked personal cost, e.g., in work and social functioning and must not be caused by drugs or alcohol, psychiatric or medical conditions, or environmental factors.[14] Herbal hypnotics and sedatives usually work via modulation of the adenosine receptors, melatoninergic effects, or via GABAergic activity.[10]

The significance of depression, anxiety, and insomnia being covered in the chapter concerns a common comorbidity between them. Mechanisms of action to treat these disorders, while varied, still interface with each other, and often when certain underlying neurological, endocrine, or circadian factors are addressed, a beneficial effect may occur on the nervous system as a whole. This may impact the treatment of other comorbid psychiatric disorders, e.g., if depression is treated then anxiety may resolve, or if insomnia is addressed depression may be relieved. In the following section, the current evidence base of botanical medicines in the treatment of these major psychiatric disorders is outlined. Mechanisms of action are detailed in addition to a review of major clinical evidence and suggestions for clinical potential application.

BOTANICAL ANXIOLYTICS AND ANXIETY

As evidenced in a study by Wittchen,[15] GAD was found in 22% of primary care patients who complained of anxiety problems. Consistent with the DSM-IV manual's description of the 1 year prevalence of GAD as approximately 3%,[14] a sample of 10,641 Australians interviewed in 1997 had a 1 month prevalence of 2.8%, and a 12 month prevalence of 3.6%.[16] Lifetime prevalence of GAD is approximated at 5%–6%.[15] The socioeconomic burden of GAD is immense, with sufferers more likely than any other patient group to make frequent medical appointments and utilize medical resources.[17] As in major depressive disorder, only about 40% of sufferers seek treatment, and only 60% achieve full or partial remission for over 5 years.[18]

The pathophysiology of GAD is still being unraveled, although current evidence indicates that the neurobiological influence involves abnormalities of serotonergic, noradrenergic, and GABA transmission.[19] The involvement of these pathways is reflected in the efficacy of selective serotonin re-uptake inhibitors (SSRIs), selective serotonin and noradrenalin re-uptake inhibitors (SNRIs), and benzodiazepines.[18] The main neuro-circuitry in the panic, fear, or anxiety responses in humans involves the prefrontal cortex, hippocampus, and amygdala.[20] Psychological determinants may also exist, such as a cognitive bias to increased attention and misinterpretation of ambiguous stimuli, which are perceived as threatening.[18]

As detailed in the introduction, plant medicines that possess anxiolytic properties usually have effects on GABA pathways either via direct receptor binding, ionic channel modulation or effects on the cell membranes of the cells. The subsequent increased GABA neurotransmission has a damping effect of stimulatory pathways, which ultimately provides a psychologically calming effect. In Table 14.2, eight herbal medicines with known anxiolytic effects are detailed. The mechanisms of action of these phytomedicines have been detailed, as elucidated via in vitro and in vivo studies. As outlined in the "clinical applications" section, aside from treating anxiety disorders, many of these anxiolytic plant medicines have additional applications. These include improving mood (such as *Melissa officinalis* [lemon balm] or *Piper methysticum* [kava]), providing a sedative or hypnotic action for insomnia (e.g., *Passiflora incarnata* [passionflower] or *Scutellaria lateriflora* [scullcap]), reducing muscle tension or pain (e.g., *Eschscholzia californica* [Californian poppy]), or enhancing cognition via nootropic activities (e.g., *Bacopa monniera* [Bacopa] or *Ginkgo biloba* [Ginkgo]).[10,12] Herbal medicines (such as *Withania somnifera* [Withania]) may also provide an adaptogenic effect applicable in cases of comorbid fatigue.

BOTANICAL ANTIDEPRESSANTS AND DEPRESSION

It is estimated that by the year 2020, depression is projected to cause the second greatest increase in morbidity after cardiovascular disease, presenting a significant socioeconomic burden.[52] The lifetime prevalence of depressive disorders varies depending on the country, age, sex, and socioeconomic group, and approximates about one in six people.[53,54] The 12 month prevalence of clinical depression (also known as major

TABLE 14.2
Botanical Anxiolytics: Mechanisms of Action and Clinical Applications

Botanical Medicine	Mechanisms of Action	Potential Clinical Application
Bacopa[21–25] (*Bacopa monniera*)	Metal chelation/β-amyloid protection Cholinesterase inhibition 5HT-2c modulation Antioxidant effects Antidepressant effects in forced swim and learned helplessness animal models	Cognitive impairment Anxiety Depression Nervous exhaustion
California poppy[26–30] (*Eschscholzia californica*)	Binding affinity with GABA receptors (flumazenil antagonist) Anxiolysis in animal models (familiar environment test and anti-conflict tests)	Anxiety Insomnia Pain
Ginkgo[31,32] (*Ginkgo biloba*)	Modulation of cholinergic and monoamine pathways Antioxidant, anti-PAF, anti-inflammatory effects GABAergic effects Nitric oxide activity	Cognitive impairment Anxiety Depression
Kava[33–38] (*Piper methysticum*)	GABA channel modulation (lipid membrane structure and sodium channel function) Weak GABA binding (increased synergistic effect of [3H]muscimol binding to GABA-α receptors) β-adrenergic downregulation MAO-B inhibition Re-uptake inhibition of noradrenaline in the prefrontal cortex	Anxiety Comorbid depression Anxious insomnia ADHD Pain
Lemonbalm[39–41] (*Melissa officinalis*)	Potent in vitro inhibitor of rat brain GABA transaminase (GABA-T) MAO-A inhibition Acute dosing caused a significant increase in self-rated calmness in human stress tests	Acute stress Anxiety Depression
Passionflower[42–47] (*Passiflora* spp.)	GABA-system mediated anxiolysis Benzodiazepine receptor partial agonist Animal behavioral models have shown non-sedative anxiolytic effects (elevated-plus maze, light/dark box choice tests)	Anxiety Insomnia
Scullcap[48] (*Scutellaria lateriflora*)	Posited GABA-α binding affinity Anxiolysis in animal maze-test model (increasing the number of unprotected head dips and entries into open-field area and arms)	Anxiety Nervous exhaustion Insomnia

(continued)

TABLE 14.2 (continued)

Botanical Anxiolytics: Mechanisms of Action and Clinical Applications

Botanical Medicine	Mechanisms of Action	Potential Clinical Application
Withania[49–51] (*Withania somnifera*)	GABA-mimetic activity (enhanced flunitrazepam binding) Anxiolytic effect comparable to that produced by lorazepam in animal models (elevated plus-maze, social interaction and feeding latency in an unfamiliar environment, tests)	Anxiety Insomnia Fatigue Nervous exhaustion

depressive disorder [MDD]) is approximately 5%–8%, with women being approximately twice as likely as men to experience an episode.[53,54]

The pathophysiology of MDD is complex, and it appears that a variety of biological causations exist.[55] The main premise concerning the biopathophysiology of MDD has in the last several decades focused on monoamine impairment (dysfunction in monoamine expression and receptor activity, lowering of monoamine production, or secondary messenger system malfunction [e.g., G proteins or cyclic AMP]).[56,57] In recent years, added attention has focused on neuroendocrinological abnormality concerning the HPA-axis, cortisol production and brain-derived neurotropic factor (this interface affects neurogenesis), impaired endogenous opioid function, changes in GABAergic and/or glutamatergic transmission, cytokine or steroidal alterations, and abnormal circadian rhythm.[56–60] From a psychological perspective, cognitive and behavioral causations (or manifestations) of MDD include negative or erroneous thought patterns or schemas, impaired self-efficacy, challenged social roles, and depressogenic behaviors or lifestyle choices.[61–63]

As a recent review by Fournier et al.[64] in *JAMA* details, emerging evidence has revealed that synthetic antidepressants (such as SSRIs, tricyclics, and MAOIs) have limited efficacy in persons with milder forms of depression. Furthermore, clinical guidelines often do not endorse antidepressants as the primary first-line intervention for milder forms of MDD, and are often regarded as widely over-prescribed.[65] Furthermore, only 30%–40% of people achieve a satisfactory response to first-line antidepressant prescriptions thus approximately 40% do not achieve remission after several antidepressant prescriptions, thus further pharmacotherapeutic developments are required.[66,67] Many herbal medicines have been revealed to provide antidepressant activity (see Table 14.3). Some plant medicines provide strong thymoleptic effects (as in the case of *E. coca*), however, due to pronounced dopaminergic effects (which may cause addiction) are not viable clinical options.[10] Others, however, such as *Hypericum perforatum* (St. John's wort) have thymoleptic effects that are mediated primarily via modulation of monoamine transmission. It should be noted that as in the case of most phytomedicines *H. perforatum*'s antidepressant mechanism of action is not as clearly

TABLE 14.3
Botanical Antidepressants: Mechanisms of Action and Clinical Applications

Botanical Medicine	Mechanisms of Action	Potential Application
Iranian borage[72] (*Echium amoenum*)	Anxiolysis shown in an animal model (elevated plus maze test) Antidepressant mechanism currently unknown	Depression Anxiety
Lavender[73–77] (*Lavandula* spp.)	GABA modulation (based on volatile constituents) Anxiolysis shown in animal models (elevated plus maze and open field tests)	Depression Anxiety Somatic tension
Korean ginseng[78–83] (*Panax ginseng*)	HPA-axis modulation Monoamine modulation (dopamine, serotonin) Anti-inflammatory, antioxidant effects Nitric oxide synthase inhibition	Fatigue Depression Poor cognition
Mimosa[84–86] (*Albizia julibrissin*)	5-HT$_{1A}$ serotonergic receptor binding Antidepressant and anxiolytic effects in animal models (elevated plus maze and tail suspension tests)	Depression Anxiety Insomnia
Rhodiola[87–91] (*Rhodiola rosea*)	HPA-axis modulation (inhibition of cortisol, stress-induced protein kinases, nitric oxide) Monoamine oxidase A inhibition Monoamine modulation Normalization of 5-HT and anti-stress effects in animal depression models	Fatigue Cognitive impairment Depression Anxiety
Saffron[92–94] (*Crocus sativus*)	↑Re-uptake inhibition of monoamines (dopamine, norepinephrine, serotonin) NMDA receptor antagonism GABA-α agonism Anxiolytic effects in animal models (elevated plus maze and open field test)	Depression Anxiety
St John's wort[95–99] (*Hypericum perforatum*)	Modulation of monoamine transmission via Na+ channel Nonselective inhibition of re-uptake of serotonin, dopamine, noradrenalin Decreased degradation of neurochemicals Increased binding/sensitivity/density to 5-HT$_{1A,B}$ Dopaminergic activity (prefrontal cortex) HPA-axis modulation Antidepressant and anxiolytic activity in animal models	Depression Bipolar depression

defined as SSRIs, having a multitude of biological effects on re-uptake and receptor binding of various monoamines, in addition to HPA-axis modulation.

Comorbidity between MDD and anxiety disorders is the rule, not the exception.[68] Several herbal medicines with mood elevating effects (such as *Rhodiola rosea* [rhodiola] or *Crocus sativus* [saffron]) may also have anxiolytic effects. This may be due to modulation of neurological pathways that have both antidepressant and anxiolytic effects (e.g., GABA, serotonin, noradrenaline, or dopamine systems), or this may be due to a "halo effect" whereby when depression is successfully treated, anxiety may also be reduced.[69,70] This was found in the case of a recent RCT involving participants with generalized anxiety, which found that *P. methysticum* (an established anxiolytic), in addition to anxiety reduction, also provided a statistically significant reduction of comorbid depression on the Montgomery–Asberg depression rating scale.[71]

BOTANICAL HYPNOTICS AND INSOMNIA

The prevalence of general sleep disturbance experienced by people over a year is estimated at approximately 85%, while the estimate of diagnosed chronic insomnia is estimated at around 10%.[100] As in the case of other psychiatric disorders, women have a slightly higher incidence of primary insomnia.[100] The pathophysiology behind sleep disorders appears to involve "hyperarousal" of the neuroendocrine system caused by abnormalities in circadian rhythm (involving clock genes, melatonin secretion, and adenosine receptors), GABA pathways, endocrine factors (HPA/cortisol hyperactivation), and excitory pathways involving glutamate and aspartate.[101,102]

Herbal hypnotics usually affect somnolence via modulation of GABA pathways, adenosine, or the circadian rhythym.[10] Anxiolysis that occurs from GABA modulation may also have follow-on soporific effects via relaxing the mind. This may occur due to shared common pathways via a general downregulation of neurological stimulatory activity. Due to this, plant medicines such as *Zizyphus jujuba* (sour date) and *Valeriana officinalis* (valerian), while used in practice commonly for insomnia, both can also be used potentially to treat anxiety. This is reflected in the continuum that exists with sedating agents. As Spinella[10] outlines, at one end of the sedation spectrum, substances that cause mild sedation will cause relaxation and anxiolysis. As the strength of the downregulation of biological arousal increases, marked sedation occurs, followed by somnolence, coma, and then death. While this continuum may reflect plants such as *P. somniferum*, this is not reflective of many herbal anxiolytic/hypnotics that, while exerting a dose-dependent response, do not cause pronounced sedation. While only four hypnotic herbal medicines are detailed in Table 14.4, several other herbal medicines that are potentially beneficial for insomnia also exist and are detailed in Table 14.2. These include *P. incarnata, E. californica, P. methysticum*, and *S. lateriflora*.

CURRENT CLINICAL EVIDENCE

Table 14.5 details 10 individual plant medicines that have varying levels of human clinical trial evidence in reducing anxiety, depression, and sleeplessness. Three phytomedicines displayed level A evidence; *P. methysticum* for generalized anxiety

TABLE 14.4
Botanical Hypnotics: Mechanisms of Action and Clinical Applications

Botanical Medicine	Mechanisms of Action	Potential Applications
Chaste tree[103] (*Vitex agnus castus*)	Circadian rhythm modulation via increased melatonin secretion (dose-dependent effect that may benefit sleep latency insomnia)	Insomnia Menstrual disorders (menstrual dysphoria)
Hops[104–106] (*Humulus lupus*)	Melatonin receptor modulation (binding affinity to M_1 and M_2 receptors) Hypothermic activity	Insomnia
Sour date[107–110] (*Zizyphus jujuba*)	Inhibits glutamate-mediated pathways in the hippocampus	Insomnia
	Animal models using Suanzaoren (a TCM formula containing *Z. jujuba* as the principle herb) have found modulation of central monoamines and limbic system interaction	Anxiety
Valerian[111 116] (*Valeriana spp.*)	Adenosine (A_1 receptor) interactions GABA modulation (decreased re-uptake and degradation of GABA) Valerenic acid from valerian has demonstrated GABA-A receptor ($\beta 3$ subunit) agonism 5-HT5a partial agonism	Insomnia Anxiety Somatic tension CNS stimulant withdrawal

and *H. perforatum* and *C. sativus* for unipolar depression. The majority of phytomedicines (seven) had level B evidence, denoting that many trials have not been replicated, and thus more RCTs are required to make firm conclusions. Among those with level B evidence in potentially ameliorating depressive symptoms, *R. rosea* and *Echium amoenum* (Iranian borage) appear to have promise in reducing depression. It should be noted however that *E. amoenum* has been restricted in Australia due to concerns over pyrrolizidine alkaloids (potentially hepatotoxic). Promising anxiolytics with emerging evidence include traditionally used herbal relaxants *S. lateriflora* and *P. incarnata*. *Ginkgo biloba* has shown in one study to possess significant anxiolytic activity over placebo, and in line with cognitive enhancing properties may be appropriate in cognitive insufficiency with comorbid anxiety. Two phytomedicines were found to have level C grade of evidence: *R. rosea* for GAD and *V. officinalis* for insomnia. *Rhodiola rosea* revealed encouraging positive results, however the study used a small sample and an open label design, thus inhibiting confidence in the results. While positive studies exist for *V. officinalis*, the results of many studies are equivocal to placebo, thereby precluding firm recommendation in the treatment of insomnia.

As shown in Tables 14.2 through 14.5, several herbal psychotropics with positive animal model evidence have not currently been trialed in humans in depression, anxiety, or insomnia. Herbal medicines such as *Albizia julibrissin* (Mimosa),

TABLE 14.5
Botanical Psychopharmacology: Clinical Studies

Botanical Medicine	Major Studies	Methodology	Results[a]	Evidence Level
Iranian borage (*Echium amoenum*)	Sayyah et al.[117]	*Depression*: 6 week, double-blind, parallel group trial; 375 mg of borage vs. placebo (n = 35)	Statistically significant reduction versus placebo on HAMD, but no significant effect on HAMA	B
Ginkgo (*Ginkgo biloba*)	Woelk et al.[31]	*Anxiety*: 4 week RCT (n = 107) 240 mg, 480 mg EGb761 vs. placebo	Dose-dependent significant reduction of anxiety -14.3 points vs. 7.8 points on HAMA (480 mg EGb 716 vs. placebo)	B
Kava (*Piper methysticum*)	Pittler et al.[118]	*Anxiety*: Cochrane Review of 11 RCTs (N = 645) and a meta-analysis of 6 RCTs (N = 345)	Significantly greater anxiolysis from Kava than placebo; 5.0-point reduction compared to placebo on HAMA (95% CI: 1.1–8.8)	A
	Witte et al.[119]	*Anxiety*: Meta-analysis. Kava WS1490 extract 6 RCTs included	Odds ratio in favor of Kava = 3.3 (95% CI: 2.09–5.22)	
Lavender (*Lavandula* spp.)	Akhondzadeh et al.[120]	*Depression*: 4 week RCT (n = 45) using Lavender tincture (1:5 50% alcohol, 60 drops) vs. imipramine, or the combination	The addition of Lavender to imipramine was more effective in reducing HAMD rated depression than imipramine alone, indicating a synergistic effect	B[b]
Passionflower (*Passiflora incarnata*)	Akhondzadeh et al.[121]	*Anxiety*: 4 week RCT (n = 36) using 45 drops of passionflower vs. 30 mg of oxazepam	Passionflower was as effective (with less side effects) as oxazepam in reducing anxiety	B
	Movafegh et al.[122]	*Anxiety*: Acute study RCT (n = 60) using 500 mg of passionflower vs. placebo for pre-surgical anxiety	Anxiety scores were significantly lower in the passionflower group than in the control group on a numerical rating scale	

Botanical	Reference	Study	Findings	Level
Rhodiola (*Rhodiola rosea*)	Darbinyan et al.[123]	*Depression*: 6 week 3 arm RCT (n = 89) comparing 340 mg vs. 680 mg of standardized Rhodiola vs. placebo	Both Rhodiola groups showed significant reduction of HAMD symptoms and in insomnia, somatization and emotional instability subscale outcome measures	B
	Bystritsky et al.[124]	*Anxiety*: 10 week open label (n = 10) using 340 mg of standardized Rhodiola	Rhodiola significantly endpoint anxiety on HAMA (although no placebo comparator was employed)	C
Saffron (*Crocus sativus*)	Akhondzadeh et al.[125,126], Moshiri et al.[127], Noorbala et al.[128]	*Depression*: 4 RCTs: Two trials using 30–90 mg of saffron vs. placebo; two vs. synthetic antidepressants	Significant improvement in reducing depression compared to placebo on HAMD. An equivalent therapeutic response vs. imipramine and fluoxetine	A
Scullcap (*Scutellaria lateriflora*)	Wolfson et al.[129]	*Anxiety*: Acute cross-over RCT (n = 19) using scullcap vs. placebo	Scullcap dose-dependently reduced symptoms of anxiety and tension after acute administration compared with placebo	B
St John's wort: SJW (*Hypericum perforatum*)	Linde et al.[130]	*Depression*: Meta-analyses SJW vs. placebo 9 larger RCTs (N = 2044) 9 smaller RCTs (N = 1020) SJW vs. SSRIs 12 (N = 1794) SJW vs. Tri/tetracyclics (N = 1016)	SJW showed significant effect on HAMD vs. placebo, comparable to synthetics RR 1.28 (1.10–1.49) RR 1.87 (1.22–2.37) RR 1.00 (0.90–1.15) RR 1.02 (0.90–1.15)	A
Valerian (*Valeriana* spp.)	Bent et al.[131]	*Insomnia*: Systematic review and meta-analysis. 16 RCTs (N = 1093). Valerian vs. placebo or vs. active controls such as antihistamines	Meta-analysis revealed that valerian significantly improved sleep quality compared to placebo (RR of improved sleep = 1.8, 95% CI:1.2, 2.9). However, the review revealed 9/16 studies were not supportive of valerian as a soporific	C

Level A, meta-analyses or replicated RCTs with positive results; level B, one or more RCTs, mainly positive results; level C, One or more clinical trials with poor methodology or mixed evidence from clinical trials; RR, relative risk. HAMD = Hamilton Depression Rating Scale; HAMA = Hamilton Anxiety Rating Scale.

a "Significant" is *p* < 0.05.

b Evidence-based on Lavender as an adjunctive rather than as a monotherapy.

M. officinalis, *Z. jujuba*, *B. monniera*, and *E. californica* have been researched over the past three decades in animal and in vitro studies with positive results, however surprisingly these have not been studied as monotherapies in the treatment of psychiatric disorders (excepting *B. monniera* in the treatment of cognitive decline). Preclinical evidence in these phytomedicines reveals an array of monoamine and neuropeptide modulatory activities that have potential to benefit sufferers of depression, anxiety, and insomnia.

FUTURE RESEARCH

As revealed in the comparison between the mechanism of action and clinical evidence tables, several herbal medicines that have solid in vitro or in vivo evidence of pharmacodynamic effects have not been studied as monotherapies in human clinical trials. Although positive in vitro or animal models do not always translate into clinical efficacy in humans, this information does provide a guidepost to potential future research. Promising anxiolytics that to date have not been tested by RCTs for anxiety disorders include *B. monniera*, *E. californica*, *M. officinalis*, and *W. somnifera*. Promising antidepressants include *Panax ginseng* (Korean ginseng) and *A. julibrissin*. It is interesting to note that while these plant medicines in modern phytotherapeutic practice are used to treat anxiety, depression, and insomnia, some have specific applications that may be utilized for comorbid anxiety presenting with specific conditions. For example, *B. monniera* is considered beneficial in treating cognitive insufficiency; *E. californica* has been used for insomnia and pain; *M. officinalis* is used for gastrointestinal complaints, e.g., nervous dyspepsia; *W. somnifera* and *P. ginseng* are regarded as a "tonics" that via HPA-axis modulation may allay fatigue. While RCTs are encouraged to assess these phytomedicines, in clinical practice a "one size fits all" approach may not be valid, and thus each may be specifically more effective in precise applications.

Research on promising psychotropic herbal medicines is in its infancy. This is evidenced by only three herbal medicines having multiple replicated RCTs: *H. perforatum* for depression, *P. methysticum* for anxiety, and *V. officinalis* for insomnia. Positive results from the first two listed phytomedicines should encourage further research on herbs with promising in vitro or in vivo studies. Surprising omissions of clinical evidence, with a deficit of RCTs being found in a database search, include *Humulus lupulus* and *Z. jujuba* in the treatment of insomnia; *E. californica*, *M. officinalis*, and *B. monniera* in anxiety; and *A. julibrissin* and *P. ginseng* for depressive disorders.

Adequate preclinical data exist to encourage further research of the previously listed phytomedicines in a range of psychiatric disorders. In addition to exploration of promising individual psychotropics for psychiatric disorders, future research could center on combinations. While "gold standard" methodology will usually be in the form of an RCT, combination formulations may also be of merit. Such an example is found in a 2003 open, practice-orientated study using a combination of *H. perforatum* and *V. officinalis* to treat depression co-occurring with anxiety (anxious depression).[132] The trial demonstrated marked success, with the herbal combination ameliorating anxiety more effectively than *H. perforatum* mono-therapy.

It should be noted however that not all combination studies are positive. A 4 week 2009 RCT using a combination of *H. perforatum* and *P. methysticum* for the treatment of MDD with comorbid anxiety revealed mixed results.[133] Although significant results occurred in self-reported depression on the Beck Depression Inventory (BDI-II) over placebo in the first controlled phase, no effects were found on the pooled analysis or on anxiety outcomes. Regardless, other potential combinations exist that may be beneficial in the treatment of depression. Formulations involving combinations of *R. rosea, C. sativus, A. julibrissin,* and *H. perforatum* may provide increased synergistic antidepressant effects for treatment of depression. In the treatment of anxiety, novel formulations including *P. incarnata* and *S. lateriflora*, in combination with *P. methysticum* may prove beneficial. While, further research using *Z. jujuba* and *Vitex agnus castus* in combination with other hypnotics may also provide benefits for the treatment of insomnia. These combinations do exist commonly in over-the-counter herbal medicine products throughout the world, however at present a paucity of research has been conducted to validate claims of efficacy. This remains a research area of great potential.

REFERENCES

1. Holmes P. *The Energetics of Western Herbs.* Vol. 1. Berkeley, CA: NatTrop Publishing, 1993.
2. Maciocia G. *The Foundations of Chinese Medicine.* Singapore: Churchill Livingstone, 1989.
3. Freud S. Mourning and melancholia. *J Nerv Ment Dis.* 1922;56 (5):543–545.
4. Kessler RC, Soukup J, Davis RB et al. The use of complementary and alternative therapies to treat anxiety and depression in the United States. *Am J Psychiatry.* February 2001;158(2):289–294.
5. Elkins G, Rajab MH, Marcus J. Complementary and alternative medicine use by psychiatric inpatients. *Psychol Rep.* February 2005;96(1):163–166.
6. Alderman C, Kiepfer B. Complementary medicine use by psychiatry patients of an Australian hospital. *Ann Pharmacother.* 2003;37(12):1779–1784.
7. García-García P, López-Muñoz F, Rubio G, Martín-Agueda B, Alamo C. Phytotherapy and psychiatry: Bibliometric study of the scientific literature from the last 20 years. *Phytomedicine.* 2008;15(8):566–576.
8. Pengelly A. *The Constituents of Medicinal Plants.* Vol. 2. Glebe New South Wales, Australia: Fast Books, 1997.
9. Kumar V. Potential medicinal plants for CNS disorders: An overview. *Phytother Res.* 2006;20(12):1023–1035.
10. Spinella M. *The Psychopharmacology of Herbal Medicine: Plant Drugs that Alter Mind, Brain and Behavior.* Cambridge, MA: MIT Press, 2001.
11. Sarris J. Herbal medicines in the treatment of psychiatric disorders: A systematic review. *Phytother Res.* August 2007;21(8):703–716.
12. Mills S, Bone K. *Principles and Practice of Phytotherapy.* London, U.K.: Churchill Livingstone, 2000.
13. Kessler RC, Chiu WT, Demler O, Merikangas KR, Walters EE. Prevalence, severity, and comorbidity of 12-month DSM-IV disorders in the National Comorbidity Survey Replication. *Arch Gen Psychiatry.* June 2005;62(6):617–627.
14. American Psychiatric Association. *Diagnostic and Statistical Manual of Mental Disorders.* 4th 'Text Revision' edn. Arlington, TX: American Psychiatric Association, 2000.

15. Wittchen HU. Generalized anxiety disorder: Prevalence, burden, and cost to society. *Depress Anxiety*. 2002;16(4):162–171.
16. Hunt C, Issakidis C, Andrews G. DSM-IV generalized anxiety disorder in the Australian National Survey of Mental Health and Well-Being. *Psychol Med*. May 2002;32(4):649–659.
17. Rynn MA, Brawman-Mintzer O. Generalized anxiety disorder: Acute and chronic treatment. *CNS Spectr*. October 2004;9(10):716–723.
18. Tyrer P, Baldwin D. Generalised anxiety disorder. *Lancet*. December 16 2006; 368(9553):2156–2166.
19. Nutt DJ, Ballenger JC, Sheehan D, Wittchen HU. Generalized anxiety disorder: Comorbidity, comparative biology and treatment. *Int J Neuropsychopharmacol*. December 2002;5(4):315–325.
20. Bremner J. Brain Imaging in anxiety disorders. *Future Drugs*. 2004;4(2):275–284.
21. Sairam K, Dorababu M, Goel RK, Bhattacharya SK. Antidepressant activity of standardized extract of *Bacopa monniera* in experimental models of depression in rats. *Phytomedicine*. April 2002;9(3):207–211.
22. Stough C, Lloyd J, Clarke J et al. The chronic effects of an extract of *Bacopa monniera* (Brahmi) on cognitive function in healthy human subjects. *Psychopharmacology (Berl)*. August 2001;156(4):481–484.
23. Tripathi YB, Chaurasia S, Tripathi E, Upadhyay A, Dubey GP. *Bacopa monniera* Linn. as an antioxidant: Mechanism of action. *Indian J Exp Biol*. June 1996;34(6):523–526.
24. Limpeanchob N, Jaipan S, Rattanakaruna S, Phrompittayarat W, Ingkaninan K. Neuroprotective effect of *Bacopa monnieri* on beta-amyloid-induced cell death in primary cortical culture. *J Ethnopharmacol*. October 30 2008;120(1):112–117.
25. Krishnakumar A, Nandhu MS, Paulose CS. Upregulation of 5-HT2C receptors in hippocampus of pilocarpine-induced epileptic rats: Antagonism by *Bacopa monnieri*. *Epilepsy Behav*. October 2009;16(2):225–230.
26. Kleber E, Schneider W, Schafer HL, Elstner EF. Modulation of key reactions of the catecholamine metabolism by extracts from *Eschscholtzia californica* and *Corydalis cava*. *Arzneimittelforschung*. February 1995;45(2):127–131.
27. Schafer HL, Schafer H, Schneider W, Elstner EF. Sedative action of extract combinations of *Eschscholtzia californica* and *Corydalis cava*. *Arzneimittelforschung*. February 1995;45(2):124–126.
28. Hanus M, Lafon J, Mathieu M. Double-blind, randomised, placebo-controlled study to evaluate the efficacy and safety of a fixed combination containing two plant extracts (*Crataegus oxyacantha* and *Eschscholtzia californica*) and magnesium in mild-to-moderate anxiety disorders. *Curr Med Res Opin*. January 2004;20(1):63–71.
29. Rolland A, Fleurentin J, Lanhers MC et al. Behavioural effects of the American traditional plant *Eschscholzia californica*: Sedative and anxiolytic properties. *Planta Med*. June 1991;57(3):212–216.
30. Rolland A, Fleurentin J, Lanhers MC, Misslin R, Mortier F. Neurophysiological effects of an extract of *Eschscholzia californica* Cham. (Papaveraceae). *Phytother Res*. August 2001;15(5):377–381.
31. Woelk H, Arnoldt K, Kieser M, Hoerr R. Ginkgo biloba special extract EGb 761 in generalized anxiety disorder and adjustment disorder with anxious mood: A randomized, double-blind, placebo-controlled trial. *J Psychiatr Res*. 2007;41(6):472–480.
32. Di Renzo G. *Ginkgo biloba* and the central nervous system. *Fitoterapia*. August 2000;71 (Suppl 1):S43–S47.
33. Davies LP, Drew CA, Duffield P, Johnston GA, Jamieson DD. Kava pyrones and resin: Studies on GABAA, GABAB and benzodiazepine binding sites in rodent brain. *Pharmacol Toxicol*. August 1992;71(2):120–126.

34. Jussofie A, Schmiz A, Hiemke C. Kavapyrone enriched extract from *Piper methysticum* as modulator of the GABA binding site in different regions of rat brain. *Psychopharmacology (Berl).* December 1994;116(4):469–474.

35. Gleitz J, Beile A, Peters T. (+/−)-Kavain inhibits veratridine-activated voltage-dependent Na(+)-channels in synaptosomes prepared from rat cerebral cortex. *Neuropharmacology.* September 1995;34(9):1133–1138.

36. Uebelhack R, Franke L, Schewe HJ. Inhibition of platelet MAO-B by kava pyrone-enriched extract from *Piper methysticum* Forster (kava-kava). *Pharmacopsychiatry.* September 1998;31(5):187–192.

37. Baum SS, Hill R, Rommelspacher H. Effect of kava extract and individual kavapyrones on neurotransmitter levels in the nucleus accumbens of rats. *Prog Neuropsychopharmacol Biol Psychiatry.* October 1998;22(7):1105–1120.

38. Seitz U, Schule A, Gleitz J. [3H]-monoamine uptake inhibition properties of kava pyrones. *Planta Med.* December 1997;63(6):548–549.

39. Kennedy DO, Scholey AB, Tildesley NT, Perry EK, Wesnes KA. Modulation of mood and cognitive performance following acute administration of *Melissa officinalis* (lemon balm). *Pharmacol Biochem Behav.* July 2002;72(4):953–964.

40. Kennedy DO, Little W, Scholey AB. Attenuation of laboratory-induced stress in humans after acute administration of *Melissa officinalis* (Lemon Balm). *Psychosom Med.* July–August 2004;66(4):607–613.

41. Awad R, Muhammad A, Durst T, Trudeau VL, Arnason JT. Bioassay-guided fractionation of lemon balm (*Melissa officinalis* L.) using an in vitro measure of GABA transaminase activity. *Phytother Res.* August 2009;23(8):1075–1081.

42. Grundmann O, Wang J, McGregor GP, Butterweck V. Anxiolytic activity of a phytochemically characterized *Passiflora incarnata* extract is mediated via the GABAergic system. *Planta Med.* December 2008;74(15):1769–1773.

43. Dhawan K, Kumar S, Sharma A. Anxiolytic activity of aerial and underground parts of *Passiflora incarnata. Fitoterapia.* December 2001;72(8):922–926.

44. Dhawan K, Kumar S, Sharma A. Anti-anxiety studies on extracts of *Passiflora incarnata* Linneaus. *J Ethnopharmacol.* December 2001;78(2–3):165–170.

45. Dhawan K, Kumar S, Sharma A. Comparative anxiolytic activity profile of various preparations of *Passiflora incarnata* Linneaus: A comment on medicinal plants' standardization. *J Altern Complement Med.* June 2002;8(3):283–291.

46. Grundmann O, Wahling C, Staiger C, Butterweck V. Anxiolytic effects of a passion flower (*Passiflora incarnata* L.) extract in the elevated plus maze in mice. *Pharmazie.* January 2009;64(1):63–64.

47. Sena LM, Zucolotto SM, Reginatto FH, Schenkel EP, De Lima TC. Neuropharmacological activity of the pericarp of *Passiflora edulis* flavicarpa degener: Putative involvement of C-glycosylflavonoids. *Exp Biol Med (Maywood).* August 2009;234(8):967–975.

48. Awad R, Arnason JT, Trudeau V et al. Phytochemical and biological analysis of skullcap (*Scutellaria lateriflora* L.): A medicinal plant with anxiolytic properties. *Phytomedicine.* November 2003;10(8):640–649.

49. Mehta AK, Binkley P, Gandhi SS, Ticku MK. Pharmacological effects of *Withania somnifera* root extract on GABAA receptor complex. *Indian J Med Res.* August 1991;94:312–315.

50. Bhattacharya SK, Bhattacharya A, Sairam K, Ghosal S. Anxiolytic-antidepressant activity of *Withania somnifera* glycowithanolides: An experimental study. *Phytomedicine.* December 2000;7(6):463–469.

51. Bhattacharya SK, Muruganandam AV. Adaptogenic activity of *Withania somnifera*: An experimental study using a rat model of chronic stress. *Pharmacol Biochem Behav.* June 2003;75(3):547–555.

52. WHO. Mental and Neurological disorders 'Depression'. 2006. http://www.who.int/ mental_health/management/depression/definition/en/.

53. Kessler RC, Berglund P, Demler O et al. The epidemiology of major depressive disorder: Results from the National Comorbidity Survey Replication (NCS-R). *JAMA*. June 18 2003;289(23):3095–3105.

54. Alonso J, Angermeyer MC, Bernert S, Bruffaerts R, Brugha TS, Bryson H. Prevalence of mental disorders in Europe: Results from the European Study of the Epidemiology of Mental Disorders (ESEMeD) project. *Acta Psychiatr Scand Suppl*. 2004;420(420):21–27.

55. Belmaker RH, Agam G. Major depressive disorder. *N Engl J Med*. Jan 3 2008; 358(1):55–68.

56. Ressler KJ, Nemeroff CB. Role of serotonergic and noradrenergic systems in the pathophysiology of depression and anxiety disorders. *Depress Anxiety*. 2000;12 (Suppl 1):2–19.

57. Hindmarch I. Expanding the horizons of depression: Beyond the monoamine hypothesis. *Hum Psychopharmacol*. April 2001;16(3):203–218.

58. Raison CL, Capuron L, Miller AH. Cytokines sing the blues: Inflammation and the pathogenesis of depression. *Trends Immunol*. January 2006;27(1):24–31.

59. Plotsky PM, Owens MJ, Nemeroff CB. Psychoneuroendocrinology of depression. Hypothalamic-pituitary-adrenal axis. *Psychiatr Clin North Am*. June 1998; 21(2): 293–307.

60. Antonijevic IA. Depressive disorders—Is it time to endorse different pathophysiologies? *Psychoneuroendocrinology*. January 2006;31(1):1–15.

61. Molina J. Understanding the biopsychosocial model. *Int J Psychiatry Med*. 1983; 13(1):29–36.

62. Southwick SM, Vythilingam M, Charney DS. The psychobiology of depression and resilience to stress: Implications for prevention and treatment. *Annu Rev Clin Psychol*. 2005;1:255–291.

63. Haeffel GJ, Grigorenko EL. Cognitive vulnerability to depression: Exploring risk and resilience. *Child Adolesc Psychiatr Clin N Am*. Apr 2007;16(2):435–448, x.

64. Fournier J, DeRubeis R, Hollon S et al. Antidepressant drug effects and depression severity: A patient-level meta-analysis. *JAMA*. January 2010;303(1):47–53.

65. Jureidini J, Tonkin A. Overuse of antidepressant drugs for the treatment of depression. *CNS Drugs*. 2006;20(8):623–632.

66. Berton O, Nestler EJ. New approaches to antidepressant drug discovery: Beyond monoamines. *Nat Rev Neurosci*. February 2006;7(2):137–151.

67. Warden D, Rush AJ, Trivedi MH, Fava M, Wisniewski SR. The STAR*D Project results: A comprehensive review of findings. *Curr Psychiatry Rep*. December 2007;9(6):449–459.

68. Kessler RC, Gruber M, Hettema JM, Hwang I, Sampson N, Yonkers KA. Co-morbid major depression and generalized anxiety disorders in the National Comorbidity Survey follow-up. *Psychol Med*. March 2008;38(3):365–374.

69. Brady KT, Verduin ML. Pharmacotherapy of comorbid mood, anxiety, and substance use disorders. *Subst Use Misuse*. 2005;40(13–14):2021–2041, 2043–2028.

70. Nierenberg AA. Current perspectives on the diagnosis and treatment of major depressive disorder. *Am J Manag Care*. September 2001;7(11 Suppl):S353–S366.

71. Sarris J, Kavanagh D, Byrne G, Bone K, Adams J, Deed G. The Kava Anxiety Depression Spectrum Study (KADSS): A randomized, placebo-controlled, cross-over trial using an aqueous extract of *Piper methysticum*. *Psychopharmacology (Berl)*. 2009;205(3):399–407.

72. Rabbani M, Sajjadi SE, Vaseghi G, Jafarian A. Anxiolytic effects of *Echium amoenum* on the elevated plus-maze model of anxiety in mice. *Fitoterapia*. July 2004;75(5):457–464.

73. Perry N, Perry E. Aromatherapy in the management of psychiatric disorders. *CNS Drugs*. 2006;20(4):257–280.

74. Bradley BF, Starkey NJ, Brown SL, Lea RW. Anxiolytic effects of *Lavandula angustifolia* odour on the *Mongolian gerbil* elevated plus maze. *J Ethnopharmacol.* May 22 2007;111(3):517–525.

75. Atsumi T, Tonosaki K. Smelling lavender and rosemary increases free radical scavenging activity and decreases cortisol level in saliva. *Psychiatry Res.* February 28 2007;150(1):89–96.

76. Shaw D, Annett JM, Doherty B, Leslie JC. Anxiolytic effects of lavender oil inhalation on open-field behaviour in rats. *Phytomedicine.* Sep 2007;14(9):613–620.

77. Toda M, Morimoto K. Effect of lavender aroma on salivary endocrinological stress markers. *Arch Oral Biol.* October 2008;53(10):964–968.

78. Park JH, Cha HY, Seo JJ, Hong JT, Han K, Oh KW. Anxiolytic-like effects of ginseng in the elevated plus-maze model: Comparison of red ginseng and sun ginseng. *Prog Neuropsychopharmacol Biol Psychiatry.* July 2005;29(6):895–900.

79. Bhattacharya SK, Mitra SK. Anxiolytic activity of *Panax ginseng* roots: An experimental study. *J Ethnopharmacol.* August 1991;34(1):87–92.

80. Kim DH, Moon YS, Jung JS et al. Effects of ginseng saponin administered intraperitoneally on the hypothalamo-pituitary-adrenal axis in mice. *Neurosci Lett.* May 29 2003;343(1):62–66.

81. Joo SS, Won TJ, Lee DI. Reciprocal activity of ginsenosides in the production of proinflammatory repertoire, and their potential roles in neuroprotection in vivo. *Planta Med.* May 2005;71(5):476–481.

82. Dang H, Chen Y, Liu X et al. Antidepressant effects of ginseng total saponins in the forced swimming test and chronic mild stress models of depression. *Prog Neuropsychopharmacol Biol Psychiatry.* November 13 2009;33(8):1417–1424.

83. Chen X. Cardiovascular protection by ginsenosides and their nitric oxide releasing action. *Clin Exp Pharmacol Physiol.* August 1996;23(8):728–732.

84. Kim JH, Kim SY, Lee SY, Jang CG. Antidepressant-like effects of *Albizzia julibrissin* in mice: Involvement of the 5-HT1A receptor system. *Pharmacol Biochem Behav.* May 2007;87(1):41–47.

85. Kim WK, Jung JW, Ahn NY et al. Anxiolytic-like effects of extracts from *Albizzia julibrissin* bark in the elevated plus-maze in rats. *Life Sci.* October 22 2004; 75(23):2787–2795.

86. Jung JW, Cho JH, Ahn NY et al. Effect of chronic *Albizzia julibrissin* treatment on 5-hydroxytryptamine1A receptors in rat brain. *Pharmacol Biochem Behav.* May 2005;81(1):205–210.

87. Kucinskaite A, Briedis V, Savickas A. Experimental analysis of therapeutic properties of *Rhodiola rosea* L. and its possible application in medicine. *Medicina (Kaunas).* 2004;40(7):614–619.

88. Perfumi M, Mattioli L. Adaptogenic and central nervous system effects of single doses of 3% rosavin and 1% salidroside *Rhodiola rosea* L. extract in mice. *Phytother Res.* January 2007;21(1):37–43.

89. Chen QG, Zeng YS, Qu ZQ et al. The effects of *Rhodiola rosea* extract on 5-HT level, cell proliferation and quantity of neurons at cerebral hippocampus of depressive rats. *Phytomedicine.* September 2009;16(9):830–838.

90. van Diermen D, Marston A, Bravo J, Reist M, Carrupt PA, Hostettmann K. Monoamine oxidase inhibition by *Rhodiola rosea* L. roots. *J Ethnopharmacol.* March 18 2009;122(2):397–401.

91. Mattioli L, Funari C, Perfumi M. Effects of *Rhodiola rosea* L. extract on behavioural and physiological alterations induced by chronic mild stress in female rats. *J Psychopharmacol.* March 2009;23(2):130–142.

92. Schmidt M, Betti G, Hensel A. Saffron in phytotherapy: Pharmacology and clinical uses. *Wien Med Wochenschr.* 2007;157(13–14):315–319.

93. Lechtenberg M, Schepmann D, Niehues M, Hellenbrand N, Wunsch B, Hensel A. Quality and functionality of saffron: Quality control, species assortment and affinity of extract and isolated saffron compounds to NMDA and sigma1 (sigma-1) receptors. *Planta Med.* June 2008;74(7):764–772.

94. Hosseinzadeh H, Noraei NB. Anxiolytic and hypnotic effect of *Crocus sativus* aqueous extract and its constituents, crocin and safranal, in mice. *Phytother Res.* June 2009;23(6):768–774.

95. Muller WE, Rossol R. Effects of hypericum extract on the expression of serotonin receptors. *J Geriatr Psychiatry Neurol.* October 1994;7 (Suppl 1):S63–S64.

96. Singer A, Wonnemann M, Muller WE. Hyperforin, a major antidepressant constituent of St. John's Wort, inhibits serotonin uptake by elevating free intracellular Na+1. *J Pharmacol Exp Ther.* September 1999;290(3):1363–1368.

97. Franklin M, Hafizi S, Reed A, Hockney R, Murck H. Effect of sub-chronic treatment with Jarsin (extract of St John's wort, *Hypericum perforatum*) at two dose levels on evening salivary melatonin and cortisol concentrations in healthy male volunteers. *Pharmacopsychiatry.* January 2006;39(1):13–15.

98. Yoshitake T, Iizuka R, Yoshitake S et al. *Hypericum perforatum* L (St John's wort) preferentially increases extracellular dopamine levels in the rat prefrontal cortex. *Br J Pharmacol.* June 2004;142(3):414–418.

99. Butterweck V. Mechanism of action of St John's wort in depression: What is known? *CNS Drugs.* 2003;17(8):539–562.

100. Roth T, Roehrs T. Insomnia: Epidemiology, characteristics, and consequences. *Clin Cornerstone Chronic Insomnia.* 2003;5(3):5–15.

101. Roth T, Roehrs T, Pies R. Insomnia: Pathophysiology and implications for treatment. *Sleep Med Rev.* February 2007;11(1):71–79.

102. Sateia MJ, Nowell PD. Insomnia. *Lancet.* November–December 2004;364(9449): 1959–1973.

103. Dericks-Tan JS, Schwinn P, Hildt C. Dose-dependent stimulation of melatonin secretion after administration of *Agnus castus. Exp Clin Endocrinol Diabetes.* February 2003;111(1):44–46.

104. Abourashed EA, Koetter U, Brattstrom A. in vitro binding experiments with a Valerian, hops and their fixed combination extract (Ze91019) to selected central nervous system receptors. *Phytomedicine.* November 2004;11(7–8):633–638.

105. Butterweck V, Brattstrom A, Grundmann O, Koetter U. Hypothermic effects of hops are antagonized with the competitive melatonin receptor antagonist luzindole in mice. *J Pharm Pharmacol.* 2007;59:549–552.

106. Brattstrom A. Scientific evidence for a fixed extract combination (Ze 91019) from valerian and hops traditionally used as a sleep-inducing aid. *Wien Med Wochenschr.* 2007;157(13–14):367–370.

107. Morishita S, Mishima Y, Hirai Y, Saito T, Shoji M. Pharmacological studies of water extract of the Zizyphus seed and the Zizyphus seed containing drug. *Gen Pharmacol.* 1987;18(6):637–641.

108. Hsieh MT, Chen HC, Kao HC, Shibuya T. Suanzaorentang, and anxiolytic Chinese medicine, affects the central adrenergic and serotonergic systems in rats. *Proc Natl Sci Counc Repub China B.* October 1986;10(4):263–268.

109. Hsieh MT, Chen HC, Hsu PH, Shibuya T. Effects of Suanzaorentang on behavior changes and central monoamines. *Proc Natl Sci Counc Repub China B.* January 1986;10(1):43–48.

110. Chen HC, Hsieh MT, Lai E. Studies on the suanzaorentang in the treatment of anxiety. *Psychopharmacology (Berl).* 1985;85(4):486–487.

111. Benke D, Barberis A, Kopp S et al. GABA(A) receptors as in vivo substrate for the anxiolytic action of valerenic acid, a major constituent of valerian root extracts. *Neuropharmacology.* 2009;56(1):174–181.

112. Murphy K, Kubin ZJ, Shepherd JN, Ettinger RH. *Valeriana officinalis* root extracts have potent anxiolytic effects in laboratory rats. *Phytomedicine.* December 2009;17(8–9):674–678.

113. Trauner G, Khom S, Baburin I, Benedek B, Hering S, Kopp B. Modulation of GABAA receptors by valerian extracts is related to the content of valerenic acid. *Planta Med.* January 2008;74(1):19–24.

114. Sichardt K, Vissiennon Z, Koetter U, Brattstrom A, Nieber K. Modulation of postsynaptic potentials in rat cortical neurons by valerian extracts macerated with different alcohols: Involvement of adenosine A(1)- and GABA(A)-receptors. *Phytother Res.* October 2007;21(10):932–937.

115. Ortiz JG, Nieves-Natal J, Chavez P. Effects of *Valeriana officinalis* extracts on [3H] flunitrazepam binding, synaptosomal [3H]GABA uptake, and hippocampal [3H]GABA release. *Neurochem Res.* November 1999;24(11):1373–1378.

116. Dietz BM, Mahady GB, Pauli GF, Farnsworth NR. Valerian extract and valerenic acid are partial agonists of the 5-HT5a receptor in vitro. *Brain Res Mol Brain Res.* Aug 18 2005;138(2):191–197.

117. Sayyah M, Sayyah M, Kamalinejad M. A preliminary randomized double blind clinical trial on the efficacy of aqueous extract of *Echium amoenum* in the treatment of mild to moderate major depression. *Prog Neuropsychopharmacol Biol Psychiatry.* January 2006;30(1):166–169.

118. Pittler MH, Ernst E. Kava extract for treating anxiety. *Cochrane Database Syst Rev.* 2003(1):CD003383.

119. Witte S, Loew D, Gaus W. Meta-analysis of the efficacy of the acetonic kava-kava extract WS1490 in patients with non-psychotic anxiety disorders. *Phytother Res.* March 2005;19(3):183–188.

120. Akhondzadeh S, Kashani L, Fotouhi A et al. Comparison of *Lavandula angustifolia* Mill. tincture and imipramine in the treatment of mild to moderate depression: A double-blind, randomized trial. *Prog Neuropsychopharmacol Biol Psychiatry.* February 2003;27(1):123–127.

121. Akhondzadeh S, Naghavi HR, Vazirian M, Shayeganpour A, Rashidi H, Khani M. Passionflower in the treatment of generalized anxiety: A pilot double-blind randomized controlled trial with oxazepam. *J Clin Pharm Ther.* October 2001;26(5):363–367.

122. Movafegh A, Alizadeh R, Hajimohamadi F, Esfehani F, Nejatfar M. Preoperative oral *Passiflora incarnata* reduces anxiety in ambulatory surgery patients: A double-blind, placebo-controlled study. *Anesth Analg.* June 2008;106(6):1728–1732.

123. Darbinyan V, Aslanyan G, Amroyan E, Gabrielyan E, Malmstrom C, Panossian A. Clinical trial of *Rhodiola rosea* L. extract SHR-5 in the treatment of mild to moderate depression. *Nord J Psychiatry.* 2007;61(5):343–348.

124. Bystritsky A, Kerwin L, Feusner JD. A pilot study of *Rhodiola rosea* (Rhodax) for generalized anxiety disorder (GAD). *J Altern Complement Med.* March 2008;14(2):175–180.

125. Akhondzadeh S, Fallah-Pour H, Afkham K, Jamshidi AH, Khalighi-Cigaroudi F. Comparison of *Crocus sativus* L. and imipramine in the treatment of mild to moderate depression: A pilot double-blind randomized trial [ISRCTN45683816]. *BMC Complement Altern Med.* September 2 2004;4:12.

126. Akhondzadeh S, Tahmacebi-Pour N, Noorbala AA et al. *Crocus sativus* L. in the treatment of mild to moderate depression: A double-blind, randomized and placebo-controlled trial. *Phytother Res.* February 2005;19(2):148–151.

127. Moshiri E, Basti AA, Noorbala AA, Jamshidi AH, Hesameddin Abbasi S, Akhondzadeh S. *Crocus sativus* L. (petal) in the treatment of mild-to-moderate depression: A double-blind, randomized and placebo-controlled trial. *Phytomedicine.* 2006;13(9–10):607–611.

128. Noorbala AA, Akhondzadeh S, Tahmacebi-Pour N, Jamshidi AH. Hydro-alcoholic extract of *Crocus sativus* L. versus fluoxetine in the treatment of mild to moderate depression: A double-blind, randomized pilot trial. *J Ethnopharmacol.* February 2005;97(2):281–284.

129. Wolfson P, Hoffmann D. An investigation into the efficacy of *Scutellaria lateriflora* in healthy volunteers. *Altern Ther Health Med.* March–April 2003;9(2):74.

130. Linde K, Berner M, Kriston L. St John's wort for major depression. *Cochrane Database Syst Rev.* 2008(4):CD000448.

131. Bent S, Padula A, Moore D, Patterson M, Mehling W. Valerian for sleep: A systematic review and meta-analysis. *Am J Med.* 2006;119:1005–1012.

132. Muller D, Pfeil T, von den Driesch V. Treating depression comorbid with anxiety—Results of an open, practice-oriented study with St John's wort WS 5572 and valerian extract in high doses. *Phytomedicine.* 2003;10 (Suppl 4):25–30.

133. Sarris J, Kavanagh DJ, Deed G, Bone KM. St. John's wort and Kava in treating major depressive disorder with comorbid anxiety: A randomised double-blind placebo-controlled pilot trial. *Hum Psychopharmacol.* January 2009;24(1):41–48.

134. Sarris J., in *Clinical Naturopathy: An Evidence Guide to Practice* (eds. Sarris, J. and Wardle, L.) Clinical Depression Sydney, New South Wales, Australia: Elsevier; 2010.

15 Huperzine A
An Update on the Psychopharmacology of a Putative Disease-Modifying Alkaloid of Interest in Alzheimer's Disease and Other Neurological Conditions

Andrea Zangara

CONTENTS

INTRODUCTION

Huperzine A (HupA) is an alkaloid originally isolated by Chinese scientists from a Chinese club moss, *Huperzia serrata*, in 1986 (Liu et al., 1986a). *Qian Ceng Ta* (the traditional name of the plant) has been used for over 1000 years in China for the treatment of a number of ailments, including contusions, strains, swellings, schizophrenia, myasthenia gravis, and more recently organophosphate poisoning (College, 1985; Liu et al., 1986a; Ma and Gang, 2004; Ma et al., 2007). It acts as a potent, highly specific, and reversible inhibitor of acetylcholinesterase (AChE) that crosses the blood–brain barrier (Tang et al., 1989; Wang et al., 2006; Zhang et al., 2008). HupA is similar or superior in potency and duration of AChE inhibition to the AChEIs physostigmine, galantamine, donepezil, and tacrine; the latter three were approved for the treatment of Alzheimer's disease (AD) in the United States and some European countries (Saxena et al., 1994; Rocha et al., 1998; Lallement et al., 2002b; Gordon et al., 2005; Liu and Sun, 2005; Eckert et al., 2006).

Memory enhancement has been demonstrated with HupA in both animal and clinical trials. HupA was approved for the treatment of AD in China in 1994, and has been widely used to improve the memory deficits in elderly people and patients with age-associated memory impairment (AAMI), AD, and vascular dementia (Kelley and Knopman, 2008; Little et al., 2008). There is also considerable interest in the West, where HupA is currently sold as over-the-counter memory supplement (often with poor control over safety and efficacy), to establish the clinical effectiveness of HupA. Increasing evidence indicates that HupA exerts multiple neuroprotective effects that are not the result of its specific inhibition on AChE, and given the multifactorial etiology of AD, this is of great clinical interest. A previous review of the same author (Zangara, 2003) provided a comprehensive survey of preclinical and clinical studies with HupA; the present update expands to more recent findings.

PHYTOBIOLOGY

HupA is extracted from the Chinese club moss known as *H. serrata* (from a new botanical classification), or *Lycopodium serrata*. Natural AChEIs have been obtained from other plants as well, e.g., the alkaloids physostigmine and galantamine, found in the plants *Physostigma venenosum* and *Galanthus nivalis*, respectively.

The whole herb contains triterpenoids (probably relevant for the traditional uses of the plant) and various alkaloids (0.2% of the total content), including lycodoline, lycoclavine, serratinine, and huperzines. Huperzine occurs in different chemical species, with similar properties but different strengths. The (–)-A species is about 10 times as strong as the B form. While the B form demonstrates lower anti-AChE

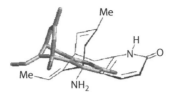

FIGURE 15.1   **(See color insert.)** Chemical structure of HupA and a 3-D view superimposed. (From Zangara, A., *Pharmacol. Biochem. Behav.*, 75(3), 675, 2003.)

activity than HupA, it has a longer duration of action and exhibits a higher therapeutic index; and like HupA, shows a relative lack of toxicity (Xu and Tang, 1987; Yan et al., 1987). The huperzines contain a nitrogen within one of the rings and an NH_2 group attached to the ring structure (Figure 15.1).

The average content of HupA in plants is 0.011% (Liu et al., 1986b). The low content of this alkaloid in nature brought successful attempts to synthesize this molecule or analogs (Kozikowsky and Xia, 1989; Qian and Ji, 1989; Kozikowsky et al., 1998; Mulzer et al., 2001). In particular, Huprine X is a new hybrid drug that combines huperzine A with tacrine, and has one of the highest affinities yet reported (Ki of 26 nM) for human AChE, 180 times that of huperzine A, 1200 times that of tacrine, and 40 times that of donepezil. Another synthetic derivative of HupA named ZT-1 is being developed in Europe and China as a new treatment for AD. ZT-1 possesses AChEl activity similar to HupA, but displayed less toxicity in mice than HupA (Ma and Gang, 2004).

BIOCHEMISTRY

HupA is an unsaturated sesquiterpene alkaloid with a pyridone moiety and primary amino group, whose empirical formula is $C_{15}H_{18}N_2O$, and molecular weight is 242. HupA is 9-amino-13-ethylidene-11-methyl-4-azatricyclo[7.3.1.0(3.8)] trideca-3(8),6,11-trien-5-one.

The compound is optically active and in the plant is present only in its (–)-enantiomer. It has a white-crystal appearance, is soluble in aqueous acid and CHCL3, and is a very stable molecule (Geib et al., 1991).

PHARMACOLOGY

ANTICHOLINESTERASE ACTION

HupA is a potent reversible inhibitor of AChE (Ki = 20–40 nM) that binds with aromatic residues in the active site gorge of AChE, localizing between Trp86 and Tyr337 in the enzyme (Ved et al., 1997). Butyrylcholinesterase (BuChE) is inhibited 1000-fold less than AChE (Ashani et al., 1992). HupA works at nanomolar concentrations and with a stereoselective mechanism, being (–) HupA, the naturally occurring form the more potent enantiomer, while (+) HupA inhibits the enzyme 38-fold less potently (Hanin et al., 1993). A 3-D computer image of AChE–HupA binding revealed how the HupA blocks the enzyme by sliding smoothly into the active site of AChE where acetylcholine is broken down, and latches onto this site via a large number of subtle chemical links (Raves et al., 1997).

In Vitro

HupA is a more effective inhibitor of AChE activity than tacrine and galantamine, but less effective than donepezil. However, HupA inhibited BuChE at a higher concentration than needed for AChE as compared to donepezil (Bai et al., 2000). The Ki values (inhibition constants, in nM) revealed that HupA was more potent than tacrine and galantamine, but less potent than donepezil by about twofold (Table 15.1). The weaker inhibition of HupA of human serum BuChE as compared with AChE, similar to galantamine, might suggest a better profile of side effects (Scott and Gao, 2000; Giacobini, 2004). The prolongation of incubation with HupA in vitro did not cause AChE activity to decrease, and after being washed five times the AChE activity returned to 94% of the control, demonstrating that the inhibitory action was of reversible type (Wang et al., 1986).

In Vivo

Following the administrations of oral HupA at doses of 0.12–0.5 mg/kg (Tang et al., 1989; Cheng and Tang, 1998), a marked dose-dependent inhibition of AChE was demonstrated in the brains of rats. In contrast to the AChE inhibition in vitro, the relative inhibitory effect of oral HupA over AChE was found to be about 24-fold more potent than donepezil and 180-fold more potent than tacrine on an equimolar basis. In rats, while the intraperitoneal (i.p.) injection of tacrine and donepezil showed greater inhibition of both AChE activity and serum BuChE than following oral administration, the i.p. administration of HupA exhibited similar efficacy of AChE inhibition (Wang and Tang, 1998a). After intraventricular injection, the inhibitory action of HupA on brain AChE was less than that of donepezil, but more than that of tacrine (Cheng and Tang, 1998). Following oral administration of 0.36 mg/kg HupA, maximal AChE inhibition was reached at 30–60 min and maintained for 360 min in rat cortex and whole brain (Tang et al., 1989, 1994a; Wang and Tang, 1998a). In contrast to donepezil and tacrine, it was the oral administration of HupA

TABLE 15.1

Cholinesterase Power of HupA, Donepezil, Tacrine, and Galantamine (See Text for Details)

| | IC_{50} µM | | |
AChEIs	AChE (Rat Cortex)	BuChE (Rat Serum)	Ki (nM)[a]
HupA	0.082	74.43	24.9
Donepezil	0.010	5.01	12.5
Tacrine	0.093	0.074	105.0
Galantamine	1.995	12.59	210.0

Source: Bai, D.L. et al., *Curr. Med. Chem.*, 7(3), 355, 2000.
[a] Rat erythrocyte membrane AChE.

TABLE 15.2
AChE Inhibition of Oral HupA, Donepezil, and Tacrine in Rats

		AChE Inhibition (%), n = 6		
	mg/kg	Cortex	Hippocampus	Striatum
HupA	0.36	20 ± 6^a	17 ± 3^a	18 ± 4^a
	0.24	16 ± 6^a	15 ± 3^a	16 ± 8^a
	0.12	10 ± 6^a	8 ± 7	13 ± 10^b
Donepezil	6.66	18 ± 6^a	12 ± 5^a	12 ± 8
	5.00	11 ± 6^a	10 ± 4^a	10 ± 6
	3.33	9 ± 11	6 ± 8	8 ± 6
Tacrine	28.2	20 ± 6^a	11 ± 10^b	11 ± 10^b
	21.1	8 ± 6^a	9 ± 6	8 ± 41^a
	14.1	7 ± 7	2 ± 2	2 ± 5

Source: Tang, X.C. and Han, Y.F., *CNS Drug. Rev.*, 5, 281, 1999.

Values expressed as percent inhibition ± standard deviation.

[a] $p < 0.01$.

[b] $p < 0.05$ versus saline group.

that produced the greatest AChE inhibition (Table 15.2), which indicated that it penetrates the blood–brain barrier more easily and has greater bio-availability (Wang et al., 2006).

Repeated doses of HupA showed no significant decline in AChE inhibition in comparison to that of single dose, demonstrating no occurrence of tolerance to HupA (Laganiere et al., 1991).

EFFECTS ON NEUROTRANSMITTERS

HupA significantly increased ACh levels in rat brain. Rats treated with HupA at doses of 0.3, 0.5, or 2 mg/kg demonstrated increased brain ACh 6 h after administration (Tang et al., 1989, 1994b; Zhu and Giacobini, 1995). A more prolonged increase of ACh levels in whole brain following HupA was observed as compared to tacrine, heptylphysostigmine, physostigmine, and metrifonate (De Sarno et al., 1989). The degree of ACh elevation was regionally selective with the maximal increase at 60 min in frontal and parietal cortex, intermediate at 30 min in the hippocampus and at 5 min in the medulla oblongata, and only slight increases at 30 min in the striatum (Tang et al., 1989, 1994a; Wang and Tang, 1998a). Considering that the level of ACh is particularly low in the cerebral cortex of patients with AD (Perry et al., 1978; Bowen et al., 1983), this specificity may represent a therapeutic advantage. HupA has been found to have little direct effect on nicotinic receptors compared to other AChEIs such as galantamine and tacrine, in studies on displacement of 3H-QNB and 3H-(–) nicotine binding (De Sarno et al., 1989; Tang et al., 1989). HupA also showed no effect on muscarinic receptors. Huprine X, however, a hybrid between tacrine and HupA, exhibited micromolar activity at M(1) and M(2) receptors that was probably agonistic (Roman et al., 2002), and could provide therapeutic advantages in the treatment of dementia (Fisher et al., 2002). Brain norepinephrine (NE) and dopamine (DA) levels increased significantly in a dose-dependent manner, following either systemic administration of HupA or local administration of HupA through microdialysis probe into the rat cortex (Zhu and Giacobini, 1995); NE and DA levels were increased more than 100% after the 0.3 and 0.5 mg/kg, but 5-HT level was not affected. Another trial (Liang and Tang, 2006) found that HupA had similar potency on increasing medial prefrontal cortex (mPFC) ACh and DA levels as compared to the 11- and 2-fold dosages of donepezil and rivastigmine, respectively, and had longer lasting effects after oral dosing; however, NE and 5-HT levels in the mPFC and hippocampus were not affected by any of the three treatments. These effects, if confirmed, might contribute to explain the cognitive enhancing properties of HupA, as the clinical effect of ChEIs has been related to the stimulation of both cholinergic and monoaminergic systems (Decker and McGaugh, 1991; Alhainen et al., 1993).

PROTECTIVE PROPERTIES

Other interesting properties of HupA pharmacology concern an impressive span of protective actions, which could represent a way to target dementia, a multifactorial illness, from several facets, and possibly to have a role in the prevention of neurological diseases. It possesses the ability to protect cells against glutamate, experimentally induced cytotoxicity and apoptosis, oxidative stress, β-amyloid peptide (Abeta), hydrogen peroxide, ischemia. These protective effects are related to its ability to reduce oxidative stress, regulate the expression of apoptotic proteins, protect mitochondria, modulate nerve growth factor and its receptors, and interfere with amyloid precursor protein metabolism. It also has antagonizing effects on N-methyl-D-aspartate receptors and potassium currents, which may contribute to its neuroprotection. In addition, HupA can be used as a protective (prophylactive) agent against organophosphate (OP) poisoning.

GLUTAMATE TOXICITY

It has been proposed that HupA functions as a noncompetitive antagonist of N-methyl-D-aspartate (NMDA) receptors, and is neuroprotective against glutamate toxicity. Excitatory aminoacid (EAA)-induced over stimulation is related to many acute and chronic neurodegenerative disorders, and is caused by an excess of glutamate that activates NMDA receptors and increases the flux of calcium ions in the neurons. Calcium at toxic levels can kill neuronal cells. Pretreatment of rat primary neuron cultures with HupA reduced glutamate-induced calcium mobilization and cytotoxicity (Wang et al., 1999). Additionally, HupA showed an age-dependent neuroprotection: more mature neuronal cells showed a greater sensitivity as well as neuroprotection to NMDA-induced toxicity. By reducing glutamate-related toxicity, HupA could be used to treat dementia and as a preventive agent for AD similarly to memantine, a drug that protects the brain against the excess of glutamate observed in AD.

Oxidative Stress

Another pathway by which HupA could be a disease-modifying agent in dementia and a preventive treatment is through protection of the brain against oxidative stress. An increase in oxidative stress, resulting from free radical damage to cellular function, can lead to AD and to brain lesions called tangles and plaques, the latest caused by the deposition of Abeta. In animal trials, HupA demonstrated a significant reduction in blood markers of oxidative stress (including erythrocyte and plasma lipoperoxides), thus suggesting a systemic antioxidant effect; interestingly, levels of lipid peroxidation and superoxide dismutase were lowered in the hippocampus, cerebral cortex, and serum of aged rats (Shang et al., 1999). A reduction in the plasma and erythrocyte oxygen free radicals was also demonstrated in a clinical study (Xu et al., 1999). Huperzine B showed similar neuroprotective properties to HupA, and to other AChEIs (donepezil, galantamine, tacrine), attenuating the hydrogen peroxide–induced injury (Zhang and Tang, 2000). HupA also exerted neuroprotection in an oxygen-glucose deprivation-induced injury model in C6 rat glioma cells (Zhao and Li, 1999).

Abeta Toxicity

HupA protected cultivated cortical neurons against Abeta toxicity, reducing the apoptosis (programmed cell death) that follows Abeta injection, and reducing the level of abnormal free radicals. HupA halted the suppressive effect of Abeta on long-term potentiation in rat hippocampal slices (Ye and Qiao, 1999). It appears that HupA treatment modifies the processing of the amyloid precursor protein by increasing α-secretase activity, and reducing the generation of amyloidogenic fragments (Peng et al., 2006). This effect could reduce Abeta toxicity and fibrillar amyloid accumulation in AD brain.

Hypoxic-Ischemic Brain Injury

The effect on brain metabolism, demonstrated by the protective effect of HupA on hypoxic-ischemic (HI) brain injury on neonatal rats with experimental brain damage, also enhances the value of HupA in the treatment of dementia, especially in the vascular type. HupA administrated daily to neonatal rats (Wang et al., 2002), at the dose of 0.1 mg/kg i.p. for 5 weeks after HI injury, produced significant protection from damage subsequent to HI injury on behavior (decreased escape latency in water maze) and on neuropathology (less extensive brain injury). Administration of subchronical oral doses of HupA (0.1 mg/kg, twice daily for 14 days) following 5 min of global ischemia in gerbils also showed similar neuroprotection (Zhou et al., 2001).

Nerve Gas Poisoning

The protective actions of HupA pharmacology have proved to be effective also against soman and other poisonous nerve gases used as chemical weapons. Pretreatment with HupA-protected rats and primates against soman, without the cholinergic side effects of other pretreatments (Grunwald et al., 1994; Lallement et al., 1997, 2002a). This peculiar protective property appears related both to the selectivity of HupA for red cell AChE, preserving the scavenger capacity of plasma butyrylcholinesterases for OP agents, and to the protection conferred on cerebral AChE (Lallement et al., 2002b).

NERVE GROWTH FACTOR

Nerve growth factor (NGF) promotes the survival and outgrowth of central cholinergic neurons. Growing evidence suggest that cholinergic mechanisms are involved in the expression and secretion of NGF synthesis and release (Shigeta et al., 2002). HupA has been shown to increase neurite outgrowth from undifferentiated PC12 cells and to promote the expression and secretion of NGF (Tang et al., 2005). This effect suggests the fascinating possibility that HupA might increase the NGF-induced enhancement of neuron survival and function improvement, preserving endangered neurons in neurodegenerative disease.

TOXICOLOGY

A series of toxicological studies conducted in different animal species indicated less severe undesirable side effects associated with cholinergic activation for HupA than for other ChEIs such as physostigmine and tacrine (Yan et al., 1987; Wang and Tang, 1998a). In mice, the LD_{50} doses were 4.6 mg (p.o.), 3.0 mg (s.c.), 1.8 mg (i.p.), and 0.63 mg (i.v.). Histopathological examinations after administration of HupA for 180 days showed no changes in liver, kidney, heart, lung, and brain in dogs (0.6 mg/kg i.m.) and in rats (1.5 mg/kg p.o.). No mutagenicity was found in rats (Zenghong and Meiying, 1990), and no teratogenic effect in mice or rabbits. Acute administration, as compared to tacrine, did not induce histopathological changes in the liver (Ma et al., 2003a).

COGNITION ENHANCEMENT

PRECLINICAL

HupA has been found to be an effective cognition enhancer with a superior safety/efficacy ratio (Tang and Han, 1999) when compared with other AChEIs in a number of different animal species (Table 15.3). Beneficial effects on learning and memory performance were seen in rodents on various tasks including spatial discrimination of radial arm maze (Xiong and Tang, 1995), water maze (Liu et al., 1998), and passive footshock avoidance (Zhu and Tang, 1988). Reversal of the effects of both aging (Lu et al., 1988; Ye et al., 1999) and experimental cognitive impairment such as cholinergic lesions (Xiong et al., 1998), scopolamine (Wang and Tang, 1998b; Ye et al., 1999; Gao et al., 2000), electroshock, cycloheximide, $NaNO_2$, CO_2 (Lu et al., 1988), reserpine, and yohimbine (Ou et al., 2001). The dose–response curve for HupA is of the inverted U-shape type, similarly to other cognitive enhancers. The effects on learning and memory retention lasted longer than those obtained with the other

TABLE 15.3

Comparison of Efficacy/Toxicity of Cholinesterase Inhibitors in Mice[a]

ChEIs	Memory Enhancement (µM/kg, p.o.)	Acute LD5 (µM/kg, p.o.)
HupA	0.83	17.31
Physostigmine	1.09	6.14
Galantamine	5.43	71.96
Tacrine	68.17	199.83

Source: Tang, X.C. and Han, Y.F., *CNS Drug. Rev.*, 5, 281, 1999.

[a] Memory enhancement (retention memory) assessed by step-down passive avoidance performance.

AChEIs, physostigmine, galantamine, and tacrine (Tang et al., 1994a). Further, after repeated daily administrations of HupA, no significant tolerance in cognitive improvement was seen (Xiong and Tang, 1995).

The dose-related deficit induced by scopolamine (0.01, 0.02, and 0.03 mg/kg) in young adult monkeys was reversed significantly up to 24 h with various doses of HupA (0.001, 0.01, 0.1, and 0.2 mg/kg) (Ye et al., 1999) using a delayed-response task. HupA (0.01–0.1 mg/kg, i.m.) improved the memory deficits induced by scopolamine in young adult monkeys. In aged monkeys, HupA (0.001–0.01 mg/kg, i.m.) significantly increased choice accuracy in delayed-response performance. The beneficial effect lasted for 24 h in both groups, and no adverse signs were observed, even at the highest doses.

CLINICAL TRIALS

Several randomized controlled clinical trials in China indicate that HupA is safe and effective in enhancing cognitive function. One early study with HupA, and the only known study of HupA for myasthenia gravis, demonstrated an improvement in muscle weakness in 128 patients (Cheng et al., 1986). HupA 0.4 mg i.m. daily was administered to 59 patients for 10 days, whereas the control group received neostigmine 0.5 mg/day every other day and HupA 0.4 mg daily on intervening days. The duration of effect was 7 h, 3 h more than that of neostigmine. Secondary effects were mainly cholinergic, and with the exception of nausea, were significantly milder than with neostigmine. One of the first studies investigating the effects of HupA in AD patients compared acute treatment of HupA (30 µg i.m.) to a treatment of 6 mg hydergine (dihydroergotoxine). In a group of 100 aged subjects, 17 with AD and the others reporting memory problems, HupA had a positive effect in comparison with dihydroergotoxine, on the memory of the subjects between 1 and 4 h after administration, with no remarkable side effects (Zhang, 1986).

A multicenter, randomized, double-blind, placebo-controlled study (Zhang et al., 1991), evaluated HupA for the treatment of memory disorders in 56 patients with multi-infarct dementia or AD. Patients were given 0.05 mg of HupA i.m. or placebo twice daily for 1 month. A further 104 patients with presenile or senile memory disorders were given 0.03 mg i.m. of HupA or placebo twice daily for 2 weeks. Therapeutic effects were measured with the Wechsler Memory Scale (WMS). Patients treated with HupA showed significant improvement in their memory quotient, and only few side effects were reported (mainly nausea and dizziness).

In a randomized, double-blind placebo-controlled study in 103 patients, enrolled after a diagnosis of AD according to the *Diagnostic and Statistical Manual of Mental Disorders-third revision* (DSM III-R) criteria, patients were given 200 μg of HupA or placebo orally and twice daily for 8 weeks (Xu et al., 1995). Their conditions were evaluated with the WMS, Hasegawa dementia scale (HDS), mini-mental state examination (MMSE), and activities of daily living scale (ADL). The improvement over placebo was still significant when measured after 8 weeks in the cognitive scales. Side effects were principally cholinergic and mild.

The efficacy and safety of HupA for treating AD were tested on 60 patients using multicenter, double-blind, placebo-controlled, parallel, positive controlled, and randomized methods (Xu et al., 1999). Patients were divided into two equal groups, one receiving four capsules of HupA (each containing 50 μg) and four tablets of placebo, while the other received four tablets of HupA (each containing 50 μg) and four capsules with placebo. Evaluation of cognitive and behavioral functions was based on monthly administration of MMSE, HDS-revised (HDS-R), IADL (instrumental activity of daily living), GBS-SDS (Gottfries-Bräne-Steen scale for dementia syndromes), and TESS (treatment emergent symptoms scale). Measures of ECG, EEG, and WMS were taken at the start and the end of the trial. There were significant differences on all the psychological evaluations taken between "before" and "after" 60 days of HupA treatment, but there was no significant difference between the two groups. The changes of oxygen free radicals showed marked improvement in the plasma and erythrocytes, The incidence of peripheral cholinergic side effects (mild to moderate nausea and mild to moderate insomnia) of HupA revealed with TESS score was up to 33%, and they diminished or disappeared by the end of the trial. The efficacy and safety of HupA in tablets and capsules appeared therefore to be equal.

Zhang and colleagues (Zhang et al., 2002) have evaluated the use of HupA in patients meeting the DSM-4 criteria for possible or probable AD. Two-hundred and two patients aged between 50 and 80 years were enrolled from 15 centers in 5 cities in China. The study was a multicenter, double-blind, randomized, and placebo-controlled; and the testing period was 12 weeks. The initial dose of HupA was 100 μg twice daily. The dose was increased to 150 μg twice daily from week 2 to week 3, and up to 200 μg twice daily from week 4 to week 12. The dosage was adjusted according to the reaction of the patients. The placebo group followed the same scheme. Both groups were also given as a basic treatment an oral dose of Vitamin E 100 mg twice daily. There was significant improvement (4.6 points compared with baseline) in the HupA group on the ADAS-Cog (cognitive portion of the Alzheimer's Disease Assessment Scale) as well as improvement on the MMSE (2.7 points), behavior and mood (1.5 points on the ADAS-Noncog) and on the activities of daily living scale. Four-point improvement on the ADAS-Cog was scored by 56%

of the patients in the active group, and 12.5% in the placebo group. HupA treatment was also associated with significant improvement on a global measure (the CIBIC-plus: clinicians interview-based impression of change plus caregiver input) compared with placebo. The percentage of mild and transient adverse events was 3% (insomnia and bilateral ankle edema), similar to the placebo group. The authors concluded that while HupA appears to be a safe and effective treatment for AD, this needs to be confirmed in larger and longer clinical trials, using different doses (400 µg is not the maximum dose).

A large clinical study reported by Wang was conducted in 819 patients who met the AD criteria of National Institute for Communicative Disorders and Stroke-Alzheimer's Disease and Related Disorders Association (NINCDS-ADRDA) and Diagnostic and Statistic Manual of Mental Disorders-Third Edition-Revised (DSM-III-R) at 39 mental hospitals in China (Wang et al., 2006). After treatment with HupA at a dose of 0.03–0.4 mcg/d, patients showed improvement in their memory, cognitive skills, and ability in their daily life. No severe side effects were found.

In the most recent Chinese placebo-controlled trial (Zhang et al., 2006), the efficacy of 0.5 mcg HupA was evaluated in 120 patients with AD. Patients in the treatment group had significantly greater improvement ($p < 0.01$) compared with those in the control group after 18 weeks of treatment, as evaluated by the ADAS-Cog. Adverse events in the HupA group included transient gastrointestinal dysfunction and elevations of ALT.

The superiority of combined therapy over HupA alone was suggested (Zhou et al., 2004) by significant favorable differences in MMSE, activity of daily living (ADL), and clinical dementia rating (CDR) scores after AD patients were treated with HupA plus nicergoline, conjugated estrogen, and nilestriol (for female AD patients).

In the United States the safety and efficacy of HupA were evaluated in 26 patients meeting the DSM-IVR and the NINCDS–ADRDA criteria for uncomplicated AD and possible or probable AD (Mazurek, 1999). This study (office-based) lasted 3 months and was open-label. HupA was very well tolerated at doses up to 200 µg bid, and effective in enhancing cognition as measured by the MMSE. This study demonstrated that the addition of HupA 100 µg bid to prior treatment regimens (including donepezil and tacrine) resulted in improvement on MMSE of 1.5, 1.75, and 2.2 points at 1, 2, and 3 months, respectively.

Two unpublished U.S. Phase I studies of HupA have recently been completed to determine the safety and tolerability of HupA in healthy elderly volunteers (age 65–70); the first evaluated doses up to 200 µg bid, while the second escalated doses up to 400 µg bid for 1 week. Another recent U.S. Phase 2 clinical trial run in collaboration with the National Institute on Aging and the Alzheimer's Disease Cooperative Study Group (ADCS) was completed in November 2007 (Rafii et al., 2011). The study was a multicenter, randomized, double-blind, placebo-controlled trial in 210 patients with mild to moderate AD. The trial compared the safety, tolerability, and efficacy of either 200 or 400 µg of HupA administered orally twice a day for 16 weeks versus placebo. Measures were cognitive function, activities of daily living, and behavior. Of the 210 patients enrolled, nearly half received concomitant treatment with Namenda, an FDA-approved drug for AD. HupA 200 µg bid did not influence change in ADAS-Cog at 16 weeks. In secondary analyses, however, HupA 400 μg bid showed a 2.27-point improvement in ADAS-Cog at 11 weeks versus 0.29-point decline in the placebo group (p = 0.001), and a 1.92-point improvement versus 0.34-point improvement in the placebo arm (p = 0.07) at week 16. Changes in clinical global impression of change, NPI, and activities of daily living were not significant at either dose. HupA was safe and well tolerated. Overall, the incidence of adverse events during the study was similar between both doses of HupA and placebo.

DISCUSSION OF CLINICAL TRIALS

Efficacy of AChEIs is limited, increasing MMSE score by 1–1.5 points (Lleo et al., 2006). The increase in MMSE score correlates with a reduction in the ADAS-Cog of 2.5–3.75. The ADAS-Cog is the reference standard for evaluating the efficacy of AD drugs, but MMSE is often used because more practical and brief.

Clinical trials evaluating HupA have demonstrated an improvement in MMSE score by 1–5 points in trials lasting 8 weeks at doses ranging from 100 to 500 μg daily (from Desilets et al., 2009); its effects on MMSE have shown a continued benefit for up to 36 weeks of treatment (Table 15.4). These results are very promising; however, it should be taken into account inconsistent methods of evaluation and reported outcomes, small study populations, weak trial design, short study duration, and lack of standardized product, dose, and dosing frequency. The majority of the published AD trials with HupA have been conducted in China, and only two Chinese studies were published in the English literature (Xu et al., 1995, 1999). A systematic Cochrane review (Li et al., 2008) considered 6 trials (454 patients) from 17 available, but only one (Zhang, et al. 2002) was considered relatively good for methodological quality. The review of the evidence presented in the six trials conducted in China concluded that HupA had some beneficial effects and no serious adverse effects for patients with AD. Among the four trials considered in one analysis based on the measure of MMSE, in two trials HupA produced statistically significant results over placebo, and two showed no statistical difference; but combining evidence quantitatively yielded a positive result. In terms of the measure of ADAS-Cog, a single trial with more than 200 patients found that HupA was superior to placebo statistically.

The review concluded that as only one study was of adequate quality and sample size, the overall evidence was inadequate to support its clinical use at present.

The authors of an updated meta-analysis of placebo-controlled randomized clinical trials of HupA in patients with AD (Wang et al., 2009) pointed that the Cochrane review results were limited by the low quality of individual studies and improper inclusion of one report. The authors searched for randomized trials comparing HupA with placebo in the treatment of AD with primary outcome measures the MMSE and ADL. Data were extracted from four randomized clinical trials, from an original selection of eleven articles, and analyzed using standard meta-analysis and meta-regression methods. The results of this meta-analysis indicated that oral administration of HupA for 8–24 weeks (300–500 μg daily) might lead to a significant improvement in the cognitive function and ADLs of patients with AD, and longer duration of treatment might produce better results.

TABLE 15.4

Effects of Huperzine A in Patients with Alzheimer's Disease

References	Study Design	Intervention	Results (Change from Baseline)	Adverse Events
Xu et al. (1995)	MC, DB, PC 8 weeks N = 103	Huperzine A 0.2 mg bid (n = 50) Placebo bid (n = 53)	MQ: +12.0 (p < 0.01 vs. placebo) MMSE: +3.0 (p < 0.01 vs. placebo) HDS: +4.0 (p < 0.01 vs. placebo) ADL: −4.0 MQ: +4.0 MMSE: +1.0 HDS: −1.0 ADL: +0.1	No significant difference between groups
Ma et al. (1998)	DB N = 80 (55 AD, 25 VD)	Huperzine A 0.1 mg bid (n = 40) control (n = 40)	MQ: +9.37 (p < 0.01 vs. control) MQ: +1.90	Gastric discomfort and nausea (n = 3) and dizziness (n = 3) significantly greater in treatment group
Xu et al. (1999)	MC, DB, DM, RC 60 days N = 60	Huperzine A 200 μg capsules bid (n = 30) Huperzine A 200 μg tablets bid (n = 30)	MQ: +8.0 (p < 0.01 vs. baseline) MMSE: +5.0 (p < 0.01 vs. baseline) MQ: +6.0 (p < 0.01 vs. baseline) MMSE: +4.0 (p < 0.01 vs. baseline)	10 pts. (capsule: n = 6, tablet: n = 4) reported mild nausea/vomiting; 27 pts. (capsule: n = 13, tablet: n = 14) categorized as having miscellaneous adverse events, mild insomnia, 8 pts. (capsule: n = 3, tablet: n = 5)
Mazurek (1999)	OL 3 mo N = 29 (26 AD, 3 other dementia)	Huperzine A 50 μg bid (n = 22) Huperzine A 100 μg bid (n = 7)[a]	MMSE: +1.09 MMSE: +1.00	No GI adverse effects (nausea, diarrhea) reported

(continued)

TABLE 15.4 (continued)
Effects of Huperzine A in Patients with Alzheimer's Disease

References	Study Design	Intervention	Results (Change from Baseline)	Adverse Events
Zhang et al. (2002)	MC, RC, DB, PC 12 weeks N = 202	Huperzine A 400 µg/day (n = 100) placebo (n = 102)	ADAS-Cog ≥4-point improvement: 56.1% MMSE ≥4-point improvement: 37.8% ADAS-Cog, ≥4-point improvement: 12.5% (p = 0.000 between groups) MMSE, ≥4-point improvement: 10.1% (p = 0.000 between groups)	3% of patients in the huperzine A group experienced mild, transient bilateral edema of ankles and insomnia
Zhang et al. (2006)	PC 18 weeks N = 120	Huperzine A mean dose 0.5 mg/day (n = 60) placebo (n = 60)	ADAS-Cog improvement significantly greater in treatment group (p < 0.01)	Transient GI effects and increase in ALT in huperzine A group
Rafil et al. (2011)	MC, DB, PC 16 weeks N = 210	Huperzine A 200 µg bid (n = 70) Huperzine A 400 µg bid (n = 70) placebo (n = 70)	ADAS-Cog no change at week 16 with 200 µg in secondary analyses, 400 µg showed a 2.27-point improvement in ADAS-Cog at 11 weeks vs. 0.29-point decline in placebo (p = 0.001)	Safe and well tolerated. Overall the incidence of adverse events during the study was similar between both doses of HupA and placebo

Source: Updated from Desilets, A.R. et al., *Ann. Pharmacother.*, 43, 514, 2009.

AD, Alzheimer's disease; ADAS-Cog, Alzheimer's Disease Assessment Scale: Cognitive; ADL, activities of daily living; ALT, alanine aminotransferase; DB, double-blind; DM, double mimic; GI, gastrointestinal; HDS, Hasegawa Dementia Scale; MC, multicenter; MMSE, Mini-Mental State Examination; MQ, memory quotient; NR, not reported; OL, open-label; PC, placebo-controlled; RC, randomized controlled; VD, vascular dementia.

[a] Five patients increased to this dose after 1 month or longer.

[b] Results of this trial have not yet been published.

CLINICAL TRIALS WITH DEBIO 9902 (FORMERLY ZT-1)

Debio 9902 has been developed as a new drug candidate by cooperation between the Shanghai Institute of Materia Medica at the Shanghai Academy of Sciences in China and Debiopharm in Switzerland. In vitro studies have shown Debio 9902, through its biotransformation into HupA, to be a highly potent and selective AChE inhibitor that increases HupA half-life from 6 to 20 h. In vivo Debio 9902 produced a marked dose-dependent inhibition of AChE and increased acetylcholine brain cortical levels in rats and reversal of scopolamine-induced memory deficits in both rats and monkeys. In one human study in 10 healthy elderly subjects, Debio 9902 had the ability to partially reverse the scopolamine-related cognitive impairment (Zangara et al., 2006) with similar results to Donepezil, but apparent longer duration of effects and earlier onset (Figure 15.2).

Debio 9902 has recently been developed in slow-release implants that are inserted subcutaneously. Implants release for up to 28 days and show an improved side effect profile in healthy volunteers. Continuous release of an AChEI provides a treatment alternative with the potential for better tolerance through a progressive, steady release of the active compound, increasing compliance with a superior efficacy profile than daily administration of an oral product.

Based on Debio 9902 efficacy in previous phase IIa study (Orgogozo et al., 2006), a superiority design was chosen for a head-to-head comparative study with donepezil. A randomized, double-blind, double-dummy, oral donepezil controlled study on the safety and efficacy of repeated monthly subcutaneous injections of a sustained-release implant of Debio 9902 in patients with moderate AD has recently concluded, and preliminary results, as recently presented at the *12th International Conference on Alzheimer's Disease (ICAD) 2009 Meeting*, appear very promising in terms of safety and efficacy, showing the potential to be a useful alternative to current AChEIs for the treatment of AD.

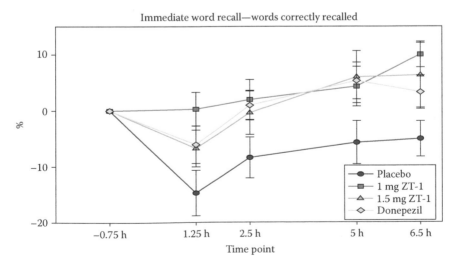

FIGURE 15.2 ZT1 versus donepezil and placebo (Zangara et al., 2006) in a scopolamine model (Cognitive Drug Research test battery, Immediate Word Recall).

TABLE 15.5

Memory Quotient in the Two Groups before and after 4 Weeks

Memory Quotient	HupA	Placebo
Baseline	92 ± 7^a	94 ± 8
4 Weeks trial	115 ± 6	104 ± 9
Odds	23 ± 7^a	11 ± 10

Source: Data from Sun, Q.Q. et al., *Acta Pharmacol. Sin.,* 20, 601, 1999.

Data are expressed as value ± SD (n = 34).

[a] $p < 0.01$.

TRIALS IN NON-AD POPULATIONS

Clinical data suggest that HupA is effective for the treatment of benign senescent forgetfulness (Ma et al., 1998) and possibly other neurological pathologies. Ma et al. (2003) have verified the effect of HupA on cognition in schizophrenic patients showing significant improvement in memory functions after treatment with HupA. Similar results were also reported by other researchers (Fang et al., 2002; Yang, 2003).

The effect of HupA on the performance of adolescent students was also studied using a double-blind, matched-pair design (Sun et al., 1999), reporting enhancement of the memory and learning performance (Table 15.5). In another study with children with language delay and other developmental conditions, treatment with 0.05 mg HupA twice daily for more than 3 months improved language delay scores by a total of 67.56% (Liao et al., 2002).

It has also been reported that HupA can improve sleeping: alternating HupA treatment with clonazepam (a benzodiazepine) treatment improves the quality of sleeping in patients with chronic insomnia, allowing for a gradual reduction of the benzodiazepine (Gao et al., 2003).

HupA SOURCE PLANTS

The original source plant of HupA, *H. serrata,* has been extensively used for over 1000 years in different areas of China as a popular traditional Chinese medicine (Ma et al., 2007). The powerful analgesic and antipyretic properties of the plant alkaloids were the main reason for its traditional use in the treatment of fever and swelling (Pilotaz and Masson, 1999). The Huperziaceae is one of the oldest existent vascular plant lineages, dating back to the late Silurian period (Ma and Gang, 2004). These plants grow very slowly and in specific habitats: from spore germination to maturity it can take 15–20 years. HupA can be chemically synthesized in laboratory (Qian and Ji, 1989); and in recent years as previously discussed, Kozikowski's group has developed methods to synthesize HupA on an industrial scale. However, as the demand for unregulated nutraceuticals containing *H. serrata* and related species continues to be sustained, plants are over harvested from the wild as they yield a very low content of HupA (Ma and Gang, 2004), signaling a possible risk to the species, or even extinction in the future.

To avoid this, alternative strategies must take place, such as controlled cultivation, identification of alternative species containing the right active alkaloids, and genetic selection.

CONCLUSIONS

Because of the significant pharmacological activities of HupA and its derivatives, Huperziaceae species have recently been a hot target of many investigations related to the chemical, biological, pharmacological, and medical properties of these plants. Results from these investigations clearly show that there is a strong relationship between the ethnopharmacological use of these plants and the medicinal properties of important compounds identified from them, such as HupA and HupB. Based on the characteristic cholinergic deficits in AD, AChE inhibitors are still the drugs of choice for the symptomatic therapy of AD. HupA, relative to other well-known AChE inhibitors, has high potency, high selectivity with respect to its AChE inhibitory effect, and marked memory-enhancing efficacy in a broad range of animal models of cognitive impairment, and in patients with AD. HupA has additional potentially disease-modifying properties, including neuroprotective neurotrophic activity and amyloid precursor protein metabolism modulation. These encouraging preclinical and clinical findings suggest that HupA is a promising candidate for the treatment of neurodegenerative diseases such as AD. More research is needed to explore further the actions of this alkaloid and its analogs and to allow sustainability of the original plant material.

REFERENCES

Alhainen K, Helkala EL, Reinikainen K, Rickkinen P. The relationship of cerebrospinal fluid monoamine metabolites with clinical response to tetrahydroaminoacridine in patients with Alzheimer's disease. *J Neural Transm Park Dis Dement* 1993; 5(3):185–192.

Ashani Y, Peggins JO, Doctor BP. Mechanism of inhibition of cholinesterase by huperzine A. *Biochem Biophys Res Commun* 1992; 184:7719–7726.

Bai DL, Tang XC, He XC. Huperzine A, a potential therapeutic agent for treatment of Alzheimer's disease. *Curr Med Chem* 2000; 7(3):355–374.

Bowen DM, Allen SJ, Benton JS, Goodhardt MJ, Haan EA, Palmer AM, Sims NR, Smith CC, Spillane JA, Esiri MM, Neary D, Snowdon JS, Wilcock GK, Davison AN. Biochemical assessment of serotonergic and cholinergic dysfunction and cerebral atrophy in Alzheimer's disease. *J Neurochem* 1983; 41:266–272.

Cheng YS, Lu CZ, Ying ZL, Ni WY, Zhang CL, Sang GW. 128 Cases of myasthenia gravis treated with huperzine A. *New Drugs Clin Remedies* 1986; 5:197–199.

Cheng DH, Tang XC. Comparative studies of huperzine A, E2020, and tacrine on behavior and cholinesterase activities. *Pharmacol Biochem Behav* 1998; 60:377–386.

College JNM. *The Dictionary of Traditional Chinese Medicine*. Shanghai Sci-Tech Press, Shanghai, China, 1985.

De Sarno P, Pomponi M, Giacobini E, Tang XC, Williams E. The effect of heptyl-physostigmine, a new cholinesterase inhibitor, on the central cholinergic system of the rat. *Neurochem Res* 1989; 14:971–977.

Decker MW, McGaugh TL. The role of interaction between cholinergic system and other neuromodulatory systems in learning and memory. *Synapse* 1991; 7:151–168.

Desilets AR, Gickas JJ, Dunican KC. Role of huperzine A in the treatment of Alzheimer's disease. *Ann Pharmacother* March 2009; 43:514–518.

Eckert S, Eyer P, Muckter H, Worek F. Kinetic analysis of the protection afforded by reversible inhibitors against irreversible inhibition of acetylcholinesterase by highly toxic organophosphorus compounds. *Biochem Pharmacol* 2006; 72:344–357.

Fang CX, Guo CR, Wu B, Jing YT. Effects of huperzine A on memory patients with schizophrenia. *Shandong Jing Shen Yi Xue* 2002; 15:39–40.

Fisher A, Brandeis R, Bar-Ner RH, Kliger-Spatz M, Natan N, Sonego H, Marcovitch I, Pittel Z. AF150(S) and AF267B: M1 muscarinic agonists as innovative therapies for Alzheimer's disease. *J Mol Neurosci* 2002; 19(1–2):145–153.

Gao Y, Tang XC, Guan LC, Kuang PZ. Huperzine A reverses scopolamine- and muscimol-induced memory deficits in chick. *Acta Pharmacol Sin* 2000; 21:1169–1173.

Gao X, Yu QP, Cao QH. A clinical trial of day and night alternated administration of huperzine A and clonazepam to treat chronic insomnia. *Chin J Nerv Ment Dis* 2003; 29:58–59.

Geib SJ, Tuckmantel W, Kozikowski AP. Huperzine A: A potent acetylcholinesterase inhibitor of use in the treatment of Alzheimer's disease. *Acta Crystallogr C* 1991; 47:824–827.

Giacobini E. Cholinesterase inhibitors: New roles and therapeutic alternatives. *Pharmacol Res* 2004; 50(4): 433–440.

Gordon RK, Haigh JR, Garcia GE, Feaster SR, Riel MA, Lenz DE, Aisen PS, Doctor BP. Oral administration of pyridostigmine bromide and huperzine A protects human whole blood cholinesterases from ex vivo exposure to soman. *Chem Biol Interact* 2005; 157–158, 239–246.

Grunwald J, Raveh L, Doctor BP, Ashani Y. Huperzine A as a pretreatment candidate drug against nerve agent toxicity. *Life Sci* 1994; 54:991–997.

Hanin I, Tang XC, Kindel GL, Kozikowski AP. Natural and synthetic huperzine A: Effect on cholinergic function in vitro and in vivo. *Ann N Y Acad Sci* 1993; 695:304–306.

Kelley BJ, Knopman DS. Alternative medicine and Alzheimer disease. *Neurologist* 2008; 14(5):299–306.

Kozikowski AP, Prakash KR, Saxena A, Doctor BP. Synthesis and biological activity of an optically pure 10-spirocyclopropyl analog of huperzine A. *J Chem Soc Chem Commun* 1998; 1287–1288.

Kozikowsky AP, Xia Y. A practical synthesis of the chinese "nootropic" agent huperzine A: A possible lead in the treatment of Alzheimer's disease. *J Am Chem Soc* 1989; 111:4116–4117.

Laganiere S, Corey J, Tang XC, Wülfert E, Hanin I. Acute and chronic studies with the anticholinesterase huperzine A: Effect on central nervous system cholinergic parameters. *Neuropharmacology* 1991; 30:763–768.

Lallement G, Baille V, Baubichon D, Carpentier P, Collombet JM, Filliat P, Foquin A, Four L, Masqueliez C, Testylier G, Tonduli L, Dorandeu F. Review of the value of huperzine as pretreatment of organophosphate poisoning. *Neurotoxicology* 2002a; 23(1):1–5.

Lallement G, Demoncheaux JP, Foquin A, Baubichon D, Galonnier M, Clarencon D, Dorandeu F. Subchronic administration of pyridostigmine or huperzine to primates: Compared efficacy against soman toxicity. *Drug Chem Toxicol* 2002b; 25:309–320.

Lallement G, Veyret J, Masqueliez C, Aubriot S, Bukhart MF, Bauichon D. Efficacy of huperzine in preventing soman-induced seizures, neuropathological changes and lethality. *Fundam Clin Pharmacol* 1997; 11:387–394.

Li J, Wu HM, Zhou RL, Liu GJ, Dong BR. Huperzine A for Alzheimer's disease. *Cochrane Database Syst Rev* 2008; (2):CD005592.

Liang Y, Tang X. Comparative studies of huperzine A, donepezil, and rivastigmine on brain acetylcholine, dopamine, norepinephrine, and 5-hydroxytryptamine levels in freely-moving rats. *Acta Pharmacol Sin* 2006; 27:1127–1136.

Liao JX, Chen L, Huang TS. Pilot trial of huperzine A to treat child language delay. *J Pediatr Pharm* 2002; 8:26–27.

Little JT, Walsh S, Aisen PS. An update on huperzine A as a treatment for Alzheimer's disease. *Expert Opin Investig Drugs* 2008; 17(2):209–215.

Liu L, Sun JX. Advances on study of organophosphate poisoning prevented by huperzine A. *Wei Sheng Yan Jiu* 2005; 34:224–226.

Liu JS, Yu CM, Zhou YZ, Han YY, Wu FW, Qi BF, Zhu YL. Study on the chemistry of huperzine A and B. *Acta Chim Sin* 1986a; 44:1035–1040.

Liu J, Zhang HY, Tang XC, Wang B, He XC, Bai DL. Effects of synthetic (–)-huperzine A on cholinesterase activities and mouse water maze performance. *Acta Pharmacol Sin* 1998; 19.413–416.

Liu JS, Zhu YL, Yu CM, Zhou YZ, Han YY, Wu FW, Qi BF. The structure of huperzine A and B, two new alkaloids exhibiting marked anticholinesterase activity. *Can J Chem* 1986b; 64:837–839.

Lleo A, Greenberg SM, Growdon JH. Current pharmacotherapy for Alzheimer's disease. *Annu Rev Med* 2006; 57:513–533.

Lu WH, Shou J, Tang XC. Improving effect of huperzine A on discrimination performance in aged rats and adult rats with experimental cognitive impairment. *Acta Pharmacol Sin* 1988; 9:11–15.

Ma X, Gang DR. The Lycopodium alkaloids. *Nat Prod Rep* 2004; 21:752–772.

Ma X, Tan C, Zhu D, Gang DR, Xiao P. HupA from Huperzia species: An ethnopharmacological review. *J Ethnopharmacol* 2007; 113:15–34.

Ma XC, Xin J, Wang HX, Zhang T, Tu ZH. Acute effects of huperzine A and tacrine on rat liver. *Acta Pharmacol Sin* 2003a; 24:247–250.

Ma JD, Zheng H, Wang YJ. Effect of huperzine A on the memory disorders of schizophrenic patients during rehabilitation period. *Chin J Health Psychol* 2003b; 11:340–341.

Ma YX, Zhu Y, Gu YD, Yu ZY, Yu SM, Ye YZ. Double-blind trial of huperzine-A (HUP) on cognitive deterioration in 314 cases of benign senescent forgetfulness, vascular dementia, and Alzheimer's disease. *Ann N Y Acad Sci* 1998; 854:506–507.

Mazurek A. An open label trial of huperzine A in the treatment of Alzheimer's disease. *Altern Ther* 1999; 5(2):97–98.

Mulzer J, Hogenauer K, Baumann K, Enz A. Synthesis and acetylcholinesterase inhibition of 5-desamino huperzine A derivatives. *Bioorg Med Chem Lett* 2001; 11:2627–2630.

Orgogozo JM, Tamches E, Wilkinson D, Todorova Yancheva S, Gagiano C, Grosgurin P, Porchet H, Scalfaro P. ZT-1 for the symptomatic treatment of mild to moderate Alzheimer's disease. *Neurobiology of Aging* 2006; 27:S16.

Ou LY, Tang XC, Cai JX. Effect of huperzine A on working memory in reserpine- or yohimbine-treated monkeys. *Eur J Pharmacol* 2001; 433:151–156.

Peng Y, Jiang L, Lee DY, Schachter SC, Ma Z, Lemere CA. Effects of huperzine A on amyloid precursor protein processing and beta-amyloid generation in human embryonic kidney 293 APP Swedish mutant cells. *J Neurosci Res* 2006; 84(4): 903–911.

Perry EK, Perry RH, Blessed G, Tomlison BE. Changes in brain cholinesterases in senile dementia of Alzheimer type. *Neuropathol Appl Neurobiol* 1978; 4(4): 273–277.

Pilotaz F, Masson P. Huperzine A: An acetylcholinesterase inhibitor with high pharmacological potential. *Ann Pharm Fr* 1999; 57:363–373.

Qian L, Ji R. A total synthesis of (±)-huperzine A. *Tetrahedron Lett* 1989; 30:2089–2090.

Rafii MS, Walsh S, Little JT, Behan K, Reynolds B, Ward, C, Jin S, Thomas R, Aisen PS. Alzheimer's disease cooperative study. A phase II trial of huperzine A in mild to moderate Alzheimer disease. *Neurology* April 2011; 76(16):1389–1394.

Raves ML, Harel M, Pang YP, Silman I, Kozikowski AP, Sussman JL. Structure of acetylcholinesterase complexed with the nootropic alkaloid, (–)-huperzine A. *Nat Struct Biol* 1997; 1:57–63.

Rocha, ES, Chebabo SR, Santos MD, Aracava Y, Albuquerque EX. An analysis of low level doses of cholinesterase inhibitors in cultured neurons and hippocampal slices of rats. *Drug Chem Toxicol* 1998; 21(Suppl 1):191–200.

Roman S, Vivas NM, Badia A, Clos MV. Interaction of a new potent anticholinesterasic compound (+/–)huprine X with muscarinic receptors in rat brain. *Neurosci Lett* 2002; 325(2):103–106.

Saxena A, Quan N, Kovach IM, Qian N, Kozikowski AP, Pang YP, Vellom DC, Radić Z, Quinn D, Taylor P, Doctor BP. Identification of amino acid residues involved in the binding of huperzine A to cholinesterase. *Protein Sci* 1994; (3):1770–1778.

Scott LJ, Gao KL. Galantamine: A review of its use in Alzheimer's disease. *Drugs* 2000; 60(5):1095–1122.

Shang YZ, Ye JW, Tang XC. Improving effects of huperzine A on abnormal lipid peroxidation and superoxide dismutase in aged rats. *Acta Pharmacol Sin* 1999; 20(9):824– 828.

Shigeta K, Ootaki K, Tatemoto H, Nakanishi T, Inada A, Muto N. Potentiation of nerve growth factor-induced neurite outgrowth in PC12 cells by a *Coptidis rhizoma* extract and protoberberine alkaloids. *Biosci Biotechnol Biochem* 2002; 66:2491–2494.

Sun QQ, Xu SS, Pan JL, Guo HM, Cao WQ. Efficacy of huperzine A capsules on memory and learning performance in 34 pairs of matched junior middle school students. *Acta Pharmacol Sin* 1999; 20:601–603.

Tang L Wang R, Tang XC. Effects of huperzine A on secretion of nerve growth factor in cultured rat cortical astrocytes and neurite outgrowth in rat PC12 cells. *Acta Pharmacol Sin* 2005; 26(6):673–678.

Tang XC, De Sarno P, Sugaya K, Giacobini E. Effect of huperzine A, a new cholinesterase inhibitor, on the central cholinergic system of the rat. *J Neurosci Res* 1989; 24:276–285.

Tang XC, Han YF. Pharmacological profile of huperzine A, a novel acetylcholinesterase inhibitor from Chinese herb. *CNS Drug Rev* 1999; 5:281–300.

Tang XC, Kindel GH, Kozikowski AP, Hanin I. Comparison of the effects of natural and synthetic huperzine A on rat brain cholinergic function in vitro and in vivo. *J Ethnopharmacol* 1994a; 44:147–155.

Tang XC, Xiong ZQ, Qian BC, Zhou ZF, Zhang CC. Cognitive improvement by oral huperzine A: a novel acetylcholinesterase inhibitor. In: Giacobini E and Becker R, eds. *Alzheimer Therapy: Therapeutic Strategies*. Boston, MA: Birkhauser 1994b; pp.113–119.

Ved HS, Koenig ML, Dave JR, Doctor BP. Huperzine A, a potential therapeutic agent for dementia, reduces neuronal cell death caused by glutamate. *Neuroreport* 1997; 8:963–968.

Wang H, Tang XC. Anticholinesterase effects of huperzine A, E2020, and tacrine in rats. *Acta Pharmacol Sin* 1998a; 19:27–30.

Wang T, Tang XC. Reversal of scopolamine-induced deficits in radial maze performance by (–)-huperzine A: Comparison with E2020 and tacrine. *Eur J Pharmacol* 1998b; 349:137–142.

Wang R, Yan H, Tang XC. Progress in studies of huperzine A, a natural cholinesterase inhibitor from Chinese herbal medicine. *Acta Pharmacol Sin* 2006; 27:1–26.

Wang YE, Yue DX, Tang XC. Anticholinesterase activity of huperzine A. *Acta Pharmacol Sin* 1986; 7:110–113.

Wang XD, Zhang JM, Yang HH, Hu GY. Modulation of NMDA receptor by huperzine A in rat cerebral cortex. *Chung Kuo Yao Li Hsueh Pao* 1999; 20:31–35.

Wang L, Zhou J, Shao X, Tang X. Huperzine A attenuates cognitive deficits and brain injury in neonatal rats after hypoxia-ischemia. *Brain Res* 2002; 949(1–2):162.

Wang BS, Wang H, Wei ZH, Song YY, Zhang L, Chen HZ. Efficacy and safety of natural acetylcholinesterase inhibitor huperzine A in the treatment of Alzheimer's disease: An updated meta-analysis. *J Neural Transm* 2009; 116(4):457–465.

Xiong ZQ, Cheng DH, Tang XC. Effects of huperzine A on nucleus basalis magnocellularis lesion-induced spatial working memory deficit. *Acta Pharmacol Sin* 1998; 19:128–132.

Xiong ZQ, Tang XC. Effects of huperzine A, a novel acetylcholinesterase inhibitor, on radial maze performance in rats. *Pharmacol Biochem Behav* 1995; 51:415–419.

Xu SS, Cai ZY, Qu ZW, Yang RM, Cai YL, Wang GQ. Huperzine-A in capsules and tablets for treating patients with Alzheimer's disease. *Acta Pharmacol Sin* 1999; 20:486–490.

Xu SS, Cai ZY, Qu ZW, Yang RM, Cai YL, Wang GQ, Su XQ, Zhong XS, Cheng RY, Xu WA, Li JX, Feng B. Efficacy of tablet huperzine-A on memory, cognition and behavior in Alzheimer's disease. *Acta Pharmacol Sin* 1995; 16:391–395.

Xu, H, Tang XC. Cholinesterase inhibition by huperzine B. *Acta Pharmacol Sin* 1987; 8:18–22.

Yan XF, Lu WH, Lou WJ, Tang XC. Effects of HupA and B on skeletal muscle and electroencephalogram. *Acta Pharmacol Sin* 1987; 8:117–123.

Yang JZ. The effects of huperzine A on the cognitive deficiency of rehabilitating schizophrenia. *Chin J Clin Rehabil* 2003; 7:1440.

Ye JW, Cai JX, Wang LM, Tang XC. Improving effects of huperzine A on spatial working memory in aged monkeys and young adult monkeys with experimental cognitive impairment. *J Pharmacol Exp Ther* 1999; 288:814–819.

Ye L, Qiao JT. Suppressive action produced by beta-amyloid peptide fragment 31–35 on long-term potentiation in rat hippocampus is N-methyl-D-aspartate receptor-independent: It's offset by (−) huperzine A. *Neurosci Lett* 1999; 275:187–190.

Zangara A. The psychopharmacology of huperzine A: An alkaloid with cognitive enhancing and neuroprotective properties of interest in the treatment of Alzheimer's disease. *Pharmacol Biochem Behav* 2003; 75(3):675–686.

Zangara A, Edgar C, Wesnes K, Scalfaro P, Porchet E. Reversal of scopolamine-related deficits in cognitive functions by ZT-1, a Huperzine A precursor. *9th International Geneva/ Springfield Symposium on Advances in Alzheimer Therapy*, April 19–22, 2006, Geneva, Switzerland.

Zenghong T, Meiying W. Mutagenicity and comutagenicity of three nootropics: Huperzine A, aniracetam and piracetam. *New Drugs Clin Remedies* 1990; 9:65–68.

Zhang SL. Therapeutics effects of huperzine A on the aged with memory impairments. *New Drugs Clin Remedies* 1986; 5:260–262.

Zhang RW, Tang XC, Han YY, Sang GW, Zhang YD, Ma YX, Zhang CL, Yang RM. Drug evaluation of huperzine A in the treatment of senile memory disorders. *Acta Pharmacol Sin* 1991; 12:250–252.

Zhang HY, Tang XC. Huperzine B, a novel acetylcholinesterase inhibitor, attenuates hydrogen peroxide induced injury in PC12 cells. *Neurosci Lett* 2000; 292(1):41–44.

Zhang Z, Wang X, Chen Q, Shu L, Wang J, Shan G. Clinical efficacy and safety of huperzine Alpha in treatment of mild to moderate Alzheimer disease, a placebo-controlled, double-blind, randomized trial. *Natl Med J China* 2002; 82(14):941–944.

Zhang HY, Yan H, Tang XC. Non-cholinergic effects of huperzine A: Beyond inhibition of acetylcholinesterase. *Cell Mol Neurobiol* 2008; 28(2):173–183.

Zhang ML, Yuan G, Yao J, Bi P, Ni S. Evaluation of clinical effect and safety of huperzine A in treating 52 Alzheimer's disease. *Chin J New Drugs Clin Remedies* 2006; 25:693–695.

Zhao HW, Li XY. Ginkgolide, A, B, and huperzine A inhibit nitric oxide production from rat C6 and human BT325 glioma cells. *Acta Pharmacol Sin* 1999; 20(10):941–943.

Zhou J, Fu Y, Tang XC. Huperzine A protects rat pheochromocytoma cells against oxygen-glucose deprivation. *Neuroreport* 2001; 12:2073–2077.

Zhou BR, Xu ZQ, Kuang YF, DengYH, Liu ZF. Effectiveness of polydrug therapy for senile dementia. *Chin J Clin Rehabil* 2004; 8:1214–1215.

Zhu XD, Giacobini E. Second generation cholinesterase inhibitors: effect of (L)-huperzine A on cortical biogenic amines. *J Neurosci Res* 1995; 41:828–835.

Zhu XD, Tang XC. Improvement of impaired memory in mice by huperzine A and huperzine B. *Acta Pharmacol Sin* 1988; 9:492–497.

16 Potential Protective Effects of Wine and the Wine-Derived Phenolic Compounds on Brain Function

Creina S. Stockley

CONTENTS

INTRODUCTION

Damage to DNA, lipids, and proteins by oxidative free radicals has been implicated in accelerated aging, degenerative diseases including cancer (Ames et al. 1995), Alzheimer's disease and other dementias (Commenges et al. 2000; Smith and Perry 1995), and Parkinson's disease (Olanow 2007), as well as cardiovascular disease. Population aging is occurring on a global scale, with faster aging projected for the coming decades than has occurred in the past (Lutz et al. 2008). Globally, the population aged 60 years and over is projected to nearly triple by 2050, while the population aged 80 years and over is projected to experience a more than fivefold increase. These diseases of old age are thus expected to increase significantly over the next few decades as people increasingly survive beyond the age of 80 years (Kelner and Marx 1996). Consequently, there is interest in identifying lifestyle factors and molecular mechanisms that can minimize the risk of these debilitating conditions, including simple dietary measures.

J-SHAPED RELATIONSHIP OF ALCOHOLIC BEVERAGES

From prospective population-based studies, there is a clear J-shaped relationship between the consumption of alcoholic beverages such as wine, and the risk of cardiovascular diseases including myocardial infarction, which has been extended to a reduced risk of certain cancers, type 2 diabetes, and ischemic stroke (Bantle et al. 2008; Barstad et al. 2005; Benedetti et al. 2006; Booyse and Parks 2001; Briggs et al. 2002; Gronbaek et al. 2000; McDougall et al. 2006; Park et al. 2009; Pedersen et al. 2003; Wannamethee et al. 2002, 2003). The moderate consumption of alcoholic beverages may reduce the risk of cardiovascular diseases, for example, by ~35%, that of type 2 diabetes by ~30%, and that of ischemic stroke by 20%–28%; the risk of hemorrhagic stroke is relatively unaffected by moderate alcohol consumption. Over the last decade, evidence has accumulated which suggests that this J-shaped relationship could also be extended to a reduced risk of cognitive dysfunction, and dementias such as Alzheimer's disease, and vascular dementia (Dufouil et al. 1997; Huang et al. 2002; Lindsay et al. 2002; Luchsinger et al. 2004; Rasmussen et al. 2006; Simons et al. 2000; Zuccala et al. 2001). Mild cognitive dysfunction or impairment is a prodrome for Alzheimer's disease.

While the literature defines consistently light-to-moderate consumption as 20–40 g ethanol per day (Jackson et al. 1992; Klatsky 2003; NHMRC 2005; Palomaki and Kaste 1993), several studies have defined moderate consumption as up to 80 g ethanol per day for men (Elias et al. 1999; Zuccala et al. 2001), which may reflect country and cultural differences in alcohol consumption; 10 g ethanol approximates one drink or 100 mL wine. Above moderate consumption, the risk of alcohol-related diseases increases dose-dependently (Corrao et al. 2004).

RELATIONSHIP OF ALCOHOLIC BEVERAGES TO COGNITIVE FUNCTION AND DEMENTIA

Cognitive function is defined as the intellectual or mental processes by which knowledge is acquired, including perception, reasoning, acts of creativity, problem solving, and possible intuition. Cognitive dysfunction or impairment is associated with increased disability and an increased need for institutionalized care. Dementia is a form of cognitive dysfunction whereby an individual loses the ability to think, remember, and reason due to physical changes in the brain.

Prior to a study by Zuccala et al. (2001), there was conflicting evidence on the relationship between alcohol consumption per se and cognitive function (Cervilla et al. 2000; Dent et al. 1997; Dufouil et al. 1997; Elias et al. 1999; Harwood et al. 1999; Hendrie et al. 1996; Leibovici et al. 1999; Teri et al. 1990). Zuccala et al. (2001) analyzed the association between alcohol consumption and cognitive impairment in 15,807 hospitalized older patients who were enrolled in an Italian multicenter pharmacoepidemiology survey. The amount of alcohol use was recorded as daily wine units (100 mL or ~10 g ethanol), because wine, particularly with meals, represents the major form of alcohol consumption in this Italian population. The probability of cognitive impairment was reduced among male patients who reported an average daily alcohol consumption of 1 L or less of wine, as compared with abstainers, but the probability increased among heavier drinkers. Among women, only the lightest-drinking category (<500 mL/day) showed a decreased probability of cognitive dysfunction when compared with abstainers, whereas heavier drinking was associated with an increased probability of cognitive impairment. The prevalence of alcohol abuse was similar among participants with cognitive impairment and those with normal cognitive functioning. The results of this study indicated that less than four drinks per day for women and less than eight drinks for men was associated with reduced probability of cognitive impairment as compared with abstinence, after adjusting for potential confounders supportive of other studies. This nonlinear association persisted when cerebrovascular diseases and Alzheimer's disease were considered separately. Such a nonlinear association might explain the conflicting results of previous studies regarding the relationship between alcohol consumption and cognitive functioning.

The observed gender difference in amount of alcohol consumption necessary for improved cognitive function confirms that observed by Elias et al. (1999), who showed that "superior" cognitive performance was found within the range of four to eight drinks per day for men but only two to four drinks per day for women, compared to abstainers.

Subsequent studies have also independently assessed the association between alcohol consumption and cognitive function, and have affirmed the observations of Zuccala et al. (2001) but have also provided more detailed data (Ganguli et al. 2005; McDougall et al. 2006; Reid et al. 2006; Stampfer et al. 2005; Wright et al. 2006). For example, both current moderate consumption and cumulative lifetime alcohol consumption are associated with better cognitive function compared to abstainers. This encompasses processing speed, which is the ability to perform tasks requiring rapid visual scanning and mental processing of information, memory such as verbal knowledge or memory including immediate and delayed recall, recognition memory, figural memory, and working memory, as well as motor speed. This has been observed for both men and women (Bond et al. 2005; Cho et al. 2000; Espeland et al. 2006; Stampfer et al. 2005).

As mentioned earlier, mild cognitive dysfunction is a prodrome for dementia, and in particular Alzheimer's disease, which is a complex, late-onset disorder characterized by the loss of memory and multiple cognitive functions. In patients with mild cognitive dysfunction, consuming up to 15 g ethanol per day now appears to also decrease the rate of progression to dementia by ~85%, while 10–30 g alcohol per day reduces the risk of Alzheimer's disease and vascular dementia (Huang et al. 2002; Mukamal et al. 2003; Ruitenberg et al. 2002).

RELATIONSHIP OF WINE TO COGNITIVE FUNCTION AND DEMENTIA

Moderate wine consumption rather than alcohol consumption per se has been specifically associated with a lower risk of developing dementia and specifically Alzheimer's disease (Deng et al. 2006; Huang et al. 2002; Larrieu et al. 2004;

Leibovici et al. 1999; Lindsay et al. 2002; Luchsinger et al. 2004; Mehlig et al. 2008; Mukamal et al. 2003; Orgogozo et al. 1997; Ruitenberg et al. 2002; Simons et al. 2006; Truelsen et al. 2002). From the PAQUID study of 3777 subjects aged 65 years and older who were followed for 3 years, in the 922 subjects drinking between 125 and 250 mL/day the odds ratio was 0.55 for Alzheimer's disease. In the 318 subjects drinking between 250 and 500 mL/day, however, the odds ratio was 0.19 for incident dementia and 0.28 for Alzheimer's disease compared to the 971 nondrinkers after adjusting for age, sex, education, occupation, and other possible confounders. In the Washington Heights Inwood-Columbia Aging Project, 980 community-dwelling individuals aged 65 and older without dementia at baseline were recruited between 1991 and 1996 and followed annually (Luchsinger et al. 2004). After 4 years of follow-up, 260 individuals developed dementia and, of these, 199 developed Alzheimer's disease. After adjusting for age, sex, apolipoprotein E (APOE)-epsilon 4 status, education, and other alcoholic beverages, only consumption of up to 33 g ethanol per day as wine was associated with a lower risk of Alzheimer's disease.

A primary difference between wine and the other alcoholic beverages is that wine contains phenolic compounds similar to those contained in fruits, vegetables, and teas, the consumption of which has also been associated a lower incidence of both mild cognitive dysfunction, dementias such as Alzheimer's disease, and other cerebrovascular/neurodegenerative diseases (Frisardi et al. 2010; Solfrizzi et al. 2011).

POTENTIAL MECHANISMS OF ACTION FOR ETHANOL AND WINE-DERIVED PHENOLIC COMPOUNDS

More than 500 compounds have been identified in *Vitis vinifera* grapes and wine to date (Rapp and Pretorius 1989; Schreier 1979). Wine typically contains alcohols such as methyl, ethyl, *n*-propyl, isopropyl, isobutyl, isoamyl, act-amyl, 2-phenethanol, *n*-hexanol as well as detectable amounts of ~18 other alcohols, where the most abundant alcohol is ethyl alcohol or ethanol (Rapp and Mandery 1986). The concentration of ethanol in "table" wine generally ranges between 8% and 15% v/v (Rankine 1989). Wine also typically contains phenolic compounds and their polymeric forms and the total amount of phenolic compounds in a 100 mL glass of red wine is ~200 mg versus 40 mg in a glass of white wine (Rankine 1989). Chemically, phenolic compounds are cyclic benzene compounds possessing one or more hydroxyl groups associated directly with an aromatic ring structure (Figure 16.1a and b). Wine-derived phenolic compounds include the nonflavonoid classes of compounds such as hydroxycinnamates, hydroxybenzoates, and stilbenes such as resveratrol, as well as the more abundant flavonoid classes of compounds: flavan-3-ols such as catechin, flavonols such as quercetin, and anthocyanins. Of these, resveratrol appears to have been the most widely examined phenolic compound over the past decade. While polymeric condensed tannins and pigmented tannins constitute the majority of red wine phenolic compounds, their large size precludes absorption and they are thus unlikely to contribute to any biological mechanism (Waterhouse 2002). Data from animal studies suggest that grape- and wine-derived phenolic compounds

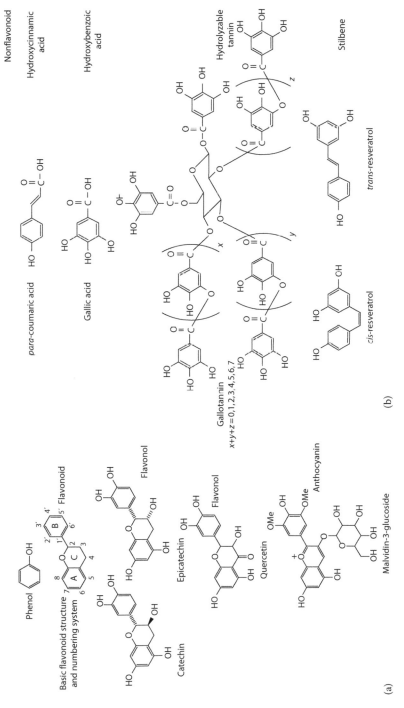

FIGURE 16.1 Examples of structure of flavonoid and non-flavonoid phenolic compounds.

are absorbed and accumulate in the brain in measurable amounts after multiple or repeated oral doses (Ferruzzi et al. 2009; Passamonti et al. 2005). Wine-derived phenolic compounds, and particularly resveratrol, have been shown to be cerebero- or neuroprotective in various models, in vitro and in vivo, and potential mechanisms have been proposed as follows. Data from similar studies using different varieties of red wines with different profiles of phenolic compounds, as well as studies comparing different phenolic compounds, suggest that the individual classes of phenolic compounds may exhibit differential effects in the brain (Hamaguchi et al. 2009; Ho et al. 2009).

HEMOSTASIS AND OXIDATIVE STRESS

The beneficial effects of alcoholic beverages such as wine on the risk of cardiovascular and cerebrovascular diseases have been partly attributed to changes in lipid and hemostatic or blood flow factors. These changes include ethanol-induced increases in the concentration of high-density lipoprotein-cholesterol, and ethanol- and phenolic-induced increases in the thrombolyic proteins tissue-type plasminogen activator activity and tissue-type plasminogen activator antigen, and induced reductions in fibrinogen, and clotting cofactors factor VII and von Willebrand factor. These changes are also associated with atherosclerosis which is the accumulation of atheromatous plaques containing cholesterol and lipids on the innermost layer of the walls of large- and medium-sized arteries. As atherosclerosis has been associated with both Alzheimer's disease and vascular dementia, it had been suggested that any beneficial effect of wine on atherosclerosis could be expected to benefit these dementias by preserving brain vasculature, consequently resulting in better cognitive function (Wright et al. 2006), however, showed that the appearance of plaque on the carotid artery which carries blood to the brain was not associated with consumption of an alcoholic beverage and associated improvements in cognitive function. This suggests then that ethanol and phenolic compounds such as resveratrol may impact cognition through a separate vascular or degenerative pathway (Kennedy et al. 2010). Vasoactive amyloid-β (Aβ), associated with Alzheimer's disease and other related neurodegenerative diseases, may also interact with cerebral blood vessels to promote free radical production and reduce local blood flow which precede other neuropathological changes in dementias, and subsequently up-regulate Aβ production (Thomas et al. 1996). Indeed, among older persons without cerebrovascular and neurodegenerative diseases, those who moderately consume alcoholic beverages such as wine have been shown to have fewer white-matter abnormalities and infarcts on magnetic resonance imaging than abstainers (Mukamal et al. 2001), where pronounced reductions in the risk of both vascular dementia and Alzheimer's disease have been shown among persons consuming one to six standard drinks per week (Mukamal et al. 2003).

A lack of heme oxygenase 1, an endogenous enzyme that is induced in neurons in response to oxidative and other stress and stimulates the degradation of prooxidant heme into free iron, carbon monoxide, and biliverdin and/or the antioxidative bilirubin, may also be associated with increased neural damage from ischemic strokes (Li et al. 2009), as well with Alzheimer's disease and Parkinson's disease

(Ma et al. 2010). Heme oxygenase 1 is dose- and time-dependently induced by res-veratrol, which may provide another cerebrovascular and neuroprotective effect for phenolic compounds.

Indeed, Parkinson's disease has been linked to increased levels of oxidative and nitrosative stress (Chung et al. 2003; Dawson and Dawson 2003) and is characterized by a progressive loss of dopaminergic neurons in the substantia nigra pars compacta region of the brain and the appearance of Lewy bodies and neurites, which comprise insoluble amyloid-like fibrils that contain the protein α-synuclein. Oxidative stress apparently promotes the aggregation of α-synuclein (Maguire-Zeiss et al. 2005). An inverse relationship between amount of wine consumed and risk has been observed where the lowest risk was observed for wine consumers of ~140–420 g/week (Fall et al. 1999). Wine-derived phenolic compounds such as catechin and epicatechin have recently been observed in vitro to inhibit the formation of α-synuclein fibrils, and to destabilize preformed fibrils (Ono et al. 2008).

NEUROTRANSMISSION

Cognition is also associated with acetylcholine. Cognitive decline associated with dementias, Huntington's disease, and Parkinson's disease, as well as with Down's syndrome and multiple sclerosis, is characterized neurochemically by a consistent deficit in cholinergic neurotransmission, in particular in the cholinergic neurons in the basal forebrain. Inhibition of acetylcholinesterase which restores cholinergic neurotransmission also appears to prevent the aggregation of Aβ peptides and for-mation of amyloid fibrillar plaques (Munoz et al. 1999).

Ethanol may stimulate the release of acetylcholine in the hippocampus leading to improved cognitive function, such that a light amount of an alcoholic beverage in normal subjects appears to improve memory for events experienced before consump-tion (Fadda and Rossetti 1998). In contrast, quercetin inhibits acetylcholinesterase (Orhan et al. 2007). Indeed, acetylcholinesterase inhibitors which reversibly bind and inactivate the enzyme that degrades acetylcholine are the primary medications prescribed associated with mild improvements in cognitive function. The impair-ment of memory performance by chronic and heavy consumption, however, parallels the reduction of acetylcholine neurotransmission.

Concerning Alzheimer's disease, which is associated with the presence of intra-cellular neurofibrillary tau tangles, extracellular Aβ peptides, synaptic failure, mitochondrial dysfunction, and depletion of acetylcholine (Anekonda and Reddy 2006), it has been suggested that ethanol may directly stimulate the release of ace-tylcholine in the hippocampus; synaptic levels of acetylcholine decrease as a result of cholinergic neuron involvement. In a rat model, a moderate concentration of etha-nol (0.8 g/kg) stimulated the release of acetylcholine while a higher concentration (2.4 g/kg) inhibited its release. The formation of amyloid fibrillar plaques is also common in diseases such as Parkinson's disease, prion diseases, Down's syndrome, and type II diabetes.

Another important intracellular signaling system involved in learning and mem-ory is protein kinase C (PKC), a family of 12 serine/threonine kinases. PKC modu-lates cell viability which protects certain neuronal cells against Aβ-induced toxicity.

Resveratrol has been observed to protect hippocampal cells against AB-induced toxicity by activating PKC (Han et al. 2004), and specific binding sites for resveratrol have been identified in the rat brain (Han et al. 2006), as have receptors for the green tea catechin gallates.

AMYLOID-β FACTORS AND ALZHEIMER'S DISEASE

Aβ is a core component of the plaque or lesion found in the neocortex and hippocampus of diseased brains. It is formed after sequential proteolytic cleavage of the amyloid precursor protein (APP), a transmembrane glycoprotein. APP can be processed by α, β, and γ secretases. Unlike α secretase which cleaves APP into nontoxic amyloid-α, the toxic amyloid-β protein is generated by successive action of the β and γ secretases. The γ secretase, which produces the C-terminal end of the Aβ peptide, cleaves within the transmembrane region of APP and can generate a number of isoforms of 39–43 amino acid residues in length. The most common isoforms are $A\beta_{40}$ and $A\beta_{42}$; the shorter form is typically produced by cleavage that occurs in the endoplasmic reticulum, while the longer form is produced by cleavage in the trans-Golgi network. The $A\beta_{40}$ form is the more common of the two, but $A\beta_{42}$ is the more fibrillogenic or polymeric and is thus associated with disease states, promoting proinflammatory responses and activating neurotoxic pathways leading to neuronal dysfunction, and the death and loss of neurons. Inhibition of the accumulation of amyloid-β peptides and the formation of Aβ fibrils/plaques from amyloid-β peptides, as well as the destabilization of preformed Aβ fibrils/plaques in the brain would, therefore, be attractive therapeutic targets for the treatment of Alzheimer's disease and other related neurodegenerative diseases.

As mutations in APP associated with early-onset Alzheimer's disease have been noted to increase the relative production of $A\beta_{42}$, a potential therapy may involve modulating the activity of β and γ secretases to produce mainly $A\beta_{40}$. Administration of a red wine to Tg2576 mice equivalent to two standard drinks, which model Alzheimer's disease Aβ neuropathology and corresponding cognitive deterioration, has been shown to promote the nonamyloidogenic processing of the APP, which acts to prevent the generation of the Aβ peptide (Wang et al. 2006). For example, administration of red wine reduced amyloidogenic $A\beta_{(1-40)}$ and $A\beta_{(1-42)}$ peptides in the neocortex and hippocampus of Tg2576 mice and correspondingly decreased the neocortical Alzheimer's disease–associated amyloid fibrils/plaque. Subsequent examination of APP processing and Aβ peptide generation increased the concentration of membrane-bound α-carboxyl terminal fragments of APP in the neocortex and α secretase activity was also increased, while there was no significant change in the neocortical concentration of β and γ carboxyl terminal fragments of APP or in β and γ secretases activity.

The typical red wine–derived phenolic compounds catechin, quercetin, epicatechin, myricetin, and tannic acid have, however, been shown in vitro to dose-dependently inhibit the formation of Aβ fibrils from fresh $A\beta_{(1-40)}$ and $A\beta_{(1-42)}$, as well as their extension, and also dose-dependently destabilized preformed Aβ fibrils (Ono et al. 2003, 2004; Roth et al. 1999). Only resveratrol, however, has been shown to decrease the level of intracellular Aβ produced by different cell lines expressing the wild type of Swedish mutant Aβ precursor protein (APP_{695}) by promoting its intracellular degradation (Marambaud et al. 2005). This mechanism was proteasome-dependent, that is, resveratrol appears to activate the proteasome involved in the degradation of Aβ, as the resveratrol-induced decrease of Aβ could be prevented by several selective proteasome inhibitors and by siRNA-directed silencing of the proteasome subunit β5. Resveratrol does not inhibit the production of Aβ because it has no effect on β and γ secretase activity.

Another potential therapy may involve preventing the aggregation of Aβ, as studies have suggested that only when aggregated in the fibrillar form Aβ is neurotoxic, although some studies alternatively suggest that the toxicity lies in soluble oligomeric intermediates rather than in the insoluble fibrils that accumulate. Grape-seed phenolic compound extract, which comprises primarily catechin and epicatechin in monomeric, oligomeric, and polymeric forms, has been shown in vitro and in animal studies to inhibit Aβ aggregation into high-molecular-weight oligomers (Wang et al. 2008). This inhibition coincided with attenuation of Alzheimer's disease–type cognitive impairment. Resveratrol and its glucoside, piceid, have, however, been shown in vitro to dose-dependently inhibit the formation of Aβ fibrils (Riviere et al. 2007, 2010); other stilbene monomers examined are less potent inhibitors. Binding may be induced by hydrophobic interactions between the phenolic rings and the hydrophobic region of Aβ, thus blocking associations between Aβ molecules and inhibiting fibril formation. These interactions may then be reinforced by the H-bond between the hydroxyl group of the phenolic rings and some donor/acceptor groups of Aβ, as observed for other peptides.

AMYLOID-β AND APOE-EPSILON 4 ALLELE

In the Washington Heights Inwood-Columbia Aging Project (1991–1996), however, a lower risk was confined to wine consumers without the APOE-epsilon 4 allele (Luchsinger et al. 2004). This allele is implicated in atherosclerosis, Alzheimer's disease, and impaired cognitive function, possibly influencing the increased deposition of Aβ in the brain as it is less effective compared to other alleles at facilitating the proteolytic breakdown of this peptide, both within and between cells (Selkoe 1997).

Indeed, the most important genetic risk factor is the ApoE genotype. ApoE is a protein that carries lipids in and out of cells. It occurs in three isoforms: ApoE2, ApoE3, and ApoE4. The gene for ApoE is on chromosome 19. One copy is inherited from each parent. The most common ApoE allele is ApoE3. Persons who are homozygous for the ApoE4 allele develop Alzheimer's disease earlier at a mean age of 70 years. The ApoE4 allele is also a risk factor for hypercholesterolemia. ApoE4 has been detected in neurofibrillary tangles and in Aβ (Luchsinger et al. 2004). This suggests that ApoE lipoproteins participate in some way in the processing of APP, perhaps by modulating APP secretases, and may also play a role in the assembly of the neuronal cytoskeleton. High cholesterol levels during midlife increase the risk of Alzheimer's disease and lipid-lowering therapies including the consumption of wine-derived phenolic compounds can lower this risk which may accordingly lower the risk of developing Alzheimer's disease (Andrade et al. 2009; Broncel et al. 2007; Franiak-Pietryga et al. 2009).

SUMMARY

Plausible biological mechanisms in various animal models in vitro and in vivo for the wine-derived phenolic compounds support a cerebro- or neuroprotective role for catechin, quercetin, and resveratrol, in particular, which go beyond their antioxidant activity and attenuation of oxidative stress. More research is required on their intracellular and molecular targets and hence protection, and it would be unwise to extrapolate these results to humans without longer-term clinical studies in patients experiencing extensive neuronal loss associated with dementias and other neurodegenerative diseases. Considering the extensive plausible biological mechanisms, these compounds provide promise as therapeutic or prophylactic agents in neurodegenerative diseases.

REFERENCES

Ames BN, Gold LS, Willett WC (1995) The causes and prevention of cancer. *Proceedings of the National Academy of Sciences of the Untied States of America* 92:5258–5265.

Andrade ACM, Cesena FHY, Consolim-Colombo FM, Coimbra SR, Benjo AM, Krieger EM, da Luz PL (2009) Short-term red wine consumption promotes differential effects on plasma levels of high-density lipoprotein cholesterol, sympathetic activity, and endothelial function in hypercholesterolemic, hypertensive, and healthy subjects. *Clinics* 64:435–442.

Anekonda TS and Reddy PH (2006) Neuronal protection by sirtuins in Alzheimer's disease. *Journal of Neurochemistry* 96:305–313.

Bantle AE, Thomas W, Bantle JP (2008) Metabolic effects of alcohol in the form of wine in persons with type 2 diabetes mellitus. *Metabolism—Clinical and Experimental* 57:241–245.

Barstad B, Sorensen TIA, Tjonneland A, Johansen D, Becker U, Andersen IB, Gronbaek M (2005) Intake of wine, beer and spirits and risk of gastric cancer. *European Journal of Cancer Prevention* 14:239–243.

Benedetti A, Parent ME, Siemiatycki J (2006) Consumption of alcoholic beverages and risk of lung cancer: Results from two case–control studies in Montreal, Canada. *Cancer Causes & Control* 17:469–480.

Bond GE, Burr RL, McCurry SM, Rice MM, Borenstein AR, Larson EB (2005) Alcohol and cognitive performance: A longitudinal study of older Japanese Americans. The Kame Project. *International Psychogeriatrics/IPA* 17:653–668.

Booyse FM, Parks DA (2001) Moderate wine and alcohol consumption: Beneficial effects on cardiovascular disease. *Thrombosis & Haemostasis* 86:517–528.

Briggs NC, Levine RS, Bobo LD, Haliburton WP, Brann EA, Hennekens CH (2002) Wine drinking and risk of non-Hodgkin's lymphoma among men in the United States: A population-based case–control study. *American Journal of Epidemiology* 156:454–462.

Broncel M, Franiak I, Koter-Michalak M, Duchnowicz P, Chojnowska-Jezierska J (2007) The comparison in vitro the effects of pravastatin and quercetin on the selected structural parameters of membrane erythrocytes from patients with hypercholesterolemia. *Pol Merkur Lekarski* 22:112–116.

Cervilla JA, Prince M, Mann A (2000) Smoking, drinking, and incident cognitive impairment: A cohort community based study included in the Gospel Oak project. *Journal of Neurology, Neurosurgery & Psychiatry* 68:622–626.

Cho E, Hankinson SE, Willett WC, Stampfer MJ, Spiegelman D, Speizer FE, Rimm EB, Seddon JM (2000) Prospective study of alcohol consumption and the risk of age-related macular degeneration. *Archives of Ophthalmology* 118:681–688.

Chung KKK, Dawson VL, Dawson TM (2003) New insights into Parkinson's disease. *Journal of Neurology* 250:15–24.

Commenges D, Scotet V, Renaud S, Jacqmin-Gadda H, Barberger-Gateau P, Dartigues JF (2000) Intake of flavonoids and risk of dementia. *European Journal of Epidemiology* 16:357–363.

Corrao G, Bagnardi V, Zambon A, La Vecchia C (2004) A meta-analysis of alcohol consumption and the risk of 15 diseases. *Preventive Medicine* 38:613–619.

Dawson TM, Dawson VL (2003) Molecular pathways of neurodegeneration in Parkinson's disease. *Science* 302:819–822.

Deng J, Zhou DH, Li J, Wang YJ, Gao C, Chen M (2006) A 2-year follow-up study of alcohol consumption and risk of dementia. *Clinical Neurology and Neurosurgery* 108:378–383.

Dent OF, Sulway MR, Broe GA, Creasey H, Kos SC, Jorm AF, Tennant C, Fairley MJ (1997) Alcohol consumption and cognitive performance in a random sample of Australian soldiers who served in the Second World War. *BMJ: British Medical Journal* 314:1655–1657.

Dufouil C, Ducimetiere P, Alperovitch A (1997) Sex differences in the association between alcohol consumption and cognitive performance. *American Journal of Epidemiology* 146:405–412.

Elias PK, Elias MF, D'Agostino RB, Silbershatz H, Wolf PA (1999) Alcohol consumption and cognitive performance in the Framingham Heart Study. *American Journal of Epidemiology* 150:580–589.

Espeland MA, Coker LH, Wallace R, Rapp SR, Resnick SM, Limacher M, Powell LH, Messina CR (2006) Association between alcohol intake and domain-specific cognitive function in older women. *Neuroepidemiology* 27:1–12.

Fadda F, Rossetti ZL (1998) Chronic ethanol consumption: From neuroadaptation to neurodegeneration. *Progress in Neurobiology* 56:385–431.

Fall PA, Fredrikson M, Axelson O, Granerus AK (1999) Nutritional and occupational factors influencing the risk of Parkinson's disease: A case–control study in southeastern Sweden. *Movement Disorders* 14:28–37.

Ferruzzi MG, Lobo JK, Janle EM, Cooper B, Simon JE, Wu QL, Welch C, Ho L, Weaver C, Pasinetti GM (2009) Bioavailability of gallic acid and catechins from grape seed polyphenol extract is improved by repeated dosing in rats: Implications for treatment in Alzheimer's disease. *Journal of Alzheimer's Disease* 18:113–124.

Franiak-Pietryga I, Koter-Michalak M, Broncel M, Duchnowicz P, Chojnowska-Jezierska J (2009) Anti-inflammatory and hypolipemic effects in vitro of simvastatin comparing to epicatechin in patients with type-2 hypercholesterolemia. *Food and Chemical Toxicology* 47:393–397.

Frisardi V, Panza F, Seripa D, Imbimbo BP, Vendemiale G, Pilotto A, Solfrizzi V (2010) Nutraceutical properties of Mediterranean diet and cognitive decline: Possible underlying mechanisms. *Journal of Alzheimer's Disease* 22:715–740.

Ganguli M, Vander Bilt J, Saxton JA, Shen C, Dodge HH (2005) Alcohol consumption and cognitive function in late life: A longitudinal community study. *Neurology* 65:1210–1217.

Gronbaek M, Becker U, Johansen D, Gottschau A, Schnohr P, Hein HO, Jensen G, Sorensen TIA (2000) Type of alcohol consumed and mortality from all causes, coronary heart disease, and cancer. *Annals of Internal Medicine* 133:411–419.

Hamaguchi T, Ono K, Murase A, Yamada M (2009) Phenolic compounds prevent Alzheimer's pathology through different effects on the amyloid-beta aggregation pathway. *American Journal of Pathology* 175:2557–2565.

Han YS, Bastianetto S, Dumont Y, Quirion R (2006) Specific plasma membrane binding sites for polyphenols, including resveratrol, in the rat brain. *Journal of Pharmacology and Experimental Therapeutics* 318:238–245.

Han YS, Zheng WH, Bastianetto S, Chabot JG, Quirion R (2004) Neuroprotective effects of resveratrol against beta-amyloid-induced neurotoxicity in rat hippocampal neurons: Involvement of protein kinase C. *British Journal of Pharmacology* 141:997–1005.

Harwood DG, Barker WW, Loewenstein DA, Ownby RL, St George-Hyslop P, Mullan M, Duara R (1999) A cross-ethnic analysis of risk factors for AD in white Hispanics and white non-Hispanics. *Neurology* 52:551–556.

Hendrie HC, Gao S, Hall KS, Hui SL, Unverzagt FW (1996) The relationship between alcohol consumption, cognitive performance, and daily functioning in an urban sample of older black Americans. *Journal of the American Geriatrics Society* 44:1158–1165.

Ho L, Chen LH, Wang J, Zhao W, Talcott ST, Ono K, Teplow D, Humala N, Cheng A, Percival SS, Ferruzzi M, Janle E, Dickstein DL, Pasinetti GM. (2009) Heterogeneity in red wine polyphenolic contents differentially influences Alzheimer's disease-type neuropathology and cognitive deterioration. *Journal of Alzheimer's Disease* 16:59–72.

Huang W, Qiu C, Winblad B, Fratiglioni L (2002) Alcohol consumption and incidence of dementia in a community sample aged 75 years and older. *Journal of Clinical Epidemiology* 55:959–964.

Jackson R, Scragg R, Beaglehole R (1992) Does recent alcohol consumption reduce the risk of acute myocardial infarction and coronary death in regular drinkers? *American Journal of Epidemiology* 136:819–824.

Kelner KL, Marx J (1996) Patterns of aging. *Science* 273:41.

Kennedy DO, Wightman EL, Reay JL, Lietz G, Okello EJ, Wilde A, Haskell CF (2010) Effects of resveratrol on cerebral blood flow variables and cognitive performance in humans: A double-blind, placebo-controlled, crossover investigation. *American Journal of Clinical Nutrition* 91:1590–1597.

Klatsky AL (2003) Drink to your health? *Scientific American* 288:75.

Larrieu S, Letenneur L, Helmer C, Dartigues JF, Barberger-Gateau P (2004) Nutritional factors and risk of incident dementia in the PAQUID longitudinal cohort. *Journal of Nutrition Health and Aging* 8:150–154.

Leibovici D, Ritchie K, Ledesert B, Touchon J (1999) The effects of wine and tobacco consumption on cognitive performance in the elderly: A longitudinal study of relative risk. *International Journal of Epidemiology* 28:77–81.

Li RC, Saleem S, Zhen GH, Cao WS, Zhuang HA, Lee J, Smith A, Altruda F, Tolosano E, Dore S (2009) Heme–hemopexin complex attenuates neuronal cell death and stroke damage. *Journal of Cerebral Blood Flow and Metabolism* 29:953–964.

Lindsay J, Laurin D, Verreault R, Hébert R, Helliwell B, Hill GB, McDowell I (2002) Risk factors for Alzheimer's disease: A prospective analysis from the Canadian Study of Health and Aging. *American Journal of Epidemiology* 156:445–453.

Luchsinger JA, Tang MX, Siddiqui M, Shea S, Mayeux R (2004) Alcohol intake and risk of dementia. *Journal of American Geriatric Society* 52:540–546.

Lutz W, Sanderson W, Scherbov S (2008) The coming acceleration of global population ageing. *Nature* 451:716–719.

Ma WW, Yuan LH, Yu HL, Ding BJ, Xi YD, Feng JF, Xiao R (2010) Genistein as a neuroprotective antioxidant attenuates redox imbalance induced by beta-amyloid peptides 25–35 in PC12 cells. *International Journal of Developmental Neuroscience* 28:289–295.

Maguire-Zeiss KA, Short DW, Federoff HJ (2005) Synuclein, dopamine and oxidative stress: Co-conspirators in Parkinson's disease? *Molecular Brain Research* 134:18–23.

Marambaud P, Zhao HT, Davies P (2005) Resveratrol promotes clearance of Alzheimer's disease amyloid-beta peptides. *Journal of Biological Chemistry* 280:37377–37382.

McDougall GJ, Becker H, Areheart KL (2006) Older males, cognitive function, and alcohol consumption. *Issues in Mental Health Nursing* 27:337–353.

Mehlig K, Skoog I, Guo X, Schutze M, Gustafson D, Waern M, Ostling S, Bjorkelund C, Lissner L (2008) Alcoholic beverages and incidence of dementia: 34-year follow-up of the prospective population study of women in Goteborg. *American Journal of Epidemiology* 167:684–691.

Mukamal KJ, Kuller LH, Fitzpatrick AL, Longstreth WT, Jr., Mittleman MA, Siscovick DS (2003) Prospective study of alcohol consumption and risk of dementia in older adults. *JAMA* 289:1405–1413.

Mukamal KJ, Longstreth WT, Jr., Mittleman MA, Crum RM, Siscovick DS, Bereczki D (2001) Alcohol consumption and subclinical findings on magnetic resonance imaging of the brain in older adults: The Cardiovascular Health Study Editorial Comment: The Cardiovascular Health Study. *Stroke* 32:1939–1946.

Munoz FJ, Aldunate R, Inestrosa NC (1999) Peripheral binding site is involved in the neurotrophic activity of acetylcholinesterase. *Neuroreport* 10:3621–3625.

NHMRC (2005) NHMRC Guidelines for the prevention, early detection and management of colorectal cancer. National Health and Medical Research Council.

Olanow C (2007) The pathogenesis of cell death in Parkinson's disease. *Movement Disorders* 22:S335–S342.

Ono K, Hasegawa K, Naiki H, Yamada M (2004) Anti-amyloidogenic activity of tannic acid and its activity to destabilize Alzheimer's beta-amyloid fibrils in vitro. *Biochimica et Biophysica Acta—Molecular Basis of Disease* 1690:193–202.

Ono K, Hirohata M, Yamada M (2008) Alpha-synuclein assembly as a therapeutic target of Parkinson's disease and related disorders. *Current Pharmaceutical Design* 14:3247–3266.

Ono K, Yoshiike Y, Takashima A, Hasegawa K, Naiki H, Yamada M (2003) Potent anti-amyloidogenic and fibril-destabilizing effects of polyphenols in vitro: Implications for the prevention and therapeutics of Alzheimer's disease. *Journal of Neurochemistry* 87:172–181.

Orgogozo JM, Dartigues JF, Lafont S, Letenneur L, Commenges D, Salamon R, Renaud S, Breteler MB (1997) Wine consumption and dementia in the elderly: A prospective community study in the Bordeaux area. *Revue Neurologique* 153:185–192.

Orhan I, Kartal M, Tosun F, Sener B (2007) Screening of various phenolic acids and flavonoid derivatives for their anticholinesterase potential. *Zeitschrift Fur Naturforschung Section C—A Journal of Biosciences* 62:829–832.

Palomaki H, Kaste M (1993) Regular light-to-moderate intake of alcohol and the risk of ischemic stroke. Is there a beneficial effect? *Stroke* 24:1828–1832.

Park JY, Mitrou PN, Dahm CC, Luben RN, Wareham NJ, Khaw KT, Rodwell SA (2009) Baseline alcohol consumption, type of alcoholic beverage and risk of colorectal cancer in the European Prospective Investigation into Cancer and Nutrition-Norfolk study. *Cancer Epidemiology, Biomarkers & Prevention* 33:347–354.

Passamonti S, Vrhovsek U, Vanzo A, Mattivi F (2005) Fast access of some grape pigments to the brain. *Journal of Agricultural and Food Chemistry* 53:7029–7034.

Pedersen A, Johansen C, Gronbaek M (2003) Relations between amount and type of alcohol and colon and rectal cancer in a Danish population based cohort study. *Gut* 52:861–867.

Rankine BC (1989) *Making Good Wine—A Manual of Winemaking Practice for Australia and New Zealand.* Sun Books, Melbourne, Victoria, Australia.

Rapp A, Mandery H (1986) Wine aroma. *Experientia* 42:873–884.

Rapp A, Pretorius PJ (1989) Flavours and off-flavours. Charalambous G (Ed.), *Proceedings of the 6th International Flavour Conference,* Rethymnon, Crete, Greece, July 5–7, 1989. Elsevier Science Publishers B.V., Amsterdam, the Netherlands, pp. 1–21.

Rasmussen HB, Bagger YZ, Tankó LB, Qin G, Christiansen C, Werge T (2006) Cognitive impairment in elderly women: The relative importance of selected genes, lifestyle factors, and comorbidities. *Neuropsychiatric Disease and Treatments* 2:227–288.

Reid MC, Van Ness PH, Hawkins KA, Towle V, Concato J, Guo Z (2006) Light to moderate alcohol consumption is associated with better cognitive function among older male veterans receiving primary care. *Journal of Geriatric Psychiatry and Neurology* 19:98–105.

Riviere C, Papastamoulis Y, Fortin PY, Delchier N, Andriamanarivo S, Waffo-Teguo P, Kapche G et al. (2010) New stilbene dimers against amyloid fibril formation. *Bioorganic & Medicinal Chemistry Letters* 20:3441–3443.

Riviere C, Richard T, Quentin L, Krisa S, Merillon JM, Monti JP (2007) Inhibitory activity of stilbenes on Alzheimer's beta-amyloid fibrils in vitro. *Bioorganic & Medicinal Chemistry* 15:1160–1167.

Roth A, Schaffner W, Hertel C (1999) Phytoestrogen kaempferol (3,4′,5,7-tetrahydroxyflavone) protects PC12 and T47D cells from beta-amyloid-induced toxicity. *Journal of Neuroscience Research* 57:399–404.

Ruitenberg A, van Swieten JC, Witteman JCM, Mehta KM, van Duijn CM, Hofman A, Breteler MMB (2002) Alcohol consumption and risk of dementia: The Rotterdam Study. *The Lancet* 359:281–286.

Schreier P (1979) Flavor composition of wines: A review. *CRC Critical Reviews in Food Science and Nutrition* 12:59–111.

Selkoe DJ (1997) Neuroscience—Alzheimer's disease: Genotypes, phenotype, and treatments. *Science* 275:630–631.

Simons LA, McCallum J, Friedlander Y, Ortiz M, Simons J (2000) Moderate alcohol intake is associated with survival in the elderly: The Dubbo Study. *The Medical Journal of Australia* 173:121–124.

Simons LA, Simons J, McCallum J, Friedlander Y (2006) Lifestyle factors and risk of dementia: Dubbo Study of the elderly. *The Medical Journal of Australia* 184:68–70.

Smith MA, Perry G (1995) Free radical damage, iron, and Alzheimer's disease. *Journal of the Neurological Sciences* 134:92–94.

Solfrizzi V, Panza F, Frisardi V, Seripa D, Logroscino G, Imbimbo BP, Pilotto A (2011) Diet and Alzheimer's disease risk factors or prevention: The current evidence. *Expert Review of Neurotherapeutics* 11:677–708.

Stampfer MJ, Kang JH, Chen J, Cherry R, Grodstein F (2005) Effects of moderate alcohol consumption on cognitive function in women. *The New England Journal of Medicine* 352:245–253.

Teri L, Hughes JP, Larson EB (1990) Cognitive deterioration in Alzheimer's disease: Behavioural and health factors. *Journal of Gerontology* 42:3–10.

Thomas T, Thomas G, McLendon C, Sutton T, Mullan M (1996) Beta-amyloid-mediated vasoactivity and vascular endothelial damage. *Nature* 380:168–171.

Truelsen T, Thudium D, Gronbaek M (2002) Amount and type of alcohol and risk of dementia: The Copenhagen City Heart Study. *Neurology* 59:1313–1319.

Wang J, Ho L, Zhao W, Ono K, Rosensweig C, Chen LH, Humala N, Teplow DB, Pasinetti GM (2008) Grape-derived polyphenolics prevent A beta oligomerization and attenuate cognitive deterioration in a mouse model of Alzheimer's disease. *Journal of Neuroscience* 28:6388–6392.

Wang J, Ho L, Zhao Z, Seror I, Humala N, Dickstein DL, Thiyagarajan M, Percival SS, Talcott ST, Pasinetti GM (2006) Moderate consumption of Cabernet Sauvignon attenuates A beta neuropathology in a mouse model of Alzheimer's disease. *FASEB J* 20:2313–2320.

Wannamethee SG, Camargo CA, Jr., Manson JE, Willett WC, Rimm EB (2003) Alcohol drinking patterns and risk of type 2 diabetes mellitus among younger women. *Archives of Internal Medicine* 163:1329–1336.

Wannamethee SG, Shaper AG, Perry IJ, Alberti KGMM (2002) Alcohol consumption and the incidence of type II diabetes. *Journal of Epidemiology & Community Health* 56:542–548.

Waterhouse AL (2002) Wine phenolics. *Annals of the New York Academy of Sciences* 957:21–36.

Wright CB, Elkind MSV, Rundek T, Boden-Albala B, Paik MC, Sacco RL (2006) Alcohol intake, carotid plaque, and cognition: The Northern Manhattan Study. *Stroke* 37:1160–1164.

Zuccala G, Onder G, Pedone C, Cesari M, Landi F, Bernabei R, Cocchi A (2001) Dose-related impact of alcohol consumption on cognitive function in advanced age: Results of a multicenter survey. *Alcoholism: Clinical and Experimental Research* 25:1743–1748.

Index

A

AA, *see* Arachidonic acid (AA)
AAMI, *see* Age-associated memory impairment (AAMI)
Acetyl-L-carnitine (ALCAR), *see* Carnitine/ALCAR
ACTH, *see* Adrenocorticotrophic hormone (ACTH)
Activities of daily living (ADL), 215, 318
Acupuncture, TCM
  animal studies
    anti-apoptotic effects, 216
    anti-dementia effects, 217
    cerebral circulation, 216
    effects, oxygen free radicals, 216
    neurotransmitter effects, 216
    vasoactive substances, effects, 216
  pain, musculoskeletal injuries and depression, 214
  treatment, 215
AD, *see* Alzheimer's disease (AD)
Adenosine triphosphate (ATP), 90, 94–95, 98
ADHD, *see* Attention-deficit hyperactivity disorder (ADHD)
ADL, *see* Activities of daily living (ADL)
Adrenocorticotrophic hormone (ACTH), 217
Age-associated cognitive decline (AACD), *see* Age-related cognitive decline (ARCD)
Age-associated memory impairment (AAMI), 8, 246, 279–280
Age-related cognitive decline (ARCD)
  AAMI and AACD, 8–9
  amphetamine, 9
  automation, tests, 5–6
  categorization, 18
  declines, normal aging, 7
  defined, 17, 242
  enhancement, 4, 6
  Ginkgo biloba and *Panax ginseng*, 10
  guidelines, 23
  interpretation, changes, 20–23
  measurement, 4–5
  neutraceutical interventions, 20
  progression, 147
  recognition, 8
  reductions, brain's gray matter, 18
  safety, 11
  tests, 19–20

Age-related neurocognitive changes
  AD, 45–46
  frontal-striatal system and medial temporal cortical regions, 45
  working and episodic memory performance, 45
ALA, *see* Alphalinolenic acid (ALA)
Alphalinolenic acid (ALA)
  deficiency symptoms, 176–177
  defined, 168
  dietary sources, 169–170
  physiological effects, 178–179
Alzheimer's disease (AD)
  age, 44
  APP, 338
  characterization, 45–46
  cholinergic deficit, 72–73
  cholinesterase inhibitors, 46
  curcumin's effect, 210
  defined, 72
  effects, huperzine A, 320–322
  free radical damage, cellular function, 315
  ginseng therapy, 211
  LA, *see* Lipoic acid (LA)
  MCI, 46
  memory deficits, 211
  multifactorial etiology, 310
  neuropathology, 338
  pathogenesis, 205–206
  proteasome-dependent, 339
Amyloid-β (Aβ) factors
  and AD, *see* Alzheimer's disease (AD)
  and Apoe-epsilon 4 allele
    hypercholesterolemia, 339
    isoforms, 339
    proteasome-dependent, 339
Amyloid precursor protein (APP)
  AD, 338
  onamyloidogenic processing, 338
  transmembrane region, 338
Antiamnesic effect
  hypoxia, 267
  scopolamine-induced amnesia, mice, 268
Antioxidants and cognition
  herbal medicine, 244
  Wechsler Memory test, 243
Antioxidant vitamins, 52–53, 57
Anxiolysis, 296

347

T - #0367 - 071024 - C372 - 234/156/16 - PB - 9780367380441 - Gloss Lamination